Advances in pig breeding and reproduction

It is widely recognised that agriculture is a significant contributor to global warming and climate change. Agriculture needs to reduce its environmental impact and adapt to current climate change whilst still feeding a growing population, i.e. become more 'climate-smart'. Burleigh Dodds Science Publishing is playing its part in achieving this by bringing together key research on making the production of the world's most important crops and livestock products more sustainable.

Based on extensive research, our publications specifically target the challenge of climate-smart agriculture. In this way we are using 'smart publishing' to help achieve climate-smart agriculture.

Burleigh Dodds Science Publishing is an independent and innovative publisher delivering high quality customer-focused agricultural science content in both print and online formats for the academic and research communities. Our aim is to build a foundation of knowledge on which researchers can build to meet the challenge of climate-smart agriculture.

For more information about Burleigh Dodds Science Publishing simply call us on +44 (0) 1223 839365, email info@bdspublishing.com or alternatively please visit our website at www.bdspublishing.com.

Related titles:

Achieving sustainable production of pig meat Volume 2
Print (ISBN 978-1-78676-092-0); Online (ISBN 978-1-78676-094-4, 978-1-78676-095-1)

Optimising pig herd health and production
Print (ISBN 978-1-78676-883-4); Online (ISBN 978-1-78676-885-8, 978-1-78676-886-5)

Understanding the behaviour and improving the welfare of pigs
Print (ISBN 978-1-78676-443-0); Online (ISBN 978-1-78676-445-4, 978-1-78676-445-4)

Chapters are available individually from our online bookshop:
https://shop.bdspublishing.com

climate•SMART•publishing
IN AGRICULTURAL SCIENCE

BURLEIGH DODDS SERIES IN AGRICULTURAL SCIENCE
NUMBER 152

Advances in pig breeding and reproduction

Edited by Professor Jason Ross, Iowa State University, USA

burleigh dodds
SCIENCE PUBLISHING

Published by Burleigh Dodds Science Publishing Limited
82 High Street, Sawston, Cambridge CB22 3HJ, UK
www.bdspublishing.com

Burleigh Dodds Science Publishing, 1518 Walnut Street, Suite 900, Philadelphia, PA 19102-3406, USA

First published 2025 by Burleigh Dodds Science Publishing Limited

Library of Congress Control Number: 2024940534

British Library Cataloguing in Publication Data
A catalogue record for this book is available from the British Library

ISBN 978-1-80146-535-9 (print)
ISBN 978-1-80146-536-6 (ePub)
ISBN 978-1-80146-537-3 (PDF)
ISSN 2059-6936 (print)
ISSN 2059-6944 (online)

DOI: 10.19103/AS.2024.0137

Typeset by Deanta Global Publishing Services, Dublin, Ireland

Contents

© Burleigh Dodds Science Publishing Limited, 2025. All rights reserved.

Series list

genome editing which restrict undesirable traits are being carried out more ethically.

Chapter 12 studies neonatal nutritional supplementation and its ability to increase growth and survival of piglets. Genetic selection for highly prolific animals has resulted in a lower survivability rate for piglets during the neonatal period. This chapter outlines the mortality risks for piglets based on their weight, showing medium weight piglets provide the greatest opportunity to improve survival. It explores different methods that can be used to support supplementation to the piglet, pre- and post-weaning to increase chances of survival. These include supplementing colostrum further with colostrum-like products that are rich in energy, protein and immunoglobulins, furthermore it compares these effects with human studies involving colostrum supplementation. It concludes by highlighting the time restraints of supplementing the nutrition of individual piglets.

The final chapter in the book, Chapter 13, addresses the challenge of feeding the prolific sow, involving proper nutrient formulation, feed delivery, and an understanding of the biological challenges that sows will undergo within their lifetimes. The chapter summarises what we know about nutrient requirements for developing gilts as well as during gestation, lactation, and the transition and post-weaning periods. The chapter covers protein, energy and fiber requirements for developing gilts and during gestation and lactation, and includes a case study on how this understanding can be used to optimise feeding levels at day 90 of gestation.

Part 1

Breeding

Chapter 1

Assessing pig genetic diversity

R. Bozzi, F. Tiezzi and M. C. Fabbri, Università degli Studi di Firenze, Italy

1 Introduction

Genetic diversity has been defined as 'any measure that quantifies the magnitude of genetic variability within a population' (Hughes et al., 2008). The degree of genetic variation is the occurrence of different genetic variants, i.e. the number of polymorphic loci (with more than two alleles) and their frequency distributions in a given population. Genetic diversity can be evaluated at different levels: among species, breeds, between lines within a breed, and even among individuals (Toro and Caballero, 2005).

The importance of preserving genetic diversity at the global level has long been recognized (United Nations Environment Programme, 1992). This is, in part, because genetic diversity provides the foundation for biodiversity, providing the basis for species diversity, which then feeds through to ecosystem diversity (Nonić and Šijačić-Nikolić, 2021). The importance of genetic diversity in pigs was highlighted in 1992 by Ollivier and Molénat (1992), who also pointed out that genetic knowledge of pig populations at that time was too limited and imprecise to identify populations or genes at risk.

The increasing availability of genetic markers has made it possible to study differences between:

- diversity (between populations); and
- variability (within populations) in pig populations.

http://dx.doi.org/10.19103/AS.2024.0137.01

Initially, measuring genetic distances between populations was considered the main tool for assessing genetic diversity. However, Ruane (1999) identified uncertainties in the reliability and relevance of such measurements. Caballero and Toro (2002) have also argued that genetic distances are inappropriate for in-species conservation. It has been suggested that other aspects need to be taken into account, such as the degree of vulnerability of the loss of a particular trait as well as focusing on traits of particular importance, whether economic, relating to historical or cultural value, resistance to biotic and abiotic stresses or traits that can improve the sustainability of production and environmental impact. Such approaches require weighting to prioritize traits, as attempted in some studies (García et al., 2002; Reist-Marti et al., 2003) using Weitzmann's theory (d'Arnoldi et al., 1998).

Another important source of genetic diversity is copy number variant (CNV) regions, segments of DNA that have different numbers of copies in different individuals (Wang et al., 2014). The number of CNVs detected varies among breeds, and they may be breed-specific, explaining e.g. certain phenotypes such as dominant white in pigs (Li et al., 2012; Rubin et al., 2012; Wang et al., 2014; Bovo et al., 2020b), as well as some differences in immune system response between Asian and European populations (Wang et al., 2014). Other important processes in pigs involving CNVs include foraging and mate recognition (Paudel et al., 2015). Studying 19 autochthonous European breeds, Bovo et al. (2020a) found that variability attributable to CNVs may contribute to phenotypic variability that distinguishes local from commercial pig breeds.

Another source of information about the genetic diversity of animal populations mediated by anthropogenic selection is selection signatures (Rubin et al., 2012; Bovo et al., 2020a). For example, a study of European local pig breeds by Muñoz et al. (2019) has shown that some breeds (e.g. Turopolje and Mora Romagnola) have a large number of specific regions, likely indicating genetic drift and limited effective population size.

3 Trends in pig breeding and genetic diversity

There are currently 16 identified subspecies of wild boar (*Sus scrofa ferus*), which are the wild relative of modern pigs (Giuffra et al., 2000). Separate ancestral populations of pigs still occur in both Europe and Asia, acting as potential genetic reservoirs (Bruford et al., 2003). Wild pigs were first domesticated around 11000–10000 BP. An important contribution to the understanding of the process of domestication was made by studies conducted using mitochondrial DNA (mtDNA) (Giuffra et al., 2000; Wu et al., 2007), as well as the study of the Y chromosome (Cliffe et al., 2010; Faria et al., 2019; Ramirez et al., 2009). By analyzing mtDNA lineages, (Bruford et al., 2003) suggested two centers of domestication, first in Anatolia (in Turkey) and subsequently in China.

A subsequent study by Larson (2005) has suggested at least five to six centers of domestication.

Food and Agriculture Organization (FAO) (World Food and Agriculture – Statistical Yearbook 2021, 2021) estimates that there are now almost 1 billion pigs globally, with a wide distribution except in the Near and Middle East, where consumption of pig meat is limited for religious reasons. In terms of live numbers, Asia accounts for more than 50% of the world population and has 41% of recorded breeds, while Europe and the Americas account for 40% of the global population with 47% of recorded breeds.

Of the 541 recorded breeds, 229 are from Asia, followed by Europe with 165 breeds. Most of these are specific to particular regions. Only 33 breeds of pigs are considered 'transboundary'/'cosmopolitan', i.e. found in more than one country. In Europe, 17 breeds are considered transboundary. Six of this subset of transboundary breeds (originating in Europe) dominate commercial production:

- Large White;
- Duroc;
- Landrace;
- Hampshire;
- Piétrain; and
- Berkshire.

Most commercially farmed pigs are hybrids derived from these breeds. Modern European pig lines date back to crosses between indigenous European and small Asian pig breeds imported from China and Southeast Asia from the eighteenth century onwards to improve both reproductive and production performance (Jones, 1998). A study of 13 European pig breeds by Wilkinson et al. (2013) e.g. demonstrated introgression from Asian breeds. Chinese introgression has e.g. been demonstrated for the British Tamworth breed (Cliffe et al., 2010), confirming previous results at the mitochondrial level (Fang and Andersson, 2006). Ramirez et al. (2009) has found that European pigs contributed to the origin of the paternal and maternal lineages of African and South American pigs.

After World War II, commercial companies applied selection ever more intensively, culminating in today's breeds. European breeding has placed particular emphasis on growth traits, nutritional efficiency, and reproductive performance. In Asia, more emphasis is placed on fertility, maternal traits, and fatness. Megens et al. (2008b) have found different overall patterns of pig breed diversity in Europe and China. A marked diversity between European and Chinese breeds was also found in studying SNP diversity (Ai et al., 2013), CNVs (Paudel et al., 2013), Y chromosome haplotypes (Cliffe et al., 2010), as well as with a resequencing study carried out by Bosse et al. (2014). Zhang et al. (2016) have found lower genetic variation in Asian pigs compared to European domestic pig populations. Yang et al. (2017) have found the genetics of Asian pig breeds to be much more geographically specific.

In general, however, intense selection over the last 50–100 years has generally resulted in a remarkable homogeneity in commercial breeds. This has led to a progressive reduction in genetic diversity (Amaral et al., 2011) and attempts to address this loss of diversity e.g. through backcrossing with wild boars (Ji et al., 2011).

4 Genetic diversity of local pig breeds

Local populations have continued to be bred in various countries, largely retaining their original traits given a lack of access to modern breeding programs e.g. due to retention of local breeding practices or different selection objectives. Gama et al. (2013) e.g. found a high level of genetic diversity within Iberian pig populations. Whilst genetic diversity is widespread within local breeds due to the founder effect or subsequent bottlenecks, there are also examples of localized reduction in genetic diversity as a result of anthropogenic selection events, identifiable at specific loci and linked regions (Boitard et al., 2010).

Their different selection histories are reflected in the considerable genetic diversity between local pig populations and cosmopolitan breeds (Nikolic et al., 2009). Herrero-Medrano et al. (2014) have analyzed candidate genes to highlight differences in selection pressure between local and commercial pig populations, looking at different mutations in the AZGP1 (lipid degradation), PLAG1 (stature), and TAS2R40 (bitter taste receptor) genes which are strongly expressed in commercial breeds. In contrast, local breeds share alleles with wild boar for immune-related genes such as IL12RB2 and STAB1 and bone synthesis genes (EIF2AK3).

Studies have found that the reproductive management of local breeds is often concentrated on the farm, exacerbating inbreeding and genetic drift (SanCristobal et al., 2006). Herrero-Medrano et al. (2014) have found a faster accumulation of deleterious mutations with a high level of inbreeding. A study of pig populations in the Balkans by Zorc et al. (2022) revealed a higher number of genetic groups than number of breeds, indicating different gene pools in some geographical regions. At the same time, however, the genetic diversity of these populations was lower than that of the conventional breeds, indicating a low level of heterozygosity due to inbreeding within the populations. This indicates that diversity between breeds does not automatically equate to genetic diversity. Given current levels of inbreeding and runs of homozygosity (ROH) (with resulting deleterious mutations), local breeds require a careful breeding program to maintain their diversity. However, since variability among breeds accounts for a small part of total genetic diversity, some local and isolated populations could act as important reservoirs of genetic diversity, even allowing for inbreeding (Gama et al., 2013). A key challenge for these small populations is balancing the need to maintain genetic diversity, respecting specific breed standards (often morphological) while, at the same time, improving traits of production or other importance (e.g. fat composition).

In looking at diversity within 12 traditional breeds, Herrero-Medrano et al. (2014) found English pig breeds showed higher diversity, probably due to traces of introgression of Asian pigs, but also high levels of ROH, which were comparable to levels found in other traditional breeds: Cinta Senese (Italy) and Mangalitza (originating in Hungary), both of which have suffered severe population declines in the recent past. A study of the Mangalitza breed by Bâlteanu et al. (2019) found remarkable differences between strains (red or blond) and countries (Romania and Hungary), suggesting local inbreeding. The exchange of genetic material could easily remedy this situation but this would need to be integrated into existing selection programs.

Muñoz et al. (2018, 2019) found that heterozygosity (both observed and expected) was low in some traditional breeds (e.g. 0.195 for Ho and 0.187 for He in the Turopolje breed), as is the inbreeding coefficient (FIS = 0.138 for the Apulo Calabrese breed). This lower variability affects options for breeding programs to conserve biodiversity for some breeds (e.g. Turopolje, Mora Romagnola, Casertana, and Apulo-Calabrese). Overall, only 7 out of 20 traditional breeds studied had a Ne value higher than 50, which is considered the lowest limit for an effective conservation program.

One factor contributing to the presence or absence of inbreeding in these populations is introgression of genes from other breeds, sometimes controlled but sometimes accepted, as in the case of crossbreeding with wild boar, which has also been documented by various authors (Manunza et al., 2016; Iacolina et al., 2018; Bovo et al., 2020). In investigating the influence of wild boar on European local pig breeds, Dadousis et al. (2022) found signs of introgression in at least six different breeds (Alentejana, Black Slavonian, Iberian, Nero Siciliano, Sarda, and Swallow-Bellied Mangalitza). The two Italian breeds showed a high degree of admixture, suggesting a very different ancestry. The distinctiveness of local European pig breeds shows both genetic diversity but also a high degree of admixture, which complicates potential breeding strategies.

Herrero-Medrano et al. (2013) have also noted introgression of genes from cosmopolitan pig populations into local breeds. The practice of crossing with boars of other breeds has grown in popularity to reduce inbreeding and also to take advantage of the better reproductive and production performance of cosmopolitan genotypes. This practice has diluted some local genotypes, with

that can be employed to measure the internal and skin temperature of the animals. These can be adapted to the conditions facing local pig breeds.

From a genetic standpoint, the management of the nucleus herd will move from managing the animals to managing the genome. High-throughput genotyping platforms allow us to tailor mating plans to meet breeding goals and maintain genetic diversity within intensively selected populations. It will probably reach a point where haplotype libraries are constructed and managed, with the breeding herd being a pool to draw from, depending on the needs for a particular mating. In low-input systems, the same approach will probably include local/autochthonous germplasm. Combining 'commercial and high-performance alleles' with 'local and high-meat quality alleles' will meet customers' demand for niche meats while allowing farmers to raise animals that grow efficiently.

Adaptation to new conditions (e.g. new diets or warmer climates) will need new '-omics' to be employed in breeding systems. Epigenomics and microbiomics will play a role in understanding the dynamics that regulate acclimation and adaptation to new and more challenging conditions. While the price of molecular genetics tools for specific platforms has fallen (e.g. whole-genome methylation, 16S microbiome sequencing), dry-lab analytic methods are also becoming more accessible in condensing a large amount of information now becoming available.

7 Where to look for further information

A comprehensive discussion of genetics in pigs can be found in M. F. Rothschild, and A. Ruvinsky. The genetics of the pig. 2nd Edition, CABI publishing. The best source of information for pig breeds is the DAD-IS database. The Domestic Animal Diversity Information System is maintained and developed by FAO, and it allows to analyze the diversity of livestock breeds on national, regional and global levels including the status of breeds regarding their risk of extinction (https://www.fao.org /dad-is/en/). For those wishing to explore the pig genome and the genetic diversity of the species, as well as the regulatory mechanisms of complex traits, the following databases are an important source of information:

- https://www.sanger.ac.uk/data/pig-genome/
- https://genome-euro.ucsc.edu/
- https://www.animalgenome.org/pig/genome/db/genepages
- https://useast.ensembl.org/Sus_scrofa/Info/Index
- https://pigbiobank.farmgtex.org/
- https://www.ncbi.nlm.nih.gov/genbank/

8 References

Ai, H., et al. 2013. Genetic diversity, linkage disequilibrium and selection signatures in chinese and western pigs revealed by Genome-Wide SNP markers. *PLoS One* 8, e56001. https://doi.org/10.1371/journal.pone.0056001

Amaral, A.J., et al. 2011. Genome-wide footprints of pig domestication and selection revealed through massive parallel sequencing of pooled DNA. *PLoS One* 6, e14782. https://doi.org/10.1371/journal.pone.0014782

Andersson, L., et al. 1994. Genetic Mapping of quantitative trait loci for growth and fatness in pigs. *Science* 263, 1771–1774. https://doi.org/10.1126/science.8134840

Bakoev, S., et al. 2021. Analysis of Homozygous-by-Descent (HBD) segments for purebred and crossbred pigs in Russia. *Life* 11, 861. https://doi.org/10.3390/life11080861

Bâlteanu, V.A., et al. 2019. The footprint of recent and strong demographic decline in the genomes of Mangalitza pigs. *Animal* 13, 2440–2446. https://doi.org/10.1017/S1751731119000582

Bergamaschi, M., et al. 2019. Genome-wide association study for carcass quality traits and growth in purebred and crossbred pigs1. *J. Anim. Sci.* 98 (1): skz360. https://doi.org/10.1093/jas/skz360.

Bertolini, F., et al. 2018. Exploiting phenotype diversity in a local animal genetic resource: identification of a single nucleotide polymorphism associated with the tail shape phenotype in the autochthonous Casertana pig breed. *Livest. Sci.* 216, 148–152. https://doi.org/10.1016/j.livsci.2018.08.007.

Biermann, A.D.M., et al. 2014. Implementation of genetic evaluation and mating designs for the endangered local pig breed 'Bunte Bentheimer.' *J. Anim. Breed. Genet.* 131, 36–45. https://doi.org/10.1111/jbg.12041

Blott, S., et al. 2004. Genetic diversity in European and Chinese pig breeds – the PigBioDiv project. *BSAP Occas. Publ.* 30, 321–323. https://doi.org/10.1017/S0263967X00042117

Boitard, S., et al. 2010. Genetic variability, structure and assignment of Spanish and French pig populations based on a large sampling: Variability of Spanish and French pig populations. Anim. Genet. 41, 608–618. https://doi.org/10.1111/j.1365-2052.2010.02061.x

Bosse, M., et al. 2014. Untangling the hybrid nature of modern pig genomes: a mosaic derived from biogeographically distinct and highly divergent *Sus scrofa* populations. *Mol. Ecol.* 23, 4089–4102. https://doi.org/10.1111/mec.12807

Bosse, M., et al. 2015. Using genome-wide measures of coancestry to maintain diversity and fitness in endangered and domestic pig populations. *Genome Res.* 25, 970–981. https://doi.org/10.1101/gr.187039.114

Bovo, S., et al. 2020a. Whole-genome sequencing of European autochthonous and commercial pig breeds allows the detection of signatures of selection for adaptation of genetic resources to different breeding and production systems. Genet. Sel. Evol. 52, 33. https://doi.org/10.1186/s12711-020-00553-7

Bovo, S., et al. 2020b. Genome-wide detection of copy number variants in European autochthonous and commercial pig breeds by whole-genome sequencing of DNA pools identified breed-characterising copy number states. *Anim. Genet.* 51, 541–556. https://doi.org/10.1111/age.12954

Bruford, M.W., et al. 2003. DNA markers reveal the complexity of livestock domestication. *Nat. Rev. Genet.* 4, 900–910. https://doi.org/10.1038/nrg1203

Buchanan, D.S. and Northcutt, S.L. 2011. The genetic principles of crossbreeding. Product of Extension Beef Cattle Resource Committee in Beef Cattle Handbook. www.iowabeefcenter.org

Caballero, A., et al. 2022. A comparison of marker-based estimators of inbreeding and inbreeding depression. *Genet. Sel. Evol.* 54, 82. https://doi.org/10.1186/s12711-022-00772-0

Caballero, A. and Toro, M. 2002. Analysis of genetic diversity for the management of conserved subdivided populations. *Conserv. Genet.* 3, 289–299. https://doi.org/10.1023/A:1019956205473

Casellas, J. 2018. On individual-specific prediction of hidden inbreeding depression load. *J. Anim. Breed. Genet.* 135, 37–44. https://doi.org/10.1111/jbg.12308

Casellas, J., et al. 2019. Inbreeding depression load for litter size in Entrepelado and Retinto Iberian pig varieties1. *J. Anim. Sci.* 97, 1979–1986. https://doi.org/10.1093/jas/skz084

Cliffe, K.M., et al. 2010. Analysis of the non-recombining Y chromosome defines polymorphisms in domestic pig breeds: ancestral bases identified by comparative sequencing. *Anim. Genet.* 41, 619–629. https://doi.org/10.1111/j.1365-2052.2010.02070.x

Clop, A., et al. 2004. Estimating the frequency of Asian cytochrome B haplotypes in standard European and local Spanish pig breeds. *Genet. Sel. Evol.* 36 (1): 97. https://doi.org/10.1186/1297-9686-36-1-97.

Dadousis, C., et al. 2022. Admixture and breed traceability in European indigenous pig breeds and wild boar using genome-wide SNP data. *Sci. Rep.* 12, 7346. https://doi.org/10.1038/s41598-022-10698-8

d'Arnoldi, C., et al. 1998. An overview of the Weitzman approach to diversity. *Genet. Sel. Evol.* 30, 149–161. https://doi.org/10.1186/1297-9686-30-2-149

Fang, M. and Andersson, L. 2006. Mitochondrial diversity in European and Chinese pigs is consistent with population expansions that occurred prior to domestication. *Proc. R. Soc. B Biol. Sci.* 273, 1803–1810. https://doi.org/10.1098/rspb.2006.3514

Faria, D.A., et al. 2019. Assessing Sus scrofa diversity among continental United States, and Pacific islands populations using molecular markers from a gene banks collection. *Sci. Rep.* 9, 3173. https://doi.org/10.1038/s41598-019-39309-9

Fontanesi, L., et al. 2010. Genetic heterogeneity and selection signature at the KIT gene in pigs showing different coat colours and patterns. *Anim. Genet.* 41 (5): 478–492. https://doi.org/10.1111/j.1365-2052.2010.02054.x

Gama, L.T., et al. 2013. Genetic structure, relationships and admixture with wild relatives in native pig breeds from Iberia and its islands. *Genet. Sel. Evol.* 45, 18. https://doi.org/10.1186/1297-9686-45-18

Ganteil, A., et al. 2021. Characterization of autozygosity in pigs in three-way crossbreeding. *Front. Genet.* 11, 584556. https://doi.org/10.3389/fgene.2020.584556

García, D., et al. 2002. *Further developments on Weitzman's approach to assess conservation of genetic diversity*. In Proceedings of the 7th World Congress on Genetics Applied to Livestock Production. Presented at the 7th World Congress on Genetics Applied to Livestock Production, Montpellier, France.

Giuffra, E., et al. 2000. The origin of the domestic pig: independent domestication and subsequent introgression. *Genetics* 154, 1785–1791. https://doi.org/10.1093/genetics/154.4.1785

González-Diéguez, D., et al. 2019. SNP-based mate allocation strategies to maximize total genetic value in pigs. *Genet. Sel. Evol.* 51, 55. https://doi.org/10.1186/s12711-019-0498-y

Gourdine, J.L., et al. 2012. There is room for selection in a small local pig breed when using optimum contribution selection: a simulation study1,2. *J. Anim. Sci.* 90, 76–84. https://doi.org/10.2527/jas.2011-3898

Groenen, M.A.M. 2016. A decade of pig genome sequencing: a window on pig domestication and evolution. *Genet. Sel. Evo* 48 (1): 23. https://doi.org/10.1186/s12711-016-0204-2.

Groenen, M.A.M., et al. 2012. Analyses of pig genomes provide insight into porcine demography and evolution. *Nature* 491, 393–398. https://doi.org/10.1038/nature11622

Groeneveld, L.F., et al. 2010. Genetic diversity in farm animals – a review. *Anim. Genet.* 41, 6–31. https://doi.org/10.1111/j.1365-2052.2010.02038.x

Gupta, S.K., et al. 2013. Cytochrome b based genetic differentiation of Indian wild pig (Sus scrofa cristatus) and domestic pig (Sus scrofa domestica) and its use in wildlife forensics. *Sci. Just.* 53 (2): 220–22. https://doi.org/10.1016/j.scijus.2012.09.005

He, J., et al. 2020. Genomic mating as sustainable breeding for Chinese indigenous Ningxiang pigs. *PLoS One* 15, e0236629. https://doi.org/10.1371/journal.pone.0236629

Herrero-Medrano, J.M., et al. 2013. Farm-by-farm analysis of microsatellite, mtDNA and SNP genotype data reveals inbreeding and crossbreeding as threats to the survival of a native Spanish pig breed. *Anim. Genet.* 44, 259–266. https://doi.org/10.1111/age.12001

Herrero-Medrano, J.M., et al. 2014. Whole-genome sequence analysis reveals differences in population management and selection of European low-input pig breeds. *BMC Genom.* 15, 601. https://doi.org/10.1186/1471-2164-15-601

Howard, D.M., et al. 2018. Selective advantage of implementing optimal contributions selection and timescales for the convergence of long-term genetic contributions. *Genet. Sel. Evol.* 50, 24. https://doi.org/10.1186/s12711-018-0392-z

Howard, J.T., et al. 2016. Characterization and management of long runs of homozygosity in parental nucleus lines and their associated crossbred progeny. *Genet. Sel. Evol.* 48, 91. https://doi.org/10.1186/s12711-016-0269-y

Howard, J.T., et al. 2017. A heuristic method to identify runs of homozygosity associated with reduced performance in livestock. *J. Anim. Sci.* 95, 4318–4332. https://doi.org/10.2527/jas2017.1664

Hughes, A.R., et al. 2008. Ecological consequences of genetic diversity. *Ecol. Lett.* 11, 609–623. https://doi.org/10.1111/j.1461-0248.2008.01179.x

Iacolina, L., et al. 2018. Hotspots of recent hybridization between pigs and wild boars in Europe. *Sci. Rep.* 8, 17372. https://doi.org/10.1038/s41598-018-35865-8

Ji, Y.-Q., et al. 2011. Multi-locus analysis reveals a different pattern of genetic diversity for mitochondrial and nuclear DNA between wild and domestic pigs in East Asia. *PLoS One* 6, e26416. https://doi.org/10.1371/journal.pone.0026416

Jones, G.F. 1998. Genetic aspects of domestication, common breeds and their origin. In M.F. Rothschild and A. Ruvinsky (Eds.), *The Genetics of the Pig* (pp. 17–50). Wallingford: CAB International.

Kadirvel, G., et al. 2023. Performance of crossbred pigs with indigenous and Hampshire inheritance under a smallholder production system in the Eastern Himalayan hill region. *Front. Genet.* 14, 1042554. https://doi.org/10.3389/fgene.2023.1042554

Kijas, J.M.H., et al. 1998. Melanocortin Receptor 1 (MC1R) mutations and coat color in pigs. *Genetics* 150, 1177–1185. https://doi.org/10.1093/genetics/150.3.1177

Kim, B., et al. 2023. Comparison of growth performance and related gene expression of muscle and fat from Landrace, Yorkshire, and Duroc and Woori black pigs. *J. Anim. Sci. Technol.* 65, 160–174. https://doi.org/10.5187/jast.2022.e93

Larson, G. 2005. Worldwide phylogeography of wild boar reveals multiple centers of pig domestication. *Science* 307, 1618–1621. https://doi.org/10.1126/science.1106927

Li, Y., et al. 2012. Identification of genome-wide copy number variations among diverse pig breeds by array CGH. *BMC Genom.* 13, 725. https://doi.org/10.1186/1471-2164-13-725

Manunza, A., et al. 2016. Romanian wild boars and Mangalitza pigs have a European ancestry and harbour genetic signatures compatible with past population bottlenecks. *Sci. Rep.* 6, 29913. https://doi.org/10.1038/srep29913

Megens, H.-J., et al. 2008a. Biodiversity of pig breeds from China and Europe estimated from pooled DNA samples: differences in microsatellite variation between two areas of domestication. *Genet. Sel. Evol.* 40, 103–128. https://doi.org/10.1051/gse:2007039

Megens, H.-J., et al. 2008b. Biodiversity of pig breeds from China and Europe estimated from pooled DNA samples: differences in microsatellite variation between two areas of domestication. *Genet. Sel. Evol.* 40, 103–128. https://doi.org/10.1051/gse:2007039

Mei, Q., et al. 2022. Impacts of additive, dominance, and inbreeding depression effects on genomic evaluation by combining two SNP chips in Canadian Yorkshire pigs bred in China. *Genet. Sel. Evol.* 54, 69. https://doi.org/10.1186/s12711-022-00760-4

Meuwissen, T.H. 1997. Maximizing the response of selection with a predefined rate of inbreeding. *J. Anim. Sci.* 75, 934. https://doi.org/10.2527/1997.754934x

Moller, M.J., et al. 1996. Pigs with the dominant white coat color phenotype carry a duplication of the KIT gene encoding the mast/stem cell growth factor receptor. *Mamm. Genome* 7 (11), 822–830. https://doi.org/10.1007/s003359900244

Muñoz, M., et al. 2018. Diversity across major and candidate genes in European local pig breeds. *PLoS One* 13, e0207475. https://doi.org/10.1371/journal.pone.0207475

Muñoz, M., et al. 2019. Genomic diversity, linkage disequilibrium and selection signatures in European local pig breeds assessed with a high density SNP chip. *Sci. Rep.* 9, 13546. https://doi.org/10.1038/s41598-019-49830-6

Nakano, H., et al. 2015. Effect of VRTN gene polymorphisms on duroc pig production and carcass traits, and their genetic relationships. *Anim. Sci. J.* 86 (2): 125–131. https://doi.org/10.1111/asj.12260

Nikolic, N., et al. 2009. What do artificial neural networks tell us about the genetic structure of populations? The example of European pig populations. *Genet. Res.* 91, 121–132. https://doi.org/10.1017/S0016672309000093

Nonić, M. and Šijačić-Nikolić, M. 2021. Genetic diversity: sources, threats, and conservation. In W. Leal Filho, et al. (Eds.), *Life on Land* (pp. 421–435). Cham: Springer.

Ollivier, L. 2009. European pig genetic diversity: a minireview. *Animal* 3, 915–924. https://doi.org/10.1017/S1751731109004297

Ollivier, L., et al. 2005. An assessment of European pig diversity using molecular markers: partitioning of diversity among breeds. *Conserv. Genet.* 6, 729–741. https://doi.org/10.1007/s10592-005-9032-6

Ollivier, L. and Molenat, M. (Institut N. de R.A., Jouy-en-Josas (France). Station de Genetique Quantitative et Appliquee). 1992. *A global review of the genetic resources of pigs*. Rome: FAO Animal Production and Health Paper (FAO).

Paudel, Y., et al. 2013. Evolutionary dynamics of copy number variation in pig genomes in the context of adaptation and domestication. *BMC Genom.* 14, 449. https://doi.org/10.1186/1471-2164-14-449

Paudel, Y., et al. 2015. Copy number variation in the speciation of pigs: a possible prominent role for olfactory receptors. *BMC Genom.* 16, 330. https://doi.org/10.1186/s12864-015-1449-9

Ramirez, O., et al. 2009. Integrating Y-chromosome, mitochondrial, and autosomal data to analyze the origin of pig breeds. *Mol. Biol. Evol.* 26, 2061–2072. https://doi.org/10.1093/molbev/msp118

Reist-Marti, S.B., et al. 2003. Weitzman's approach and conservation of breed diversity: an application to african cattle breeds. *Conserv. Biol.* 17, 1299–1311.

Roberts, K.S. and Lamberson, W.R. 2015. Relationships among and variation within rare breeds of swine1. *J. Anim. Sci.* 93, 3810–3813. https://doi.org/10.2527/jas.2015-9001

Ruane, J. 1999. A critical review of the value of genetic distance studies in conservation of animal genetic resources. *J. Anim. Breed. Genet.* 116, 317–323. https://doi.org/10.1046/j.1439-0388.1999.00205.x

Rubin, C.-J., et al. 2012. Strong signatures of selection in the domestic pig genome. *Proc. Natl. Acad. Sci.* 109, 19529–19536. https://doi.org/10.1073/pnas.1217149109

SanCristobal, M., et al. 2006. Genetic diversity within and between European pig breeds using microsatellite markers. *Anim. Genet.* 37, 189–198. https://doi.org/10.1111/j.1365-2052.2005.01385.x

Schiavo, G., et al. 2018. Taking advantage from phenotype variability in a local animal genetic resource: identification of genomic regions associated with the hairless phenotype in Casertana pigs. *Anim. Genet.* 49, 321–325. https://doi.org/10.1111/age.12665.

Schiavo, G., et al. 2020. Genome-wide association analyses for coat colour patterns in the autochthonous Nero Siciliano pig breed. *Livest. Sci.* 236, 104015. https://doi.org/10.1016/j.livsci.2020.104015

Schook, L.B., et al. 2005. Swine Genome Sequencing Consortium (SGSC): a strategic roadmap for sequencing the pig genome. *Int. J. Genom.* 6, 251–255. https://doi.org/10.1002/cfg.479.

Serrano, M.P., et al. 2008. Influence of sex and terminal sire line on performance and carcass and meat quality of Iberian pigs reared under intensive production systems. *Meat Sci.* 78, 420–428. https://doi.org/10.1016/j.meatsci.2007.07.006

Shi, L., et al. 2022. Integrating genome-wide association studies and population genomics analysis reveals the genetic architecture of growth and backfat traits in pigs. *Front. Genet.* 13, 1078696. https://www.frontiersin.org/articles/10.3389/fgene.2022.1078696

Škorput, D., et al. 2022. Managing genetic diversity in pig populations: implications of optimal contribution selection in the Black Slavonian pig. *Ital. J. Anim. Sci.* 21, 1259–1267. https://doi.org/10.1080/1828051X.2022.2104661

Szpiech, Z.A., et al. 2013. Long runs of homozygosity are enriched for deleterious variation. *Am. J. Hum. Genet.* 93, 90–102. https://doi.org/10.1016/j.ajhg.2013.05.003

Toro, M.A. and Caballero, A. 2005. Characterization and conservation of genetic diversity in subdivided populations. *Phil. Trans. R. Soc. B* 360, 1367–1378.

Uemoto, Y., et al. 2021. Genome-wide association studies for production, respiratory disease, and immune-related traits in Landrace pigs. *Sci. Rep.* 11, 15823. https://doi.org/10.1038/s41598-021-95339-2

United Nations Environment Programme. 1992. Convention on biological diversity, June 1992.

Wang, J., et al. 2016. Prediction and estimation of effective population size. *Heredity* 117, 193–206. https://doi.org/10.1038/hdy.2016.43

Wang, S., et al. 2022. Identification of homozygous regions with adverse effects on the five economic traits of duroc pigs. *Front. Vet. Sci.* 9, 855933. https://doi.org/10.3389/fvets.2022.855933

Wang, Y., et al. 2014. Analysis of genome-wide copy number variations in chinese indigenous and western pig breeds by 60 K SNP genotyping arrays. *PLoS One* 9, e106780. https://doi.org/10.1371/journal.pone.0106780

Wilkinson, S., et al. 2013. Signatures of diversifying selection in European pig breeds. *PLoS Genet.* 9, e1003453. https://doi.org/10.1371/journal.pgen.1003453

World Food and Agriculture – Statistical Yearbook 2021, 2021. FAO. https://doi.org/10.4060/cb4477en

Wu, G.-S., et al. 2007. Population phylogenomic analysis of mitochondrial DNA in wild boars and domestic pigs revealed multiple domestication events in East Asia. *Genome Biol.* 8, R245. https://doi.org/10.1186/gb-2007-8-11-r245

Xiang, T., et al. 2016. Genomic evaluation by including dominance effects and inbreeding depression for purebred and crossbred performance with an application in pigs. *Genet. Sel. Evol.* 48, 92. https://doi.org/10.1186/s12711-016-0271-4

Yang, B., et al. 2017. Genome-wide SNP data unveils the globalization of domesticated pigs. *Genet. Sel. Evol.* 49, 71. https://doi.org/10.1186/s12711-017-0345-y

Zhang, J., et al. 2016. Genetic diversity in the mitochondrial DNA D-loop region of global swine (Sus scrofa) populations. *Biochem. Biophys. Res. Commun.* 473, 814–820. https://doi.org/10.1016/j.bbrc.2016.03.125

Zhang, Y., et al. 2022. Estimate of inbreeding depression on growth and reproductive traits in a Large White pig population. *G3 GenesGenomesGenetics* 12, jkac118. https://doi.org/10.1093/g3journal/jkac118

Zhao, Q., et al. 2021. Long-term impact of conventional and optimal contribution conservation methods on genetic diversity and genetic gain in local pig breeds. *Heredity* 127, 546–553. https://doi.org/10.1038/s41437-021-00484-z

Zorc, M., et al. 2022. Genetic diversity and population structure of six autochthonous pig breeds from Croatia, Serbia, and Slovenia. *Genet. Sel. Evol.* 54, 30. https://doi.org/10.1186/s12711-022-00718-6

Chapter 2

Advances in understanding and exploiting the pig genome

Martijn F. L. Derks, Wageningen University & Research/Topigs Norsvin Research Center, The Netherlands

1 Introduction

Selective animal breeding has delivered significant improvements in the productivity of farmed animals including pigs in the absence of knowledge of the genes controlling the trait(s) of interest. Quantitative genetics has provided both the theoretical framework and statistical tools for scientific genetic improvement. Advances in molecular genetics and sequencing technologies driven by the Human Genome Project have enabled the exploitation of genomic information to deliver a revolution in selective animal breeding (Hill 2014).

The essence of selective animal breeding is concerned with identifying the candidate parents of the next generation with the highest genetic merit. In the absence of knowledge of the causal genetic variants, genetic markers linked to loci with effects on the trait of interest can be used in Marker Assisted Selection (MAS) (Haley and Visscher 1998). Genomic selection is an implementation of this approach on a genome-wide scale which has enabled significant advances in pig breeding (Knol et al. 2016). However, to this day,

http://dx.doi.org/10.19103/AS.2024.0137.02

genomic selection is still a black-box approach as many of the causal variants that are selected for are unknown (Habier et al. 2013).

The first pig genomics projects were concerned with developing the necessary polymorphic genetic markers, including single nucleotide polymorphisms (SNPs) and using those markers to map quantitative trait loci (QTL) (Ramos et al. 2009; Ernst and Steibel 2013). Using this method, regions of the genome that are associated with variation in a particular trait of interest can be identified (Rothschild et al. 2007). QTL mapping involves analyzing population genomic data to identify genetic markers that are linked to variation in the trait of interest.

However, many commercial livestock populations have relatively strong linkage disequilibrium (LD) between genetic markers due to small effective population sizes causing limited haplotype diversity (Veroneze et al. 2014). This strong LD hampers accurate detection of the causal variant because often still thousands of (sequence) variants are in strong LD encompassing a QTL signal. Therefore, recent advances have been made to aid in fine-mapping of QTL regions which leverage the functional or regulatory genome (Gross et al. 2020; Liu et al. 2020). These tools are key to further fine-map the genomic region that is responsible for the trait and study the specific genes and molecular pathways involved. This chapter outlines current developments in pig genetics and genomics and provides one case study for fine-mapping QTL regions affecting meat quality.

2 Developments in pig genomics

The first efforts to sequence the pig genome were initiated by the establishment of the Swine Genome Sequencing Consortium (SGSC) at the start of this century (Archibald et al. 2010) which resulted in the publication of the draft reference genome sequence of *Sus scrofa* in 2012, derived from a female Duroc pig (Groenen et al. 2012). The SGSC has since then been instrumental in sequencing and annotating the genomes of hundreds of individual pigs belonging to hundreds of breeds around the world, providing a valuable resource for researchers in the field.

Subsequently, the Functional Annotation of Animal Genomes (FAANG) consortium was established, an international effort to generate comprehensive maps of functional elements in the genomes of a wide range of animal species, with the focus on livestock (Andersson et al. 2015). The FAANG project aims to create a comprehensive catalog of functional elements in animal genomes, including gene expression, chromatin accessibility, and DNA methylation patterns (Giuffra et al. 2019). To further understand the role of regulatory variants and gene expression, the FarmGTex consortium was established, a resource that aims to provide a comprehensive map of gene expression patterns in a range of pig tissues (Teng et al. 2022). Together, this information will be valuable for understanding the genetic basis of complex traits and diseases in animals (Clark et al. 2020).

The SGSC, FAANG, and FarmGTex represent important initiatives in pig genomics, providing valuable resources and infrastructure for researchers in the field. By advancing our understanding of the genetics and genomics of pigs, these projects drive the development of more efficient and sustainable breeding programs, improve animal health and welfare, and contribute to a more efficient and sustainable food system.

2.1 Reference genome and telomere-to-telomere genome assembly

After the establishment of third-generation sequencing technologies (producing long reads) a new pig reference genome was generated of much-improved quality in terms of continuity and gene annotation (Sscrofa11.1, (Warr et al. 2020)). This pseudohaploid genome sequence of a single Duroc sow, plus the sequence of chromosome Y of a single Duroc male is currently used as the reference for pig genomics studies. Although this reference genome is of high quality, some difficult-to-resolve regions are still not included in the reference genome (e.g. centromeres, segmental duplications, and the short arms of acrocentric chromosomes).

However, recent advances in genomic sequencing technologies have enabled researchers to create a new type of genome assembly called 'telomere-to-telomere' (T2T) (Nurk et al. 2022). Telomere-to-telomere (T2T) genome assemblies comprise the complete sequences of all chromosomes, including the telomeres, centromeres, and other difficult to resolve regions. The long-read sequencing technologies also allow for the generation of haplotype-resolved genome assemblies (Koren et al. 2018). Currently, efforts are ongoing to create a pig T2T genome assembly as well as for other important livestock species (Clark et al. 2024). A complete T2T pig genome will be a strong baseline for further genetic and genomic research in pigs and will help uncovered the most complex regions in the pig genome.

2.2 Pig pangenome and structural variation

With the availability of third-generation sequencing technologies, the availability of high-quality genomes of different individuals and breeds have become available. Together, these sequences can be formed into a pangenome to assess structural differences between individual or breed genomes (Tian et al. 2020; Derks et al. 2022; Jiang et al. 2023).

Together, the pig pangenome should give a comprehensive representation of the genetic diversity within pigs (and possibly other Suidae species). One important aspect of the pig pangenome is the identification of structural variation between breeds and individuals, which refers to variations in the number, size, and arrangement of DNA segments. Structural variation is a major source of genetic diversity in many species, and it has been implicated in a wide range of phenotypic traits and diseases in pigs (Paudel 2015; Ho et al. 2020).

Several recent studies have used high-throughput sequencing and other genomic technologies to characterize the structural variation within the pig pangenome (Derks et al. 2022; Jiang et al. 2023). These studies have identified a wide range of structural variants, including insertions, deletions, inversions, and translocations. However, linking these structural variants to phenotypes requires further genetic analysis (i.e. include structural variants in SNP panels and genome-wide association studies (GWAS)). Structural variants that are associated with desirable traits can potentially be included on the genotyping chip to develop more efficient and sustainable breeding programs that incorporate the structural variants. However, if there is sufficient LD between the structural variant and SNP variants on the chip, then imputation of structural variants is feasible (Lee et al. 2020). Together, the growing pig pangenome and catalog of structural variants will provide further information on the diversity between breeds and structural changes that are unique to specific breeds and possibly phenotypes.

3 Genomic selection and quantitative trait loci mapping

Genomic selection uses genomic data to predict which animals to use for the next generation and enables breeders to identify and select animals with desirable traits at a much earlier stage of development than traditional breeding methods (Meuwissen et al. 2001). Genomic selection has been widely adopted in the pig industry and has led to significant improvements in the accuracy and efficiency of pig breeding programs. Genomic selection relies on the use of genome-wide markers to estimate the breeding values of individual pigs (Knol et al. 2016). These markers are typically SNPs that are segregating at varying frequencies in the populations. Genomic selection relies on LD between SNPs on the chip and causal variants as the variants on the chip are most likely not causal. Hence, genomic selection works as a black box and further genomic studies are required to unravel exact causal variants and molecular mechanisms associated with traits of interest (Habier et al. 2013).

One relatively new technique to genotype animals for genomic prediction is using low-pass sequencing (Hickey 2013). With a robust reference population comprising whole-genome sequenced individuals, the imputation accuracy achieved is high (even for low-frequency variants) (Ros-Freixedes et al. 2017; Snelling et al. 2020). The declining cost of low-pass sequencing is getting close to genotyping animals on medium-density SNP chips. This cost-effectiveness enhances the scalability and feasibility of incorporating (low-pass) sequencing data into breeding programs. Furthermore, low-pass sequencing offers the potential to capture structural variations alongside SNPs and Indels, thereby providing a more comprehensive view of genomic variation within populations. As such, the integration of low-pass sequencing data in animal breeding practices will likely become a more common practice.

3.1 Quantitative trait loci mapping functional variation

As noted above, one key approach to identifying genetic factors that underlie the traits that are critical to pig productivity and health is QTL mapping. QTL mapping involves identifying regions of the genome that are associated with variation in a particular trait of interest, such as disease resistance or meat quality. However, while identifying QTLs is relatively straight forward, identifying the causal variant(s) is challenging. QTL fine-mapping typically involves a combination of genetic and genomic techniques, including linkage analysis, association mapping, and functional validation assays (Broekema et al. 2020). With the increasing volume of whole genome sequence data and functional genomic data (e.g. DNA methylation, histone modifications, ATAC-Seq) new applications for fine-mapping have been developed for application in livestock (Xiang et al. 2019; 2021). One approach is to leverage Combined Annotation Dependent Depletion (CADD) scores and population whole genome sequence data to prioritize sequence variation in QTL regions (Gross et al. 2020; Derks et al. 2021).

One key aspect of the genotype–phenotype relationship is that most of the phenotypic changes observed in a population are due to changes in gene expression and not due to changes in the coding sequence of genes (Gamazon et al. 2015). With increasing volumes of functional genomic, data researchers can predict the epigenomic states of the genome in different tissues (Pan et al. 2021). This is especially relevant to identify regulatory regions and variants that will likely enhance QTL mapping, especially in combination with expression QTL studies (eQTL). Combining QTL mapping with functional genomics approaches will provide a more comprehensive understanding of the genetic basis and the molecular mechanisms of important traits in pigs (Johnsson 2023).

3.2 Novel traits

Genomics techniques can help to identify genomic markers that affect traits that are difficult to measure or fine map. These traits often lack a proper reference population which hampers accurate prediction of breeding values (Harlizius et al. 2020). Among these 'difficult' traits are disease resistance and robustness but also traits that are only shown in the final crossbred production animals (and not in the purebred breeding lines). The use of sequence information in combination with gene expression and epigenomic data in small-scale studies can provide genetic markers that affect important traits which can subsequently be implemented in breeding. This has been successfully applied in the selection for natural antibodies (i.e. affecting disease resistance) in chicken (Berghof et al. 2017) and selection for increased resistance to Porcine Reproductive and Respiratory Syndrome (PRRS) in pigs (Dunkelberger et al. 2017).

4 The functional genome and causal variation

In this section, we will outline several key developments in the area of genomics that will enhance the discovery of functional genetic variation underlying traits of interest.

4.1 Gene expression and expression quantitative trait loci

The first application to understand the regulatory genome is to study the effect of genetic variation on gene expression. eQTL mapping involves identifying genetic variants that are associated with variation in the expression of a particular gene. eQTLs are typically classified as either cis or trans:

- Cis-eQTLs (cis-acting expression quantitative trait loci) are genetic variants that regulate gene expression levels in nearby genomic regions.
- Trans-eQTLs (trans-acting expression quantitative trait loci) are genetic variants that regulate gene expression levels in distant genomic regions, often on different chromosomes.

Cis-eQTLs tend to contain fewer false positives because they act locally, affecting genes in close proximity, which reduces the likelihood of spurious associations due to LD or other confounding factors compared to the more distant and complex regulatory interactions of trans-eQTLs.

In pigs, eQTL mapping has been used to study a wide range of traits, including meat quality (Liu et al. 2020), disease resistance (Kommadath et al. 2017), and production traits (Carmelo and Kadarmideen 2020). However, studies on the relationship between gene expression and reproductive performance are scarce. In eQTL mapping, hotspots represent genomic regions where the expression of multiple genes show significant association with the same set of genetic variants, indicating shared regulatory mechanisms (Cesar et al. 2018). Fig. 1 outlines an example of such trans-eQTL hotspots in four different pig tissues (Gòdia et al. 2022). The identification of trans-eQTL hotspots is crucial for understanding the complex regulatory networks that govern gene expression in various biological processes (Tian et al. 2016).

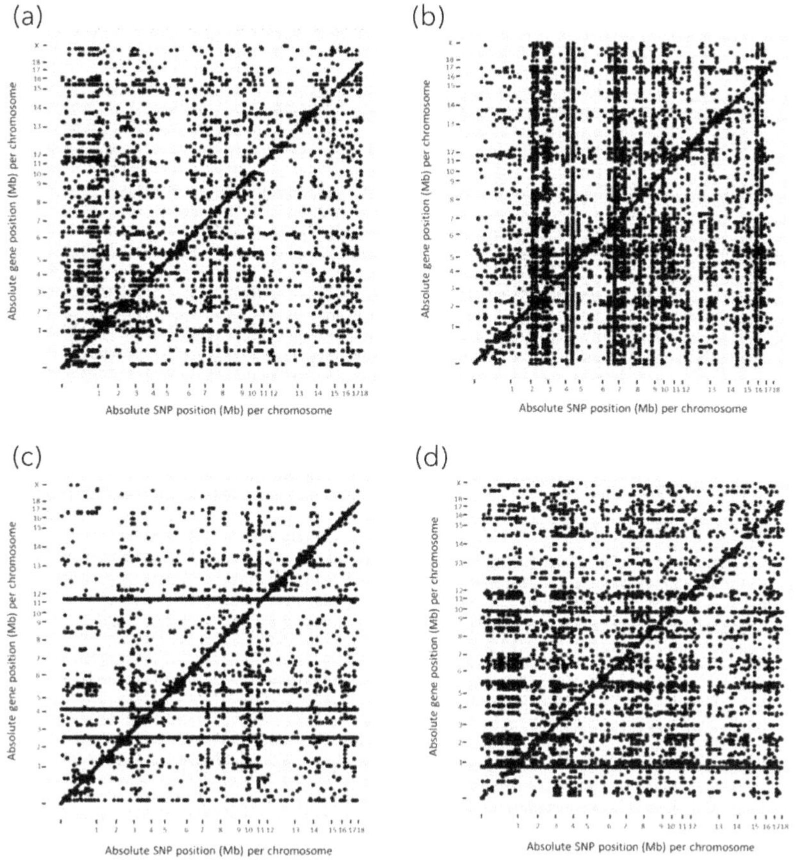

Figure 1 The absolute genomic position of target genes (y-axis) versus absolute genomic position of associated SNP-eQTLs (x-axis) in (a) liver, (b) lung, (c) spleen, and (d) muscle tissues. Each black dot represents a significant association between a target gene and a SNP-eQTL. Diagonal dots represent cis-eQTLs and off-diagonal dots represent trans-eQTLs. The presence of multiple vertical bands signifies that numerous expressed target genes are linked to a single genomic locus, indicating trans-eQTL hotspots. Source: Adapted from Gòdia et al. (2022).

Trans-eQTL hotspots have been identified in various species, including pigs (Criado-Mesas et al. 2020). These hotspots often contain key regulatory genes (e.g. transcription factors) or genetic elements that exert widespread effects on gene expression across multiple cellular pathways and tissues. However, it is worth noting that the identification and interpretation of trans-eQTL hotspots can be challenging because of the complex nature of gene regulation, the influence of environmental factors, and the potential interactions between genetic variants and genes. Thus, advanced statistical and bioinformatics methods are employed to accurately identify and interpret trans-eQTL hotspots (Nica and Dermitzakis 2013).

4.2 Functional genome (epigenomic states)

Much of the variation underlying changes in gene expression, as identified in eQTL studies, will affect regulatory sequences in the genome. By inferring the epigenomic state of the genome, we can infer the regulatory regions in the genome (Mehrmohamadi et al. 2021). These so-called epigenomic states, which are modifications to DNA or the proteins that package DNA (histones) that can affect gene expression without changing the underlying DNA sequence, will provide information on active or inactive regulatory regions including promoter regions, enhancer elements, and other regulatory sequences (Pan et al. 2021).

In pigs, Pan et al. (2021) predicted 15 different epigenomic states in 14 biologically important tissues. States include active and weak promoter regions and enhancer elements (e.g. active, weak, and poised) and open chromatin as indicated by ATAC islands. Overall, the study of functional genomic variation and epigenomic states is a rapidly evolving field (van der Velde et al. 2021). Knowing which parts of the DNA are regulatory in which tissue and/or timepoint will be crucial in identifying regulatory variants underlying phenotypic change and is especially powerful in combination with large-scale eQTL studies or the FarmGTEx resource (Teng et al. 2022).

4.3 Predicting variant consequence in (non)-coding regions

Numerous tools exist for the prediction of coding sequence variation on gene/protein function (Choudhury et al. 2022). However, predicting the impact of non-coding variation remains a challenge. One promising method to score the impact of coding as well as non-coding variants is CADD. CADD is a machine learning tool that integrates multiple annotations, including DNA sequence conservation and functional genomics data, to provide a single metric that reflects the likelihood that a given genetic variant has an impact (Rentzsch et al. 2019).

A CADD tool has been developed for pig, creating the pig CADD (pCADD) score (Gross et al. 2020). The pCADD score allows for the prediction of the functional impact of genetic variants in pigs and can be used to prioritize variants in QTL regions (Derks et al. 2021). A relatively similar approach was developed in cattle as FAETH scores (Xiang et al. 2019). However, in order to identify causal variants underlying GWAS results, a reference panel of individuals with whole genome sequence data is required to identify variants on the same haplotype (i.e. in LD) as the top SNP. Alternatively, genotype data from SNP chips can be imputed to sequence to perform a sequence-based GWAS. This GWAS in combination with variant prioritization with pCADD will provide a powerful tool to examine potential causal variants. However, although this approach provides a first step in prioritizing variants in non-coding regions, the pCADD scores need to be updated regularly with new (functional) genomic data, to improve the accuracy of the scores.

4.4 Validation of causal variants

Several strategies can be employed to validate causal regulatory variants, including functional assays and genome editing tools. These approaches aim to provide evidence that a specific genetic variant directly influences gene expression or regulatory elements. Traditional methods for studying regulatory variants involve time-consuming and expensive experiments to test the effect of each variant individually.

Recently a massively parallel reporter assay (MPRA) was developed in cattle to simultaneously perform functional testing of thousands of candidate causative regulatory variants (Littlejohn et al. 2022). MPRA allows for the high-throughput screening of thousands of variants in parallel which will accelerate the discovery of functional genomic variants. The basic principle of MPRA involves introducing candidate regulatory variants into a reporter gene, which then allows for the measurement of the effect of each variant on gene expression. This will allow for direct testing of regulatory variants and assess their impact on gene expression. This tool also holds promise in pigs to validate candidate regulatory variants at large scale.

4.5 Gene editing

Gene editing is a rapidly-advancing technology that enables precise and targeted modifications to the genome of an organism. In pigs, gene editing holds great promise for improving animal health, welfare, and productivity (Yang and Wu 2018). Gene editing with the CRISPR/Cas9 system uses a guide RNA to target specific regions of the genome and a Cas9 enzyme to make precise edits (Nidhi et al. 2021). This technology has enabled the development of highly specific and efficient gene editing tools that can be used to create precise genetic modifications in pigs.

One of the primary applications of gene editing in pigs is in the development of disease-resistant animals. Several successes have been reported including the editing of the *CD163* gene to improve PRRS resistance in pigs (Burkard et al. 2018). One factor that prevents wide usage of gene editing is maintaining an edited allele in a breeding population, without increasing inbreeding rates, requiring many thousands of edits for which the technique is currently not designed (Bastiaansen et al. 2018). In addition, gene editing is most powerful for monogenic traits in which a single edit can have a major influence on the phenotype (Mueller et al. 2021). However, most economically important traits in pig (e.g. growth or reproductive performance) are highly polygenic meaning that many genes have subtle effect on the trait. This means that, for polygenic traits, the added benefit is limited.

Despite the potential benefits of gene editing in pigs, there are also ethical and regulatory considerations that must be taken into account (de Graeff et al. 2019). Also, it is important to carefully evaluate the risks and benefits of gene editing in pigs and ensure that the technology is used in a responsible and sustainable manner. Together, gene editing represents a powerful tool for advancing the field of pig genetics and improving animal health, welfare, and productivity. However, the use of gene editing in pigs is currently still in its infancy and will require further technological advances to become a more common practice in pig breeding (Whitworth et al. 2022).

5 Case study: quantitative trait loci fine-mapping for meat quality traits identifies genes involved in muscle glycogen storage

Meat quality is an important trait in the pork industry, with factors such as flavor, tenderness, and juiciness influencing consumer preference and purchase decisions. QTL mapping has been used extensively to identify the genetic basis of meat quality traits in pigs (Hu et al. 2022). One important aspect of meat quality is glycogen storage, which can affect post-mortem metabolism and ultimately impact meat quality. Glycogen is a carbohydrate that is stored in muscle tissue and serves as an energy source during exercise and other physical activities. After slaughter, glycogen is converted into lactic acid, which can impact meat quality traits such as pH, water-holding capacity, and color (Moreno et al. 2020).

Several studies have identified major QTL affecting meat quality in pigs (Rothschild et al. 2007; Gao et al. 2021). Two examples of major QTL affecting meat quality are a missense variant in the *PRKAG3* gene (Milan et al. 2000) and a splice variant in the *PHKG1* gene (Ma et al. 2014). *PRKAG3* which regulates glycogen storage. The isoform affected by the missense mutation in *PRKAG3* has a role in the metabolic plasticity of fast-glycolytic muscle and is primarily expressed in white skeletal muscle fibers (Mahlapuu et al. 2004). Gain of function mutations in the *PRKAG3* gene are associated with increased glycogen content in skeletal muscle in pig, negatively affecting meat quality (Ciobanu et al. 2001). The *PHKG1* splice mutation causes the deficiency of glycogen breakdown, consequently leading to glycogen storage disease and acid meat pigs (Ma et al. 2014).

Further QTL studies on meat quality have identified numerous loci associated with meat quality (Derks et al. 2021). One example of a GWAS for meat quality in a synthetic boar line is provided in Fig. 2. The candidate causal

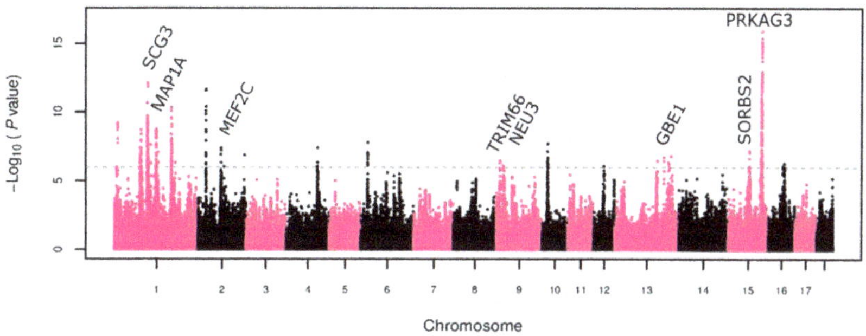

Figure 2 Manhattan plot for drip loss in a synthetic breed. The figure shows significant loci and likely causal genes identified. Source: Adapted from Derks et al. (2021).

variants and genes have shown a strong enrichment for pathways involved in glycogen synthesis and storage. Increased levels of muscle glycogen are known to lead to increased drip loss, which is considered to negatively affect meat quality. Examples discovered include regulatory variants affecting the *MEF2C*, *SCG3*, and *GBE1* genes. *MEF2C* regulates glycogen storage in the muscle, and impaired gene function leads to increased glycogen content in the muscle (Anderson et al. 2015). *GBE1* codes for a glycogen branching enzyme associated with glycogen storage disease. Two additional coding sequence variants were identified in the *NEU3* and *MAP1A* genes. *NEU3* overexpression increases glycogen deposition (Yoshizumi et al. 2007).

Meat quality provides an interesting case study to link the genotype with phenotype including in-depth understanding of the molecular pathways associated. Most variants found to be associated with meat quality affect genes that are directly or indirectly involved in glycogen deposition in the muscle. In addition, for meat quality a few QTLs were described with a major impact (Milan et al. 2000; Ma et al. 2014), which is not common in many other polygenic traits. Therefore, meat quality is one of the key traits in which direct knowledge from molecular pathways will enhance breeding for improved meat quality (Larzul 2021).

6 Mapping of deleterious variation

One major interest of breeders is to understand the relationship between inbreeding and performance. Although inbreeding rates can be defined relatively accurately using genomic tools in livestock (Doekes et al. 2021), the exact mutations contributing to inbreeding depression remain largely unknown. In the next section, the work on deleterious alleles in pigs is outlined with a focus on recessive lethal variants and other deleterious recessive alleles.

6.1 High-impact deleterious variants

The current resources of genotyped, sequenced, and phenotyped animals in pig breeding allow for the identification of deleterious variants (Bianco et al. 2015). The most severe mutations cause embryonic or pre- and postnatal lethality. Recently, many such variants have been mapped in pigs, the majority leading to embryonic lethality resulting in smaller litters in carrier-by-carrier matings (Derks et al. 2018; 2019a). However, genetic syndromes causing postnatal lethality have also been identified (Derks et al. 2019b; Matika et al. 2019).

The majority of the lethal variants are not only caused by single nucleotide changes or small indels but also large structural variants have been discovered (Derks et al. 2018). The lethal recessives can be identified by examining the deficiency of homozygous haplotypes in the mature population based on genotype data from >10 000 individuals per population. However, also direct predictions from sequence can identify deleterious alleles rather efficiently. This is a consequence of the large quantities of sequenced individuals together with improved prediction tools that can also predict the consequence of non-coding variants (Gross et al. 2020). Identifying and understanding the effects of deleterious and lethal variants will allow for the development of new strategies for improving pig health and productivity and for developing more effective breeding programs that are tailored to prevent specific matings that can lead to affected animals.

6.2 Non-additive genome-wide association studies to identify deleterious variants

Traditional GWAS models assume that the genetic effects are additive, meaning that each allele contributes additively to the phenotype. However, non-additive genetic effects, such as dominance and epistasis, are also known to play an important role in determining complex traits in pigs. In addition, many deleterious variants are recessive, meaning that only homozygous individuals show an altered phenotype. Hence, non-additive models are better suited to discover recessive deleterious variants as the group of homozygous animals will be tested specifically against the group of carriers and non-carriers.

In recent years, there has been a growing interest in using non-additive GWAS models to identify deleterious variants in livestock populations (Reynolds et al. 2021). Reynolds et al. 2021 discovered several deleterious recessive variants with a major impact on the fitness of homozygous individuals. Traits that can be examined include both production and reproduction traits (Reynolds et al. 2021, 2022). Non-additive GWAS resulted in the discovery of many novel regions compared to the 'classic' additive model. Hence, non-additive GWAS models allow researchers to identify these variants, which may be missed using traditional GWAS models. Nevertheless, large numbers of genotyped (and imputed) individuals are required to identify low-frequency deleterious variants but with the current availability of hundreds of thousands of genotyped individuals in commercial livestock population this approach has become feasible.

Boshove et al. (2023) applied such non-additive GWAS in pigs by using 200 000 genotyped pigs from two breeding lines (one sire and one dam line) for which various novel deleterious variants have been mapped, including several loss-of-function variants in genes leading to impaired growth (Boshove et al. 2024). By including non-additive genetic effects in the analysis, researchers can identify deleterious variants which in homozygous state have a negative impact on the phenotype (recessive), even if those alleles are present at low frequencies. This will likely unravel a large list of novel deleterious variants that can subsequently be purged from breeding populations.

7 Conclusion and future trends

In summary, QTL mapping and functional variation analysis are important tools used in pig genetics to identify the genetic basis of various production traits, including meat quality, growth rate, and disease resistance. These techniques involve identifying regions of the genome associated with the trait of interest and narrowing down candidate causative variants using high-resolution genetic and genomic data.

Future trends in QTL mapping and functional variation analysis in pigs include the incorporation of advanced genomic technologies, such as CRISPR-Cas9 gene editing and MPRAs, to validate candidate causative variants and to improve the accuracy and efficiency of breeding strategies. Additionally, the use of pangenomic approaches and the integration of multi-omics data, including transcriptomics, epigenomics, and proteomics, are expected to provide a more comprehensive understanding of the genetic and molecular mechanisms underlying pig traits.

Furthermore, there is a growing interest in identifying and characterizing deleterious and lethal variants in pigs and developing strategies to mitigate their effects on pig health and productivity. This involves the integration of phenotypic data, genomic data, and bioinformatics tools to identify and validate these variants, and to develop more effective breeding strategies that reduce their impact on pig performance.

8 Where to look for further inforamtion

Unraveling the genetic basis of traits in pigs requires collaboration between breeders, geneticists, and researchers utilizing advanced genomics tools. For those interested in exploring pig genomics, several resources provide detailed information and datasets. The Pig QTL Database (PigQTLdb) (https://www.animalgenome.org/cgi-bin/QTLdb/SS/index) is a valuable resource for locating quantitative trait loci (QTLs) linked to various phenotypes in pigs, offering insights into traits related to growth, reproduction, and disease resistance. The NCBI Genome Data Viewer (https://www.ncbi.nlm.nih.gov/genome/) and Ensembl Genome Browser (https://www.ensembl.org/Sus_scrofa/Info/Index) are excellent platforms for accessing annotated pig genomes. They provide tools to explore genes, variants, and genomic regions in depth. For those interested in pig transcriptomics, the PigGTEx database (https://piggtex.ipiginc.com/) is an important resource. It offers gene expression profiles across various tissues, helping researchers link gene function to specific traits and biological processes in pigs.

9 References

Anderson, C.M., et al. 2015. Myocyte enhancer factor 2C function in skeletal muscle is required for normal growth and glucose metabolism in mice. *Skelet Muscle* 5: 7.

Andersson, L., et al. 2015. Coordinated international action to accelerate genome-to-phenome with FAANG, the Functional Annotation of Animal Genomes project. *Genome Biol* 16: 57.

Archibald, A.L., et al. 2010. Pig genome sequence–analysis and publication strategy. *BMC Genom* 11: 438.

Bastiaansen, J.W.M., et al. 2018. The impact of genome editing on the introduction of monogenic traits in livestock. *Genet Sel Evol* 50: 18.

Berghof, T.V.L., et al. 2017. Genomic region containing toll-like receptor genes has a major impact on total IgM antibodies including KLH-binding IgM natural antibodies in Chickens. *Front Immunol* 8: 1879.

Bianco, E., et al. 2015. A deep catalog of autosomal single nucleotide variation in the pig. *PLoS One* 10: e0118867.

Boshove, A., et al. 2023. Imputation to sequence and GWAS to identify rare and deleterious alleles in commercial pig breeds. In *Plant and Animal Genomics (PAG)*, San Diego, USA.

Boshove, A., et al. 2024. Large scale sequence-based screen for recessive variants allows for identification and monitoring of rare deleterious variants in pigs. *PLoS Genet* 20: e1011034.

Broekema, R.V., et al. 2020. A practical view of fine-mapping and gene prioritization in the post-genome-wide association era. *Open Biol* 10: 190221.

Burkard, C., et al. 2018. Pigs lacking the scavenger receptor cysteine-rich domain 5 of CD163 are resistant to porcine reproductive and respiratory syndrome virus 1 infection. *J Virol* 92. https://doi.org/10.1128/jvi.00415-18

Carmelo, V.A.O. and Kadarmideen, H.N. 2020. Genetic variations (eQTLs) in muscle transcriptome and mitochondrial genes, and trans-eQTL molecular pathways in feed efficiency from Danish breeding pigs. *PLoS One* 15: e0239143.

Cesar, A.S.M., et al. 2018. Identification of putative regulatory regions and transcription factors associated with intramuscular fat content traits. *Bmc Genom* 19: 499.
Choudhury, A., et al. 2022. Comparative analysis of web-based programs for single amino acid substitutions in proteins. *PLoS One* 17: e0267084.
Ciobanu, D., et al. 2001. Evidence for new alleles in the protein kinase adenosine monophosphate-activated gamma(3)-subunit gene associated with low glycogen content in pig skeletal muscle and improved meat quality. *Genetics* 159: 1151–1162.
Clark, E., et al. 2024. Generating annotated telomere to telomere genome assemblies using a trio-binning approach for a cross of two production-relevant domestic pig breeds. In *Plant and Animal Genome XXXI Conference*, San Diego (CA), USA.
Clark, E.L., et al. 2020. From FAANG to fork: application of highly annotated genomes to improve farmed animal production. *Genome Biol* 21: 285.
Criado-Mesas, L., et al. 2020. Identification of eQTLs associated with lipid metabolism in Longissimus dorsi muscle of pigs with different genetic backgrounds. *Sci Rep-Uk* 10: 9845.
de Graeff, N., et al. 2019. The ethics of genome editing in non-human animals: a systematic review of reasons reported in the academic literature. *Philos Trans R Soc Lond B Biol Sci* 374: 20180106.
Derks, M.F.L., et al. 2018. Balancing selection on a recessive lethal deletion with pleiotropic effects on two neighboring genes in the porcine genome. *PLoS Genet* 14: e1007661.
Derks, M.F.L., et al. 2019a. Loss of function mutations in essential genes cause embryonic lethality in pigs. *PLoS Genet* 15: e1008055.
Derks, M.F.L., et al.. 2019b. Detection of a frameshift deletion in the SPTBN4 gene leads to prevention of severe myopathy and postnatal mortality in pigs. *Front Genet* 10: 1226.
Derks, M.F.L., et al. 2021. Accelerated discovery of functional genomic variation in pigs. *Genomics* 113: 2229–2239.
Derks, M.F.L., et al. 2022. 564. A pan-genome of commercial pig breeds. In *Proceedings of 12th World Congress on Genetics Applied to Livestock Production (WCGALP)*, pp. 2337–2341. https://doi.org/10.3920/978-90-8686-940-4_564
Doekes, H.P., et al. 2021. How depressing is inbreeding? A meta-analysis of 30 years of research on the effects of inbreeding in livestock. *Genes* 12. https://doi.org/10.3390/genes12060926
Dunkelberger, J.R., et al. 2017. Effect of a major quantitative trait locus for porcine reproductive and respiratory syndrome (PRRS) resistance on response to coinfection with PRRS virus and porcine circovirus type 2b (PCV2b) in commercial pigs, with or without prior vaccination for PRRS. *J Anim Sci* 95: 584–598.
Ernst, C.W. and Steibel, J.P. 2013. Molecular advances in QTL discovery and application in pig breeding. *Trends Genet* 29: 215–224.
Gamazon, E.R., et al. 2015. A gene-based association method for mapping traits using reference transcriptome data. *Nat Genet* 47: 1091-+.
Gao, G., et al. 2021. Genome-wide association study of meat quality traits in a three-way crossbred commercial pig population. *Front Genet* 12: 614087.
Giuffra, E., et al. 2019. Functional Annotation of Animal Genomes (FAANG): current achievements and roadmap. *Annu Rev Anim Biosci* 7: 65–88.
Gòdia, M., et al. 2022. 558. Unravelling regulatory variants affecting gene expression in four porcine tissues. In *Proceedings of 12th World Congress on Genetics Applied to*

Livestock Production (WCGALP). Wageningen Academic Publishers, pp. 2313–2316. https://doi.org/10.3920/978-90-8686-940-4_558

Groenen, M.A., et al. 2012. Analyses of pig genomes provide insight into porcine demography and evolution. *Nature* 491: 393–398.

Gross, C., et al. 2020. pCADD: SNV prioritisation in Sus scrofa. *Genet Sel Evol* 52: 4.

Habier, D., et al. 2013. Genomic BLUP decoded: a look into the black box of genomic prediction. *Genetics* 194: 597-+.

Haley, C.S. and Visscher, P.M. 1998. Strategies to utilize marker-quantitative trait loci associations. *J Dairy Sci* 81(Suppl 2): 85–97.

Harlizius, B., et al. 2020. Breeding for resilience: new opportunities in a modern pig breeding program. *J Anim Sci* 98: S150–S154.

Hickey, J.M. 2013. Sequencing millions of animals for genomic selection 2.0. *J Anim Breed Genet* 130: 331–332.

Hill, W.G. 2014. Applications of population genetics to animal breeding, from wright, fisher and lush to genomic prediction. *Genetics* 196: 1–16.

Ho, S.S., et al. 2020. Structural variation in the sequencing era. *Nat Rev Genet* 21: 171–189.

Hu, Z.-L., et al. 2022. Bringing the animal QTLdb and CorrDB into the future: meeting new challenges and providing updated services. *Nucleic Acids Res* 50: D956–D961.

Jiang, Y.F., et al. 2023. Pangenome obtained by long-read sequencing of 11 genomes reveal hidden functional structural variants in pigs. *iScience* 26: 106119.

Johnsson, M. 2023. The big challenge for livestock genomics is to make sequence data pay. *arXiv preprint arXiv:230201140*.

Knol, E.F., et al. 2016. Genomic selection in commercial pig breeding. *Anim Front* 6: 15–22.

Kommadath, A., et al. 2017. Genetic architecture of gene expression underlying variation in host response to porcine reproductive and respiratory syndrome virus infection. *Sci Rep-Uk* 7. https://doi.org/10.1038/srep46203

Koren, S., et al. 2018. De novo assembly of haplotype-resolved genomes with trio binning. *Nat Biotechnol*. https://doi.org/10.1038/nbt.4277.

Larzul, C. 2021. How to improve meat quality and welfare in entire male pigs by genetics. *Animals* 11. https://doi.org/10.3390/ani11030699.

Lee, Y.L., et al. 2020. Functional and population genetic features of copy number variations in two dairy cattle populations. *BMC Genom* 21: 89.

Littlejohn, M.D., et al. 2022. 534. A massively parallel reporter assay to screen bovine regulatory variants. In *Proceedings of 12th World Congress on Genetics Applied to Livestock Production (WCGALP)*, pp. 2216–2219. https://doi.org/10.3920/978-90-8686-940-4_534

Liu, Y., et al. 2020. Genome-wide analysis of expression QTL (eQTL) and allele-specific expression (ASE) in pig muscle identifies candidate genes for meat quality traits. *Genet Sel Evol* 52: 59.

Ma, J., et al. 2014. A splice mutation in the PHKG1 gene causes high glycogen content and low meat quality in pig skeletal muscle. *PLoS Genet* 10: e1004710.

Mahlapuu, M., et al. 2004. Expression profiling of the gamma-subunit isoforms of AMP-activated protein kinase suggests a major role for gamma3 in white skeletal muscle. *Am J Physiol Endocrinol Metab* 286: E194–E200.

Matika, O., et al. 2019. Balancing selection at a premature stop mutation in the myostatin gene underlies a recessive leg weakness syndrome in pigs. *PLoS Genet* 15: e1007759.

Mehrmohamadi, M., et al. 2021. A comparative overview of epigenomic profiling methods. *Front Cell Dev Biol* 9: 714687.

Meuwissen, T.H.E., et al. 2001. Prediction of total genetic value using genome-wide dense marker maps. *Genetics* 157: 1819–1829.

Milan, D., et al. 2000. A mutation in PRKAG3 associated with excess glycogen content in pig skeletal muscle. *Science* 288: 1248–1251.

Moreno, I., et al. 2020. Glycogen and lactate contents, pH and meat quality and gene expression in muscle Longissimus dorsi from iberian pigs under different rearing conditions. *Livest Sci* 240: 104167.

Mueller, M.L., et al. 2021. Comparison of gene editing versus conventional breeding to introgress the POLLED allele into the tropically adapted australian beef cattle population. *Front Genet* 12. https://doi.org/10.3389/fgene.2021.593154

Nica, A.C. and Dermitzakis, E.T. 2013. Expression quantitative trait loci: present and future. *Philos Trans R Soc Lond B Biol Sci* 368: 20120362.

Nidhi, S., et al. 2021. Novel CRISPR-Cas systems: an updated review of the current achievements, applications, and future research perspectives. *Int J Mol Sci* 22: 3327.

Nurk, S., et al. 2022. The complete sequence of a human genome. *Science* 376: 44–53.

Pan, Z.Y., et al. 2021. Pig genome functional annotation enhances the biological interpretation of complex traits and human disease. *Nat Commun* 12: 5848.

Paudel, Y. 2015. *Structural variations in pig genomes*. Wageningen University, Wageningen.

Ramos, A.M., et al. 2009. Design of a high density SNP genotyping assay in the pig using SNPs identified and characterized by next generation sequencing technology. *PLoS One* 4: e6524.

Rentzsch, P., et al. 2019. CADD: predicting the deleteriousness of variants throughout the human genome. *Nucleic Acids Res* 47: D886–D894.

Reynolds, E.G.M., et al. 2021. Non-additive association analysis using proxy phenotypes identifies novel cattle syndromes. *Nat Genet* 53: 949–954.

Reynolds, E.G.M., et al. 2022. Non-additive QTL mapping of lactation traits in 124,000 cattle reveals novel recessive loci. *Genet Sel Evol* 54: 5.

Ros-Freixedes, R., et al. 2017. A method for allocating low-coverage sequencing resources by targeting haplotypes rather than individuals. *Genet Sel Evol* 49: 78.

Rothschild, M.F., et al. 2007. Advances in QTL mapping in pigs. *Int J Biol Sci* 3: 192–197.

Snelling, W.M., et al. 2020. Assessment of imputation from low-pass sequencing to predict merit of beef steers. *Genes (Basel)* 11: 1312.

Teng, J., et al. 2022. A compendium of genetic regulatory effects across pig tissues. *bioRxiv*. https://doi.org/10.1101/2022.11.11.516073

Tian, J.A., et al. 2016. The dissection of expression quantitative trait locus hotspots. *Genetics* 202: 1563-+.

Tian, X., et al. 2020. Building a sequence map of the pig pan-genome from multiple de novo assemblies and Hi-C data. *Sci China Life Sci* 63: 750–763.

van der Velde, A., et al. 2021. Annotation of chromatin states in 66 complete mouse epigenomes during development. *Commun Biol* 4: 239.

Veroneze, R., et al. 2014. Linkage disequilibrium patterns and persistence of phase in purebred and crossbred pig (Sus scrofa) populations. *BMC Genet* 15: 126.

Warr, A., et al. 2020. An improved pig reference genome sequence to enable pig genetics and genomics research. *Gigascience* 9. https://doi.org/10.1093/gigascience/giaa051

Whitworth, K.M., et al. 2022. Improvements in pig agriculture through gene editing. *CABI Agri Biosci* 3: 41.

Xiang, R., et al. 2019. Quantifying the contribution of sequence variants with regulatory and evolutionary significance to 34 bovine complex traits. *Proc Natl Acad Sci U S A* 116: 19398–19408.

Xiang, R., et al. 2021. Genome-wide fine-mapping identifies pleiotropic and functional variants that predict many traits across global cattle populations. *Nat Commun* 12: 860.

Yang, H. and Wu, Z. 2018. Genome editing of pigs for agriculture and biomedicine. *Front Genet* 9: 360.

Yoshizumi, S., et al. 2007. Increased hepatic expression of ganglioside-specific sialidase, NEU3, improves insulin sensitivity and glucose tolerance in mice. *Metabolism* 56(3): 420–429.

Chapter 3

Advances in genomic selection in pig breeding

Xiaolei Liu, Lilin Yin, Dong Yin, Xinyun Li and Shuhong Zhao, Huazhong Agricultural University/Hubei Hongshan Laboratory – Wuhan, China

1 Introduction

Before the theory of genetics was proposed, people selected outstanding pigs based on their empirical experience from recorded observations of phenotypic characteristics. Pig breeders had to wait for all individuals to have their own production test scores before selection, making genetic progress in pig breeding extremely slow and imprecise. In the mid-twentieth century, with the emergence of concepts such as heritability, breeders sought to dissect phenotypic values into breeding values and environmental residuals. At the same time, Charles Roy Henderson developed the BLUP (Best Linear Unbiased Prediction) method (Henderson, 1949, 1950, 1975; Henderson et al., 1959). He used a mixed linear model (MLM) to account for unbalanced phenotypic observations from individuals in different production units to estimate breeding values. This revolutionized the field of prediction and selection for pig economic traits. The breeding value of an individual could be estimated from a mixed linear model by using the information from phenotypic observations, environmental records, and a relationship matrix derived from pedigrees. The goal of pig breeders essentially was to model all the available data to derive breeding values from phenotypes accurately, which were used to select superior pigs rather than simply relying on the original phenotypes.

http://dx.doi.org/10.19103/AS.2024.0137.03

In the early twenty-first century, Meuwissen et al. (2001) proposed the concept of genomic selection, which utilized genetic markers across the whole genome to predict the genetic merits of each individual. In addition to pedigree records, genomic information was also used to construct a genomic relationship matrix (GRM) to extend the applicability of the BLUP method (VanRaden, 2008). Compared with the relationship matrix that was derived from pedigree data only, the relationship between each pair of individuals calculated using genomic data was more accurate because the Mendelian sampling error was reduced (Habier et al., 2007; Muir, 2007). Owing to its high predictive efficiency, genomic selection has become a key technique in cattle, pig, and other livestock breeding (Georges et al., 2019), and it has also been applied in other areas, e.g. to predict the risk of human disease (Chatterjee et al., 2016).

With the development of sequencing technology, both sample size and marker density for pig genomic selection have increased rapidly. Various statistical methods and computation technologies have been developed to accommodate the growing volume of different types of data. However, how to efficiently analyze huge data sets to achieve more accurate genomic prediction is still a major challenge.

2 Improving genomic prediction accuracy: enhancing classical algorithms

Genomic prediction strategies vary among species with distinct types of reproduction, but statistical methods and genetic theories are consistent across species. The breeding values of economically important traits can be estimated by incorporating information from phenotypic records and using a relationship matrix to measure the distance among individuals which can then be incorporated into a linear mixed model. Before high-throughput genetic markers could be detected, the relationship matrix was derived directly from pedigree records (known as the 'A matrix'), a method known as PBLUP (Pedigree-based Best Linear Unbiased Prediction). However, the predicted breeding values of full sibs of new born animals were identical.

To reduce the Mendelian sampling error and to define the relationships among individuals precisely, the A matrix was replaced by a genomic relationship matrix (G matrix), constructed from genetic markers across the whole genome, which provided the basis for what became known as Genomic Best Linear Unbiased Prediction (GBLUP). Unlike the practice of obtaining breeding values directly, based on relationship information in GBLUP, breeding values could also be estimated using a 'two-step' strategy, which first estimated the effect of each marker and then multiplied the numeric coded genotype matrix by the vector of marker effects to get the breeding value for each individual. This ridge regression BLUP (RRBLUP) model assumed that the effects of all markers were from one normal distribution with a mean of zero and the same variance (Whittaker et al., 2000). Theoretically, it was identical to GBLUP (Goddard, 2009; Hayes et al., 2009).

However, the prediction accuracy of RRBLUP/GBLUP was limited by the basic assumption that marker effects followed the same normal distribution. Bayesian models allow different sets of genetic markers to have different variances that follow specific distributions. To make use of this functionality, Meuwissen et al. (2001) proposed two hypotheses:

- The first hypothesis assumed that all genetic markers affected the target trait, and their effects followed the normal distribution of $N\left(0,\sigma_g^2\right)$, and the variances were unequal and followed the inverse chi-square distribution of $x^{-2}\left(v,s\right)$
- the second hypothesis assumed that only a small proportion (1-π, π was the proportion of non-effect genetic markers) of genetic markers affected the trait.

Because the estimation strategy of unknown parameters was combined with Bayesian theory, the two resulting models were called BayesA and BayesB, respectively. After that, many classical Bayesian alphabet models were developed to improve the prediction accuracy by optimizing the assumption on the distribution of unknown parameters (e.g. BayesC, BayesCPi, BayesDPi, and BayesLASSO (Gianola, 2013; Habier et al., 2011; Kizilkaya et al., 2010; Yi and Xu, 2008).

By benefiting from a greater flexibility of assumptions, Bayesian models can be superior to the models under the LMM framework in terms of prediction accuracy, especially for traits associated with major genes. However, for each trait, abundant unknown parameters needed to be optimized by tens of thousands of Markov chain Monte Carlo (MCMC) iterations, which was technically challenging and time-consuming. The computational burden required has limited its application in practical breeding programs.

More recently, many models have been developed to provide flexible hypotheses for traits of interest with various types of genetic architectures (e.g. weighting the contributions of genetic markers in construction of a relationship matrix or assigning appropriate distributions for effects and variances to multiple groups of genetic markers). Unknown parameters (e.g. weight of each marker) can be estimated by improved methods using either an LMM framework or a Bayesian framework though the computational burden increase with marker density. A kinship-adjusted multiple-loci model (KAML) (Yin et al., 2020) used machine learning integrated with multiple regression, cross-validation, grid search, and bisection algorithms to optimize the fixed and random terms of LMM to find the ideal model for prediction. It provides a comparable predictive accuracy compared with Bayesian models while achieving similar computational efficiency as LMM.

3 Improving algorithms under the best linear unbiased prediction model framework

The main differences among the range of BLUP (e.g. GBLUP) models is the construction of relationship matrix. This reflects the similarity of pairs of individuals at the genome-level constructed from all available genetic markers across the whole genome. To improve the prediction accuracy for traits with a complex genetic architecture, a series of BLUP-based models were developed to provide built-in assumptions. The main strategy was to optimize the contributions of genetic markers to traits in the prediction model.

3.1 Integrating the best linear unbiased prediction model with the single-step strategy

It has always been difficult to genotype all individuals of a large population in practice. This meant that individuals with accurate phenotypic observations, but without genotypic information, could not be used for genomic evaluation. In 2010, a single-step GBLUP model (SSGBLUP) that integrated pedigree and genotypic and phenotypic data simultaneously was proposed to maximize the utilization of available information, which made it possible to connect all phenotypic observations for both genotyped and non-genotyped individuals (Aguilar et al., 2010; Christensen and Lund, 2010). Because more phenotypic observations were included in the model, the prediction accuracy of SSGBLUP for genotyped individuals was always higher than that of GBLUP model. For non-genotyped individuals, the prediction accuracy also increased compared with the pedigree-based BLUP model due to the inclusion of genotypic information.

3.2 Weighting genetic markers as fixed effect terms

For traits controlled by only a few major genes, the underlying genetic architecture undermines the basic assumption of BLUP models. Noise generated by other SNPs would weaken the contributions of SNPs that are in linkage disequilibrium (LD) with causal genes. The straightforward solution to overcome this weakness has been to integrate SNPs with major effects or QTLs (quantitative trait loci) that had been validated in other research as covariates in BLUP models. The genomic estimated breeding value (GEBV) then provides the summation of fixed and random terms. Based on this adjustment, prediction accuracy improves significantly because SNPs with major effects fully reflect genetic variance compared to phenotypic variance. Studies have indicated that the extent of improvement in prediction accuracy varies with the genetic complexity of traits (Lopes et al., 2017; Moore et al., 2017).

However, adding SNPs or QTLs as fixed effects to improve predictions was always likely to become a "double-edged sword" for a number of reasons. First, it was extremely risky to include covariates in the linear mixed model because any associated SNPs that were false positives decreased prediction accuracy significantly. Second, most of the agricultural economic traits of interest were polygenic, which meant that those traits were affected by hundreds, even thousands, of genes with small effects. Therefore, the phenotypic variance explained (PVE) of a subset of associated SNPs was rather limited, which restricted the room for improvement in prediction performance. Third, with the increasing density of SNPs for the whole genome, adding too many SNPs in the associated regions reduced the model's degree of freedom, which may result in underdetermined problems in solving the mixed-model equation. Finally, due to the LD differences in populations with various genetic backgrounds, validated QTLs in one population did not necessarily work in others, making it questionable to utilize QTLs directly across different populations.

3.3 Weighting genetic markers in random effect terms

Another strategy to optimize the assumptions of BLUP models has been to weight the SNPs in random effect terms. Unlike incorporating associated SNPs as fixed effects, weighting SNPs in random effect terms is more flexible and safer because the number of SNPs to be weighted in random effect was not limited, and the variance of random effect term was controllable. The SNPs can be classified into multiple groups and weighted to improve prediction accuracy. The top percentage of significantly associated SNPs or SNPs that passed a certain significance threshold can be selected (based on the prior calculated GWAS results) to construct the relationship matrix, instead of using all the SNPs. This helped to minimize the negative impact by removing 'noisy' SNPs to improve prediction performance. However, setting the appropriate threshold or a top percentage to select SNPs is challenging. A wider threshold would include more 'noisy' SNPs for traits with simple genetic architecture, and a stricter threshold may lose effective genetic information for complex traits. Previous studies have shown that prediction accuracy varies based on different SNP selection strategies (Tan and Ingvarsson, 2019).

To overcome this problem, Zhang et al. (2014) proposed the BLUP|GA model, where the threshold for selecting SNPs and their weights was optimized automatically in a cross-validation procedure. The results demonstrated an improved accuracy that was comparable with Bayesian regression models. Nevertheless, the large computational burden of updating the variance components cyclically for each validation unit of the cross-validation procedure was too time-consuming to apply to a large population size. Second, the optimized parameters from cross-validation were dependent on the rough average value of prediction accuracy of different prior parameter combinations. The difference in genetic architecture or LD structure between training and validation populations changed continuously as the sampling proceeded. It was unclear whether the parameter that was optimized ultimately was appropriate for predicting the individuals without phenotypic observations because, if the LD structure was mismatched, this created a potential gap in genetic background.

Speed and Balding proposed the MultiBLUP method that ingeniously integrated genomic selection and a genome-wide association study (Speed and Balding, 2014). They divided the chromosome into chunks and merged some of the chunks by a likelihood ratio test, treating the remaining SNPs as background chunks. The relationship matrices were then constructed for every chunk respectively. All the relationship matrices were assigned as multiple random effects in GBLUP. The GEBVs equaled the sum of all random effects. The number of chunks was optimized according to the genetic architecture of traits, and each group of SNPs captured more appropriate genetic variance. Compared with the cross-validation strategy-based model, the MultiBLUP model significantly decreased the computational complexity in parameter derivation processes.

As described previously, the threshold for SNP selection is important to model stability, and the prediction accuracy of MultiBLUP varied across species (Zeng and Zhou, 2017). With the increased number of random effects, it would be a great challenge in computing time and memory capacity to estimate many variance components accurately. Moreover, the phenotypic variance was only partitioned into genetic and residual variances. The model assumption of negligible covariance between random effects was met in many cases. However, if the chunks were correlated in reality, the covariance terms between these random effects were substantial. Estimation of variance components was thus biased, resulting in variations in prediction performance (Zhou et al., 2020).

3.4 Weighting genetic markers by incorporating multi-information sources

Variance in phenotypes can be divided simply into the variance of breeding values (additive genetic variance) and the residual variance. Variances, such as dominance and interaction, can be further split out from the residual and added to the genomic selection model to improve prediction accuracy. Guo et al. (2016) evaluated the dominance effect for backfat thickness and average daily gain in Danish Duroc pigs, with results indicating that incorporating both additive and dominance effects simultaneously improved the prediction accuracy. Weissbrod et al. (2016) developed the multi-kernel linear mixed model (MKLMM), which introduced the interaction effects among different genomic regions in the prediction model to improve accuracy.

In addition, information on functional annotations, gene pathways, reported QTLs, ENCODE, and multi-omics can be used to weight the contributions of genetic markers for a trait of interest. Xiang et al. (2019) ranked the contributions of genetic markers by gene regulation and evolution evidence to improve prediction accuracy (Xiang et al., 2019). With multi-omics data (e.g. genomes and transcriptomes), Ye et al. (2020) improved prediction accuracy with preselected SNPs based on the results of genome-wide association studies or transcriptome-wide association studies in *Drosophila*.

3.5 Software engineering for methods under best linear unbiased prediction framework

Commonly used software tools used with genomic selection methods under the BLUP framework are summarized in Table 1. The available functions of this software are largely identical with only minor differences. However, efficiency varies greatly with different methods because of the programming language and the strategy implemented to estimate genetic parameters. Software written in a compiled language should be much faster than those written in an interpreted language. Software that uses the V (phenotypic variance-covariance matrix) based direct algorithm to estimate genetic parameter estimation should be more efficient in handing genome relevant computation, while those using the MME-based indirect algorithm should be better in pedigree-related genetic evaluation.

4 Improving algorithms under the Bayesian model framework

Compared with the methods under the LMM framework, Bayesian models always have more parameters that need to be estimated. This improves prediction accuracy, but it also results in a heavy computational burden. A series of Bayesian alphabet methods has been developed to provide flexible prior distributions for marker effects and to estimate the abundant unknown parameters in an efficient manner.

4.1 Integrating Bayesian models with a single-step strategy

Similar to SSGBLUP, a technique called SSBR has been proposed for Bayesian models to utilize all available data of genotyped animals and non-genotyped animals (Fernando et al., 2014). Genetic markers were imputed for non-genotyped animals. Imputation errors were then added to accommodate the

Table 1 The most commonly used software for the methods under LMM framework.

Software	Language	BLUP methods	Website and reference
ASReml	*Fortran*	BLUP, PBLUP, GBLUP, SSGBLUP	https://vsni.co.uk/software/asreml (Gilmour et al., 2002)
BLUPF90	*Fortran*	BLUP, PBLUP, GBLUP, SSGBLUP	http://nce.ads.uga.edu/html/projects/programs/ (Misztal et al., 2002)
DMU	*Fortran*	BLUP, PBLUP, GBLUP, SSGBLUP	https://dmu.ghpc.au.dk/dmu/DMU/ (Madsen et al., 2006)
GCTA	*C++*	GBLUP	https://yanglab.westlake.edu.cn/software/gcta (Yang et al., 2011)
GVCBLUP	*C++ and Java*	GBLUP	https://animalgene.umn.edu/gvcblup (Wang et al., 2014)
KAML	*R and Rcpp*	GBLUP, KAML	https://github.com/YinLiLin/KAML (Yin et al., 2020)
LDAK	*C*	GBLUP, MultiBLUP	https://dougspeed.com/ldak/ (Speed et al., 2012)
lme4	*R and Rcpp*	BLUP	https://CRAN.R-project.org/package=lme4 (Bates et al., 2015)
MKLMM	*Python*	MKLMM	https://github.com/omerwe/MKLMM (Weissbrod et al., 2016)
MTG2	*Fortran*	GBLUP, SSGBLUP	https://sites.google.com/site/honglee0707/mtg2 (Lee and Van der Werf, 2016)
rrBLUP	*R*	rrBLUP	https://CRAN.R-project.org/package=rrBLUP (Endelman, 2011)
sommer	*R and Rcpp*	GBLUP	https://CRAN.R-project.org/package=sommer (Covarrubias-Pazaran, 2016)
DeepGS	*R*	DL, CNN	https://github.com/cma2015/DeepGS (Ma et al., 2018)
lme4GS	*R*	BLUP, GBLUP	https://github.com/perpdgo/lme4GS (Caamal-Pat et al., 2021)
hiblup	*C++*	BLUP, PBLUP, GBLUP, SSGBLUP	https://www.hiblup.com/ (Yin et al., 2023)

deviation between true and estimated genotypes. If the effects of all markers were assumed to follow the same normal distribution, the SSBR model was identical to SSGBLUP. However, unlike SSGBLUP, prior assumptions about distribution of marker effects can be diverse in SSBR.

4.2 Enriched hypotheses of Bayesian models with mixture distributions

Integrating the advantages of the LMM approach and Bayesian models, Zhou et al. (2013) proposed a method called BSLMM (Bayesian Sparse Linear Mixed Model) which assumed that all genetic markers had phenotype effects and could be divided into two groups:

As in RRBLUP, the marker effects of SNPs in the first group followed a normal distribution of $N\left(0, \sigma_a^2\right)$

The second group of SNPs had additional variance σ_b^2 that was not captured by the first group, and the distribution was written as $N\left(0, \sigma_a^2 + \sigma_b^2\right)$

The proportion of markers in those two groups π and 1-π was treated as a parameter that needed to be estimated. If π equaled 1, BSLMM was identical to GBLUP whilst, if σ_a^2 was zero, BSLMM yielded the same results as BVSR (Logsdon et al., 2010). Many studies have shown BSLMM to be superior to GBLUP, BVSR, and other Bayesian models.

Moser et al. (2015) developed the BayesR model which assumed a mixture of four normal distributions, with all available SNPs divided into four groups based on marker effects. By benefiting from the complex mixture distribution, BayesR outperformed other Bayesian models, and it is now well known for its high prediction accuracy across many traits of various species (Fikere et al., 2018; Hayes et al., 2019). A difference between BSLMM and BayesR is that all SNPs are assigned into a limited number of groups. The DPR (Dirichlet process regression) model assumes that marker effects follow a normal distribution with an unlimited number of components (Zeng and Zhou, 2017). DPR has been found to perform well compared to BayesR in terms of prediction accuracy.

4.3 Efficient estimation of unknown parameters

Unknown parameters have been estimated in MCMC iterations using a sampling strategy (e.g. Gibbs or MH (metropolis-hasting)). Unknown parameters are updated in each iteration until the calculation reached a stationary distribution. The estimation of marker effects required tens of thousands of iterations to reach convergence to derive the posterior means, which was typically time-consuming. Zhao et al. (2020) have proposed BayesXII which accelerates the sampling procedure by an efficient parallelized algorithm without loss of estimation accuracy. The modified Gibbs sampling for each genetic marker was carried out independently, and the computations were accelerated at each step of the Markov chain.

4.4 Software engineering for methods under the Bayesian framework

Commonly used software tools used with genomic selection methods under the Bayesian framework are summarized in Table 2. As Bayesian methods impose a huge computational burden to estimate genetic parameters iteratively, all this software is written in highly efficient programming languages. However, available methods for different software are distinctive. For example, all the software can deal with individual-level Bayesian models. However, JWAS can handle single-step Bayesian model, GCTB can process summary-level Bayesian model, and hibayes can simultaneously fit individual-level, summary-level, and single-step Bayesian models.

5 General techniques for improving prediction accuracy: data size, pedigree information and number of individuals

5.1 The influence of data size on genomic prediction accuracy

In addition to algorithms, data used to estimate breeding value is also an important factor affecting the accuracy of genomic prediction. For complex traits, prediction accuracy is affected by the properties of the available genetic markers (Zhang et al., 2019), including marker density and marker set design. Accuracy is also influenced by the proportion of individuals tested for genotypes and phenotypes in pedigrees.

We have simulated a pig population to try to investigate how data size affects genomic prediction. The whole genome resequencing data with 81.8 million SNPs of 78 Duroc pigs were obtained from ISwine database (http://iswine.iomics.pro/) (Fu et al., 2020; Xu et al., 2020). In order to be closer to a practical situation and to ensure an efficient simulation, we randomly selected 50,000 (50K) out of 81.8 million SNPs to construct the base population, i.e. the first generation. Nine generations at random mating were then simulated using the Simer package (https://github.com/xiaolei-lab/SIMER), assuming a litter size of 10, an average recombination of 50% across the genome, and a selection rate of 34%. The simulated phenotype was controlled by 300 QTNs (Quantitative Trait Nucleotides) with additive marker effects which followed a normal distribution with zero mean and a variance of 0.48, and with heritability set to 0.3. In total, a population composed of ten generations, including 21 088 individuals with 50K genetic markers were retained for the subsequent analysis. The reported prediction accuracy was measured by Pearson's correlation

Table 2 The most commonly used software for the methods under the Bayesian framework.

Software	Language	Bayesian methods	Website and reference
BayesR	*Fortran*	R	https://github.com/syntheke/bayesR (Moser et al., 2015)
BayesXII	*Julia*	RR, A, B, Bpi, C, Cpi, LASSO	https://github.com/zhaotianjing/BayesXII (Zhao et al., 2020)
BGLR	*R and Rcpp*	RR, A, B, Cpi, LASSO	https://cran.r-project.org/package=BGLR (Pérez and de Los Campos, 2014)
bWGR	*R and Rcpp*	RR, A, B, Bpi, C, Cpi, Dpi, LASSO	https://cran.r-project.org/package=bWGR (Xavier et al., 2020)
DPR	*C++*	DPR	https://github.com/biostatpzeng/DPR (Zeng and Zhou, 2017)
GCTB	*C++*	RR, A, B, Bpi, C, Cpi, S, R	https://cnsgenomics.com/software/gctb (Zeng et al., 2018)
GEMMA	*C++*	BSLMM, BVSR	https://github.com/genetics-statistics/GEMMA (Zhou and Stephens, 2012)
JWAS	*Julia*	RR, A, B, Bpi, C, Cpi, LASSO	https://reworkhow.github.io/JWAS.jl/latest/ (Cheng et al., 2018)
Hibayes	*R and Rcpp*	RR, A, B, Bpi, C, Cpi, LASSO, BSLMM, R	https://cran.r-project.org/package=hibayes (Yin et al., 2022)

coefficient between the predicted and true breeding values for the individuals of the tenth generation across 10 replicates.

5.2 How many generations are required for accurate genomic prediction?

In contrast to humans and plants, pedigree information for livestock is usually well documented. Genetic values were obtained from BLUP by using information on relationships among individuals and phenotypes. More generations in the pedigree data allowed more historical individuals with phenotypes to be used for the evaluation of breeding values in the next generation, which resulted in a higher prediction accuracy. The simulation results indicated that the depth of pedigree relationships affected prediction accuracy significantly. However, using pedigree data, the relationship between predicted generation and their ancestors beyond four generations became very weak. The accuracy of estimated breeding value then improved slowly when the depth of pedigree records exceeded five generations (Fig. 1).

5.3 How many individuals need to be tested for genotype and phenotype to achieve an accurate genomic prediction?

In addition to pedigree data, genotypic information was utilized to construct the relationship matrix. More genotyped individuals helped to locate the causal

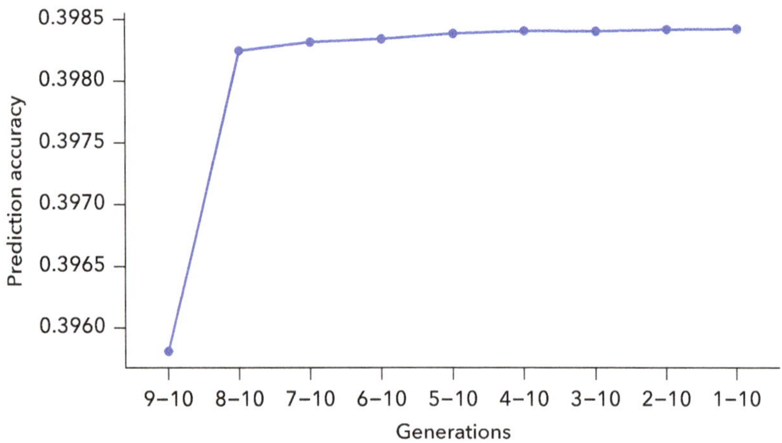

Figure 1 The prediction accuracy affected by the depth of pedigree records. (Notes: For the genetic evaluation of each point, we ran REML to estimate the variance components, which were used to solve MME to obtain individual breeding values. The prediction accuracy of the breeding values for individuals in the 10th generation improved continuously with the depth of pedigree records, and it improved negligibly when the depth of pedigree traced back over five generations).

Published by Burleigh Dodds Science Publishing Limited, 2025.

markers and to estimate the marker effects accurately, resulting in more accurate estimates of breeding value. From the simulation, we saw that the prediction accuracy of GBLUP increased rapidly with the proportion of individuals in each generation tested for genotype and phenotype (Fig. 2).

5.4 Does pedigree information help to improve the prediction accuracy when many individuals were genotyped?

To maximize the usage of individuals in a herd, the SSGBLUP method was developed to handle pedigree and genotype data simultaneously. SSGBLUP is commonly used in genomic selection of livestock animals (e.g. pigs and cattle). SSGBLUP was more effective than GBLUP for prediction when only a small proportion of individuals were genotyped, because the single-step strategy made use of the phenotype information of ungenotyped individuals (Fig. 2). With the rapid decrease in the cost of genotyping, a large proportion of animals could be genotyped. When 20–40% of the individuals were genotyped, the prediction accuracy of both GBLUP and SSGBLUP reached a plateau, but GBLUP performed slightly better than SSGBLUP when almost all individuals were genotyped. In contrast to PBLUP, when individuals from earlier generations were incorporated into the analysis, the prediction accuracy of both GBLUP and SSGBLUP continued to improve, with GBLUP improving faster than SSGBLUP.

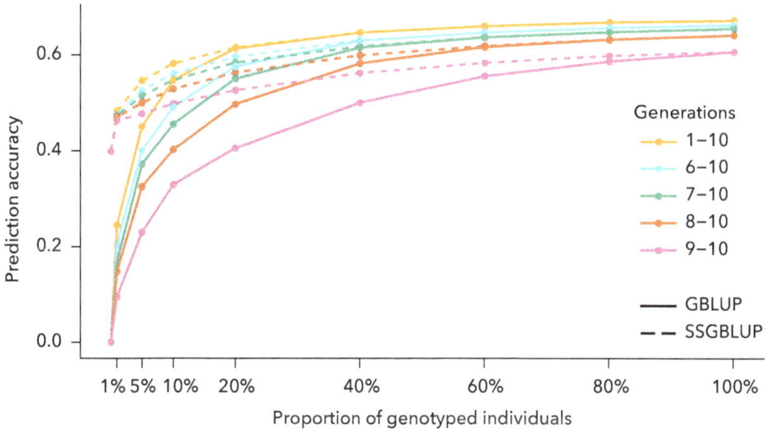

Figure 2 Prediction accuracy affected by the proportion of individuals with information on genotype and phenotype in each generation. (Notes: Both GBLUP and SSGBLUP models were conducted in this experiment and the results indicated that (1) SSGBLUP had great advantages over GBLUP when <20% of the individuals were genotyped; (2) SSGBLUP and GBLUP had a similar performance in terms of prediction accuracy when 20–40% of the individuals were genotyped; and (3) GBLUP performed slightly better than SSGBLUP when almost all the individuals were genotyped).

5.5 How many markers are required for an accurate genomic prediction?

Genomic prediction relies on the LD between available genetic markers for prediction and the underlying causal loci of the objective trait. By increasing marker density, average LD also increased, and this led to a higher prediction accuracy in theory. To further investigate the influence of marker density on the prediction accuracy of a trait, we downloaded all the resequencing data, including 642 purebred individuals from ISwine database. A phenotype with a heritability of 0.5 was simulated using a mixture of two normal terms consisting of randomly selected 110 QTNs. In the case of 100 of these, the effect size was sampled from a normal distribution with zero mean and a variance of 0.01. The effect size of the remaining 10 QTNs followed another normal distribution with zero mean and a variance of 0.1. Prediction accuracy was measured by the Pearson's correlation coefficients between the predicted and true breeding values across five replicates on five-fold cross-validation.

From Figure 3 we can see that when the number of markers was <50K, prediction accuracy improved rapidly with an increase in marker density. However, the prediction accuracy of GBLUP stopped increasing when the density of genetic markers reached a certain level. One reason is that the genomic relationship between each pair of individuals is calculated by all available markers, and it tended to be stable when the number of markers exceeded a certain level. This number varied among species due to differences in genome size and genome structure.

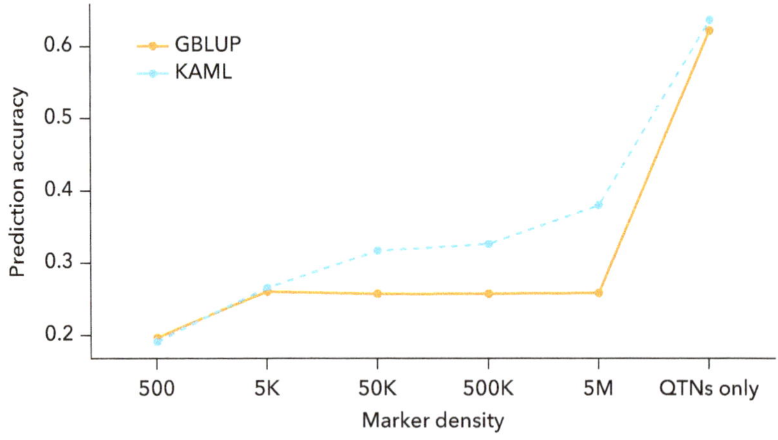

Figure 3 The prediction accuracy affected by marker density with GBLUP and KAML. (Notes: The simulation results showed that (1) the prediction accuracy of both GBLUP and KAML improved rapidly with an increase in marker density when the number of markers was <50K; (2) KAML outperformed GBLUP, and the superiority increased with an increase in marker density; and (3) prediction with QTNs only gave the highest accuracy).

However, instead of using the same genomic relationship matrix (GRM) for all traits, improved models (e.g. KAML) use a trait-specific GRM constructed from markers with different weights of genetic contributions. The optimization of the weights depended heavily on the LD between available genetic markers. A population genotyped with a higher marker density, with an appropriate weight for each marker, can potentially bring high prediction accuracy across traits. Additionally, prediction with QTNs alone gives the highest accuracy i.e. selecting functional loci and genetic markers located in a strong LD with potential QTNs for an objective trait. To sum up, when combined with improved algorithms, high-density genetic markers or an appropriate design for a marker set at a certain density can result in high genomic prediction accuracy. Improved prediction accuracy over GBLUPs models will increase for traits with genetic architectures varying from complex (i.e. polygenic) to simple (e.g. major genes).

6 References

Aguilar, I., *et al.* Hot topic: a unified approach to utilize phenotypic, full pedigree, and genomic information for genetic evaluation of Holstein final score. *J Dairy Sci* 2010;93(2):743–752.

Bates, D., *et al.* Fitting linear mixed-effects models using lme4. *J Stat Softw* 2015;67(1):1–48.

Caamal-Pat, D., *et al.* lme4gs: an r-package for genomic selection. *Front Genet* 2021;12:680569.

Chatterjee, N., Shi, J. and Garcia-Closas, M. Developing and evaluating polygenic risk prediction models for stratified disease prevention. *Nat Rev Genet* 2016;17(7):392–406.

Cheng, H., Fernando, R. and Garrick, D. JWAS: Julia implementation of whole-genome analysis software. In *Proceedings of the world congress on genetics applied to livestock production*. 2018. p. 859.

Christensen, O.F. and Lund, M.S. Genomic prediction when some animals are not genotyped. *Genet Sel Evol* 2010;42. https://doi.org/10.1186/1297-9686-42-2

Covarrubias-Pazaran, G. Genome-assisted prediction of quantitative traits using the R package sommer. *PloS One* 2016;11(6):e0156744.

Endelman, J.B. Ridge regression and other kernels for genomic selection with R package rrBLUP. *Plant Genome* 2011;4(3):250–255.

Fernando, R.L., Dekkers, J.C. and Garrick, D.J. A class of Bayesian methods to combine large numbers of genotyped and non-genotyped animals for whole-genome analyses. *Genet Sel Evol* 2014;46:50.

Fikere, M., *et al.* Genomic prediction using prior quantitative trait loci information reveals a large reservoir of underutilised blackleg resistance in diverse canola (Brassica napus L.) lines. *Plant Genome* 2018;11(2). https://doi.org/10.3835/plantgenome2017.11.0100

Fu, Y., *et al.* A gene prioritization method based on a swine multi-omics knowledgebase and a deep learning model. *Commun Biol* 2020;3(1):1–11.

Georges, M., Charlier, C. and Hayes, B. Harnessing genomic information for livestock improvement. *Nat Rev Genet* 2019;20(3):135–156.

Gianola, D. Priors in whole-genome regression: the bayesian alphabet returns. *Genetics* 2013;194(3):573–596.

Gilmour, A.R., *et al.* ASReml user guide release 1.0. 2002.

Goddard, M. Genomic selection: prediction of accuracy and maximisation of long term response. *Genetica* 2009;136(2):245–257.

Guo, X., *et al.* Genomic prediction using models with dominance and imprinting effects for backfat thickness and average daily gain in Danish Duroc pigs. *Genet Sel Evol* 2016;48(1):1–9.

Habier, D., Fernando, R.L. and Dekkers, J.C.M. The impact of genetic relationship information on genome-assisted breeding values. *Genetics* 2007;177(4):2389–2397.

Habier, D., *et al.* Extension of the bayesian alphabet for genomic selection. *BMC Bioinformatics* 2011;12:186.

Hayes, B.J., *et al.* Towards multi-breed genomic evaluations for female fertility of tropical beef cattle. *J Anim Sci* 2019;97(1):55–62.

Hayes, B.J., Visscher, P.M. and Goddard, M.E. Increased accuracy of artificial selection by using the realized relationship matrix. *Genet Res* 2009;91(1):47–60.

Henderson, C.R. Estimation of changes in herd environment. *J. Dairy Sci* 1949;32(8):706–706.

Henderson, C.R. Estimation of genetic parameters. In *Biometrics*. Hoboken, NJ, 1950, pp. 186–187.

Henderson, C.R. Best linear unbiased estimation and prediction under a selection model. *Biometrics* 1975:423–447.

Henderson, C.R., *et al.* The estimation of environmental and genetic trends from records subject to culling. *Biometrics* 1959;15(2):192–218.

Kizilkaya, K., Fernando, R. and Garrick, D. Genomic prediction of simulated multibreed and purebred performance using observed fifty thousand single nucleotide polymorphism genotypes. *J Anim Sci* 2010;88(2):544–551.

Lee, S.H. and Van der Werf, J.H. MTG2: an efficient algorithm for multivariate linear mixed model analysis based on genomic information. *Bioinformatics* 2016;32(9):1420–1422.

Logsdon, B.A., Hoffman, G.E. and Mezey, J.G. A variational Bayes algorithm for fast and accurate multiple locus genome-wide association analysis. *BMC Bioinformatics* 2010;11:58.

Lopes, M.S., *et al.* Using markers with large effect in genetic and genomic predictions. *J Anim Sci* 2017;95(1):59–71.

Ma, W., *et al.* A deep convolutional neural network approach for predicting phenotypes from genotypes. *Planta* 2018;248:1307–1318.

Madsen, P., Sørensen, P., Su, G, Damgaard, L.H., Thomsen, H., and Labouriau, R. DMU – a package for analyzing multivariate mixed models. In Book of abstracts: CD communication 27-11, 2 pp. WCGALP. 2006. p. 247.

Madsen, P., *et al.* DMU-a package for analyzing multivariate mixed models. In *8th world congress on genetics applied to livestock production*. Belo Horizonte; 2006.

Meuwissen, T.H., Hayes, B.J. and Goddard, M.E. Prediction of total genetic value using genome-wide dense marker maps. *Genetics* 2001;157(4):1819–1829.

Misztal, I., *et al.* BLUPF90 and related programs (BGF90). In *Proceedings of the 7th world congress on genetics applied to livestock production*. Montpellier; 2002.

Moore, J.K., *et al.* Improving genomic prediction for pre-Harvest sprouting tolerance in wheat by weighting large-effect quantitative trait loci. *Crop Sci* 2017;57(3):1315–1324.

Moser, G., *et al.* Simultaneous discovery, estimation and prediction analysis of complex traits using a bayesian mixture model. *PLoS Genet* 2015;11(4). https://doi.org/10.1371/journal.pgen.1004969

Muir, W.M. Comparison of genomic and traditional BLUP-estimated breeding value accuracy and selection response under alternative trait and genomic parameters. *J Anim Breed Geneti* 2007;124(6):342–355.

Pérez, P. and de Los Campos, G. Genome-wide regression and prediction with the BGLR statistical package. *Genetics* 2014;198(2):483–495.

Speed, D. and Balding, D.J. MultiBLUP: improved SNP-based prediction for complex traits. *Genome Res* 2014;24(9):1550–1557.

Speed, D., *et al.* Improved heritability estimation from genome-wide SNPs. *Am J Hum Genet* 2012;91(6):1011–1021.

Tan, B. and Ingvarsson, P.K. Integrating genome-wide association mapping of additive and dominance genetic effects to improve genomic prediction accuracy in Eucalyptus. *BioRxiv* 2019:841049. https://doi.org/10.1002/tpg2.20208

VanRaden, P.M. Efficient methods to compute genomic predictions. *J Dairy Sci* 2008;91(11):4414–4423.

Wang, C., *et al.* GVCBLUP: a computer package for genomic prediction and variance component estimation of additive and dominance effects. *BMC Bioinformatics* 2014;15(1):1–9.

Weissbrod, O., Geiger, D. and Rosset, S. Multikernel linear mixed models for complex phenotype prediction. *Genome Res* 2016;26(7):969–979.

Whittaker, J.C., Thompson, R. and Denham, M.C. Marker-assisted selection using ridge regression. *Genetics Res* 2000;75(2):249–252.

Xavier, A., Muir, W.M. and Rainey, K.M. *bWGR: bayesian whole-genome regression*. Oxford: Oxford University Press; 2020.

Xiang, R., *et al.* Quantifying the contribution of sequence variants with regulatory and evolutionary significance to 34 bovine complex traits. *PNAS* 2019;116(39):19398–19408.

Xu, J., *et al.* Whole genome variants across 57 pig breeds enable comprehensive identification of genetic signatures that underlie breed features. *J Anim Sci Biotechno* 2020;11(1):115.

Yang, J., *et al.* GCTA: a tool for genome-wide complex trait analysis. *Am J Hum Genet* 2011;88(1):76–82.

Ye, S., Li, J. and Zhang, Z. Multi-omics-data-assisted genomic feature markers preselection improves the accuracy of genomic prediction. *J Anim Sci Biotechno* 2020;11(1):1–12.

Yi, N. and Xu, S. Bayesian LASSO for quantitative trait loci mapping. *Genetics* 2008;179(2):1045–1055.

Yin, L., *et al.* hibayes: an R package to fit individual-level, summary-level and single-step bayesian regression models for genomic prediction and genome-wide association studies. *bioRxiv* 2022.

Yin, L., *et al.* HIBLUP: an integration of statistical models on the BLUP framework for efficient genetic evaluation using big genomic data. *Nucleic Acids Res* 2023:gkad074. https://doi.org/10.1093/nar/gkad074

Yin, L.L., *et al.* KAML: improving genomic prediction accuracy of complex traits using machine learning determined parameters. *Genome Biol* 2020;21(1). https://doi.org/10.1186/s13059-020-02052-w

Zeng, J., *et al.* Signatures of negative selection in the genetic architecture of human complex traits. *Nat Genet* 2018;50(5):746–753.

Zeng, P. and Zhou, X. Non-parametric genetic prediction of complex traits with latent Dirichlet process regression models. *Nat Commun* 2017;8(1):456.

Zhang, H., *et al.* Factors affecting the accuracy of genomic selection for agricultural economic traits in maize, cattle, and pig populations. *Front Genet* 2019;10:189.

Zhang, Z., *et al.* Improving the accuracy of whole genome prediction for complex traits using the results of genome wide association studies. *PLoS One* 2014;9(3):e93017.

Zhao, T., *et al.* Fast parallelized sampling of Bayesian regression models for whole-genome prediction. *Genet Sel Evol* 2020;52(1):1–11.

Zhou, X., Carbonetto, P. and Stephens, M. Polygenic modeling with bayesian sparse linear mixed models. *PLoS Genet* 2013;9(2):e1003264.

Zhou, X., Im, H.K. and Lee, S.H. CORE GREML for estimating covariance between random effects in linear mixed models for complex trait analyses. *Nat Commun* 2020;11(1):1–11.

Zhou, X. and Stephens, M. Genome-wide efficient mixed-model analysis for association studies. *Nat Genet* 2012;44(7):821–824.

Chapter 4

Advances in understanding the genetics of and breeding for resistance to infectious disease in pigs

Jack C. M. Dekkers, Iowa State University, USA

1 Introduction

Infectious disease represents one of the largest cost components to the swine industry, incurring veterinary costs, loss of pigs due to mortality, reduced performance, and reduced animal welfare. Strategies to reduce the incidence and impact of infectious disease include biosecurity, vaccination, and routine use of medications, therapeutically or metaphylactically. However, effective vaccines are only available and/or efficacious for some pathogens, while

http://dx.doi.org/10.19103/AS.2024.0137.06

biosecurity protocols that keep most infectious pathogens out may be less effective in hog-dense regions.

Additional approaches to combat disease are therefore needed to complement vaccination, biosecurity, and eradication approaches. Genetics and genetic improvement can be one of those approaches. Over the past decades, many studies have attempted to identify the genetic basis of disease resistance, with the goal to incorporate selection for traits related to disease resistance in breeding programs. The purpose here is to review the current state of knowledge of the genetics of disease resistance in pigs and of strategies for genetic improvement of disease resistance.

2 Genetics of resistance to disease

It is well known that resistance to disease or how an animal responds to infection can have a strong genetic basis (Henryon et al., 2001; Davies et al., 2009). For example, viremia level following experimental infection of nursery pigs with the porcine reproductive and respiratory syndrome virus (PRRSv) has been found to be substantially heritable, at around 0.4 (Boddicker et al., 2012). In addition, many studies have shown that different measures of immune response, such as levels of cytokines or antibodies, have substantial heritability (Henryon et al., 2006; Clapperton et al., 2009; Roth et al., 2022).

Natural genetic variation that results in complete resistance to a pathogen is, however, an exception rather than the rule and has been identified for only a few pathogens in pigs. Examples are naturally occurring mutations in genes that confer resistance to *E. coli* F18 and F4 (or K88) (Meijerink et al., 2000; Meijerink et al., 1997; Ren et al., 2012a; Zhang et al., 2008). Both of these involve a mutation in the gene that codes for the receptor that the pathogen uses to attach or adhere to, thereby preventing the pathogen to enter or establish itself in the host.

In most cases, host resistance to infectious disease is partial and is determined by a large number of genes across the genome, although some genes may have larger effects. For example, Boddicker et al. (2012) identified a major gene on pig chromosome 4 for susceptibility to experimental PRRSv infection of nursery pigs. Although pigs that have the favorable genotype for this major gene or its linked genetic marker become infected when exposed to the virus, they have lower levels of viremia following infection and maintain a higher growth rate (Boddicker et al., 2012, 2014). The effects of this gene appear to be the result of a naturally occurring mutation in the *GBP5* gene (Koltes et al., 2015), which produces a protein that is involved in innate immune response. Because of this mutation, pigs that have the unfavorable genotype are expected to produce a truncated *GBP5* protein, which hampers their innate immune response to PRRS infection. Innate immunity refers to the first line of defense to pathogens, in contrast to adaptive immunity, which involves the production of antibodies that are specific to a pathogen. Adaptive immunity results in the 'memory' that is developed as a result of vaccination. Natural polygenic variation has also been described for susceptibility to PCV2, with two major resistance loci identified (Walker et al., 2018).

Complete resistance to a disease can, however, also be achieved by gene editing, as was demonstrated by editting the *CD163* gene to create resistance to PRRSv (Wells et al., 2017; Whitworth et al., 2016). The *CD163* gene codes for a hemoglobin scavenger protein that is expressed on the membrane of macrophages (Kristiansen et al., 2001) and is used by the PRRSv to enter macrophages (Calvert et al., 2007). Knocking the *CD163* gene out or creating more precise edits of the domain that the PRRSv interacts with (Burkhard et al., 2017) prevents the PRRSv to infect macrophages. Natural genetic variation that results in complete resistance to PRRSv infection has not been identified, including in the *CD163* gene (Johnsson et al., 2018), although natural mutations in the *CD163* gene that are associated with partial resistance have been identified (Ren et al., 2012b; Lim et al., 2018; Dong et al., 2021).

It is not clear, however, whether the *CD163* edited pig is resistant to all variants of the PRRSv, or will remain resistant to novel variants, as pathogens can evolve rapidly and could mutate or recombine into strains that evade whatever resistance mechanism the host employs. This was recently observed for resistance to infectious pancreatic necrosis (IPN) virus in farmed Atlantic Salmon, for which a major gene was identified that substantially reduces mortality to IPN virus infection (Houston et al., 2008, 2010; Moen et al., 2009). This effect was attributed to a natural mutation in the Epithelial Cadherin gene, which produces a protein that binds to the IPN virus (Moen et al., 2015). However, recently, outbreaks of IPN have been observed in Salmon that were bred to be resistant based on this natural mutation (Benkaroun et al., 2021; Hillestadt et al., 20121). Recent work in Chile has shown that these outbreaks may be the result of the IPN virus having evolved to adapt to the introduction of genetically resistant salmon and/or to the vaccination strategies that were implemented simultaneously (Godoy et al., 2022). Further, single-gene resistance, either natural or artificially created, typically involves mutations in genes that have a natural role in the host and can, therefore, have detrimental effects on the host, as has been observed in several cases (Bai and Plastow, 2022). Other considerations associated with the use of gene editing to introduce single-gene resistance and traits in livestock populations have been discussed in Bastiaansen et al. (2018). In the end, it will likely be impossible to develop pigs that are completely resistant to all major diseases, even with the use of gene editing.

Given that pigs will continue to be exposed to a range of infectious pathogens and that creating pigs that have complete resistance to this multitude of continuously evolving pathogens may not be a feasible strategy, much discussion in the past decades has focussed on what selection strategies for improved health, productivity, and welfare in livestock should target (Albers et al., 1987; Raberg et al., 2007; Bishop et al., 2002; Doeschl-Wilson and Kyriazakis, 2012; Guy et al., 2012; Knap and Doeschl-Wilson, 2020). These discussions have considered the ability of the host to prevent infection, to limit the pathogen's ability to inflict damage, and to infect other animals, as well as the potential impact of genetic improvement of these host characteristics on the evolution of existing pathogens and the risk of novel pathogens entering the population, either from other demographic regions or from other species. To discuss these issues, it is important to introduce mechanisms that hosts employ to combat infectious pathogens or to limit their effects on the host, which will be described in the following sections.

3 Mechanisms to combat pathogens: resistance versus tolerance versus resilience or robustness

The terms resistance, tolerance, resilience, and robustness have been part of the vocabulary of quantitative genetics of host responses to infectious disease for decades, with, unfortunately, considerable room for interpretation (Albers et al., 1987; Bishop, 2004; Raberg et al., 2007; Colditz and Hine, 2016; Knap and Doeschl-Wilson, 2020).

The general consensus is that **resistance** refers to the ability of an animal to prevent infection when exposed to a pathogen or to limit replication of the pathogen when infected. When resistance is incomplete, the level of resistance of an animal can be quantified by measuring the animal's pathogen load for a given level of exposure, e.g. the level of viremia in blood, and can be quantified by the increase in pathogen load as pathogen exposure increases, as illustrated in Fig. 1a. Differences in resistance between animals can be the result of more resistant animals being better able to block the entry of the pathogen to reduce replication of the pathogen or to eliminate the pathogen or a combination of these.

Tolerance refers to ability of an animal to maintain performance at a given level of infection or pathogen load. Thus, at a given pathogen load, growth and performance of tolerant animals is less affected than that of less tolerant animals. This is illustrated in Fig. 1b as the slope of performance on pathogen load. Differences in tolerance exist because tolerant animals are better able to deal with the damage that is done by the pathogen or because they are more effective at mounting an immune response, while diverting less energy away from performance.

Whereas tolerance refers to the impact on performance of a given pathogen load that is present in the animal, the concept of **resilience** refers

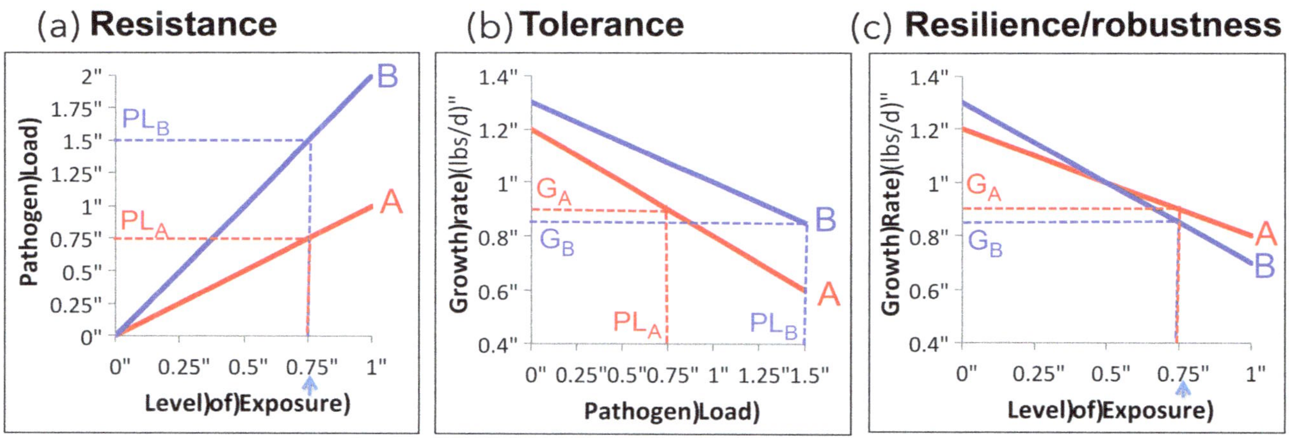

Figure 1 Illustration of the concepts of Resistance, Tolerance, and Resilience as, respectively, the inverse of the regression of pathogen load on the level of exposure (a), and the regression of performance (growth rate) on pathogen load (b) and level of exposure (c).

to the ability of an animal to maintain performance following exposure to a pathogen and is quantified as the slope of performance against the level of exposure, as illustrated in Fig. 1c. Thus, resilience combines resistance and tolerance.

To further illustrate the concepts of resistance, tolerance, and resilience, consider the relationship between the level of exposure to a pathogen and pathogen load for the progeny of two boars, A and B, as shown in Fig. 1a. The graphs show that, for a given level of exposure, progeny of boar B have a pathogen load that is twice as large as that for progeny of boar A. This reflects differences in resistance to the pathogen; progeny of boar A are more resistant. Figure 1b shows the relationship between pathogen load and growth rate. At a pathogen load of zero, i.e. in a clean environment, progeny of boar B have a 0.1 lb greater growth rate per day than progeny of boar A. In addition, as pathogen load increases, growth rate drops more quickly for progeny of boar A than for progeny of boar B; at a pathogen load of 1 (in arbitrary units), the growth rate of progeny from boar A have dropped 0.4 lb/d (from 1.2 to 0.8), compared to a drop of only 0.3 lb/d for progeny of boar B (from 1.3 to 1.0). This reflects tolerance, i.e. for a given level of pathogen load (e.g. viremia), the growth rate of the progeny of boar B is less affected, which means progeny of boar B are more tolerant. Thus, if progeny of these two boars have the same pathogen load, boar B would be preferred. The pathogen load that an animal reaches upon exposure, however, depends on its resistance. Thus, although progeny of boar A are less tolerant, their higher resistance results in their decline in performance as a function of level of exposure to be lower than for progeny of boar B. This is termed resilience and is illustrated in Fig. 1c.

Robustness is often used interchangeably with resilience but has also been described as representing a combination of a high production potential with high resilience (Knap and Doeschl-Wilson, 2020). Thus, while resilience refers to the rate of decline in performance with increasing disease exposure, and ignores differences in performance without disease, robustness also considers the level of performance without disease. If the interest is to maximize performance at a given level of exposure, both resilience and performance without disease must be considered and the preferred animal may differ depending on the level of exposure.

Colditz and Hine (2016) differentiated robustness from resilience in terms of the nature of the perturbations that the animal is exposed to (including disease and other environmental stressors), with resilience referring to the ability to maintain or quickly return to pre-challenge performance when subjected to short-term perturbations of the environment, and robustness to the ability to perform under a range of more persistent or periodic aspects of the environment. Specifically, Colditz and Hine (2016) defined robustness as the capacity to maintain productivity in a wide range of environments without compromising reproduction, health, and wellbeing. Under these definitions, they argued that robustness primarily involves innate immune and stress response mechanisms and that resilience involves not only innate but also adaptive immune and stress response mechanisms, as it can involve facultative learning.

Another important concept to introduce is infectiousness, which is the ability of an infected animal to transmit the pathogen to other animals in the herd or population (Lipschutz-Powell et al., 2012, Doeschl-Wilson et al., 2021). To this end, Doeschl-Wilson et al. (2021) emphasized the importance of herd- or population-level resilience, which is not necessarily the average of the individual resilience of animals in the herd or population.

The advantages of selecting for resistance versus tolerance versus resilience has been reviewed and discussed by several authors. Doeschl-Wilson and Kyriazakis (2012) argued that genetic improvement of resistance puts pressure on the pathogen to evolve and escape the resistance mechanisms that the host employs, unless resistance is complete and results in elimination of the pathogen from the population. In contrast, tolerance focuses on limiting or mitigating the damage done by the pathogen and likely involves less interaction between the host and the pathogen. Thus, selection to improve tolerance is less likely to result in an arms-race between the host and the pathogen (Doeschl-Wilson and Kyriazakis, 2012). A potential additional benefit of selection for resistance is that it is expected to reduce shedding, i.e. infectiousness, thereby reducing the level of pathogen exposure in the population. Mulder and Rashidi (2017) showed that selection for resilience is expected to improve both resistance and tolerance, depending on the genetic correlation between resistance and tolerance. An important benefit of selecting for resilience is that it does not require measurement of pathogen load, in contrast to selection for resistance or tolerance (Mulder and Rashidi, 2017; Knap and Doeschl-Wilson, 2020).

4 Immune competence

Bishop (2004) claimed that genetic differences in disease resistance are in large part the result of genetic differences in host immune responses to infection. Possible host molecular mechanisms associated with resistance versus tolerance were described by Glass (2012). In this regard, another important concept is that of generalized versus specific immunity. Generalized immunity (Bishop et al., 2002) or immune competence (Wilkie and Mallard, 1999) refers to the ability to respond to a range of disease challenges. This includes many components of the innate immune system but could also include the ability of an animal to mount an effective adaptive immune response. Wilkie and Mallard (1999) defined immune competence as 'the ability of the body to produce an appropriate and effective immune response when exposed to a variety of pathogens'.

In contrast, specific immunity refers to the ability to respond to a specific disease challenge. Vaccines by definition focus on developing an animal's immunity to a specific pathogen. This also means that a different vaccine is needed for each pathogen and perhaps even for different strains of the same pathogen (e.g. for heterologous PRRS virus strains). In addition, effectiveness of a vaccine may change as the pathogen evolves.

Because genetic improvement is a long-term process, the goal of genetic improvement should be to improve generalized immunity or immune competence, rather than specific immunity, potentially in conjunction with selection for resistance to important specific pathogens (Hine et al., 2014). Selection for immune competence reduces the risk that the pathogen will evolve around the enhanced resistance mechanisms, potentially improves resistance also to new infectious diseases that may enter the population in the future, and also reduces the possibility that improvement of response to one disease will be unfavorable for resistance to other diseases (Hine et al., 2014).

5 Pyramidal structure of pig breeding programs: opportunities and challenges

In order to understand the opportunities and challenges of genetic improvement of disease resilience, it is important to understand the structure of breeding programs and how genetic improvement is created in those programs. Modern genetic improvement programs for pigs typically utilize a so-called pyramid structure, as illustrated in Fig. 1. In such programs, most selection takes place in purebred nucleus breeding populations at the top of the pyramid, while commercial sows and hogs are crossbreds (Dekkers et al., 2010). In the nucleus breeding populations, animals that are estimated to have the best genetics for a comprehensive breeding goal are selected and mated to produce the next generation of purebred selection candidates, as well as germplasm that is passed on to the multiplication and commercial levels. The breeding goal typically includes growth rate, feed efficiency, backfat, and meat quality for the terminal sire line and, for the maternal lines, also female reproductive traits such as litter size.

The pyramidal breeding structure allows for multiplication of the superior genetics that are developed in the nucleus populations, while capitalizing on the benefits of crossbreeding. The latter include the ability to

- capitalize on breed complementarity, specifically by combining the high female reproductive capacity of maternal lines with the high growth potential of terminal sire lines;
- capitalize on improved performance of crossbred animals as a result of heterosis, which ranges from 0% to 30%, depending on the trait and

lines crossed (Bondoc et al., 2001; Bidanel, 2011), and results from non-additive genetic variation and differences in allele frequencies between the purebred lines;

- meet the needs of specific markets by crossing purebred lines that have been selected for specific traits (Visscher et al., 2000); and
- protect the genetic gain achieved in purebred lines from competing breeding organizations by only marketing crossbred breeding stock.

Some of the potential limits in this approach are discussed later in this chapter.

6 Phenotyping and other sources of information to estimate breeding values

Selection within the purebred nucleus populations requires availability of phenotypes that can be used to estimate the breeding value of selection candidates for traits that are included in the breeding goal or for so-called indicator traits, i.e. traits that are genetically correlated to one or more traits in the breeding goal. The latter are used if the trait in the breeding goal is difficult or expensive to measure, e.g. ultrasound backfat as an indicator trait for carcass backfat thickness.

The goal of phenotyping and genetic evaluation programs, which are important components of a breeding program, is to obtain a high accuracy of estimated breeding values (EBVs) of the purebred selection candidates at the time of selection, given the resources that are available. Here, accuracy is defined as the correlation between the true and estimated breeding value of the selection candidates and is proportional to expected response to selection (Falconer and Mackay, 1996).

The types of information that can be used to estimate breeding values of selection candidates are summarized in Fig. 2 and include phenotypes recorded on the selection candidates and on relatives. With pedigree-based Best Linear Unbiased Prediction (PBLUP) of breeding values (Henderson, 1988), breeding values of selection candidates are estimated using phenotypes collected on the selection candidates themselves and on close relatives; when using pedigree relationships, phenotypes on more distant relatives do not contribute information.

Instead of pedigree relationships, genomic prediction (GBLUP) uses genotypes at many genetic markers (typically single nucleotide polymorphisms, SNPs) across the genome for prediction of breeding values (Meuwissen et al., 2001). Ideally, these genotypes are obtained on the individuals that have been phenotyped, in addition to the selection candidates. However, to reduce genotyping costs, genotyping the parents of the phenotyped individuals is also possible (e.g. the purebred parents of commercial sows), although this

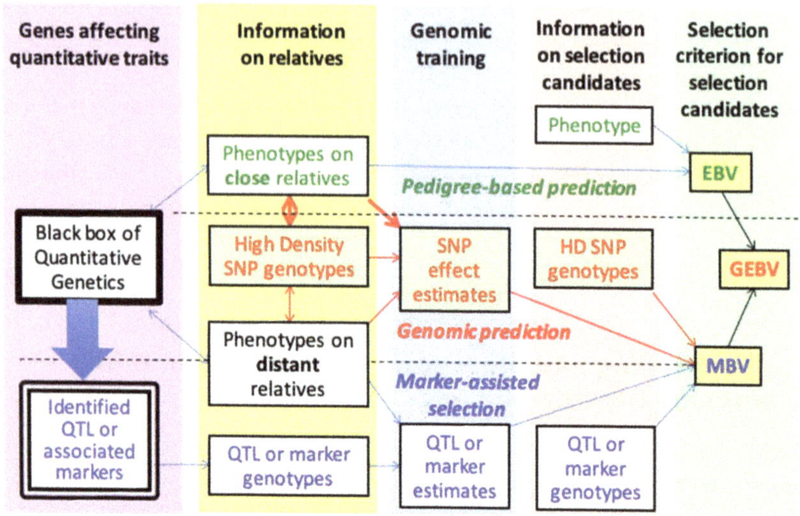

Figure 2 Sources of information that can be used to estimate the breeding value of selection candidates for a quantitative trait using pedigree-based, genomic, or marker-assisted prediction, including phenotypes and marker genotypes on the selection candidate itself and its close and distant relatives.

does result in a decline in the accuracy of the resulting EBV. In GBLUP, pedigree relationships between phenotyped individuals and selection candidates are replaced by genomic relationships (VanRaden, 2008). The latter are derived from the high-density SNP genotypes and reflect the proportion of alleles shared between individuals. So-called single-step GBLUP methods that allow for a combination of genomic- and pedigree-based relationships when not all animals have been genotyped have also been developed (Misztal et al., 2009).

The use of genomic-based instead of pedigree-based relationships result in an increase in the accuracy of EBV because genomic relationships are more differentiating than pedigree-based relationships. For example, paternal half siblings have a pedigree-based relationship of 0.25 with each other because they are expected to share 25% of their alleles. However, based on genomics, their relationship can deviate substantially from 0.25, depending on the set of alleles that they received from their common sire. Genomic relationships also allow phenotypes on distant relatives to provide information to the EBV of selection candidates. Although distant relatives have low genomic relationships with selection candidates, the number of distant relatives is large and, in contrast to pedigree-based relationships, which are essentially zero for distant relatives, they are variable between any pair of distant relatives. A recent study by Bhatia et al. (2023) showed that the use of genomic relationships resulted in higher and more accurate estimates of the heritability of pelvic organ prolapse in sows than the use of pedigree-based relationships. This increase in the heritability estimate likely stemmed from genomic relationships correcting potential pedigree errors, as well as from genomic relationships reflecting true parentage when pooled semen is used.

Another way in which GBLUP can be differentiated from PBLUP is that the GBLUP model that fits breeding values with a genomic relationship matrix has been shown to be equivalent to a model that simultaneously fits the effect of all genotyped markers (as random regression covariates on the number of copies of a given allele that the phenotyped animal carries), with the true effects assumed to originate from a normal distribution (Strandén and Garrick, 2009). This amounts to estimating the effect of each genotyped marker based on all recorded phenotypes (the training or reference population), regardless of whether they are on close or distant relatives. Thus, in contrast to PBLUP, an important feature of GBLUP is that informativeness is not limited to phenotypes on close relatives and that it allows any recorded phenotype to provide information to all selection candidates, even those that are distantly related, although the phenotype of a given animal contributes more information to close relatives.

Both PBLUP and GBLUP rely on the black box of quantitative genetics and do not require knowledge of the number, location, or effects of the loci that contribute to a phenotype. Figure 3 also illustrates the concept of marker-assisted selection, in which a limited number of quantitative trait loci (QTL) or associated genetic markers with larger effects are identified from the black box of quantitative genetics (Dekkers and Hospital, 2002). Their effects are then estimated based on a training data set of animals that have been phenotyped and genotyped for the identified markers and a so-called molecular breeding value (MBV) is estimated. This MBV, which is the EBV of the selection candidate for the identified markers, is then combined with its pedigree-based EBV to obtain a marker-assisted EBV. A similar concept also applies to genomic prediction, except that markers that contribute to the MBV are not pre-selected and, therefore, capture more of the genetic variance (Meuwissen et al., 2001).

7 Challenges to genetic improvement in pyramidal breeding programs

In pyramid breeding programs, as depicted in Fig. 3, genetic improvement is primarily created at the top of the pyramid, in the purebred lines, and is then passed on to lower layers of the pyramid through transfer of animals or germplasm. Although the target of genetic improvement is performance at the commercial crossbred level, typically, most selection within the purebred lines is based on EBV derived from phenotypes recorded on the purebred animals at the top of the pyramid. The resulting genetic improvement is primarily for additive genetics.

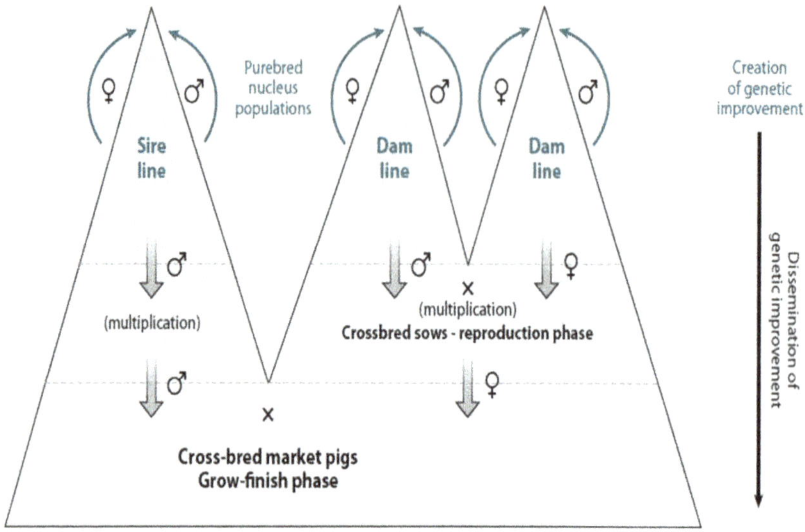

Figure 3 Pyramidal structure of a modern pig breeding program

A problem with this purebred selection strategy is that phenotypes recorded at the top of the pyramid on purebred animals may not be representative of traits that are of economic importance at the commercial level. As a result, only a portion of the genetic gain created in the purebred populations at the top of the pyramid may transfer to the commercial level. The relationship between a trait recorded on purebreds at the top of the pyramid and that same trait recorded at the commercial level on crossbreds can be quantified by the genetic correlation between the corresponding phenotypes, often referred to as the purebred–crossbred correlation (r_{PC}) (Wei and van der Steen, 1991). Through its impact on correlated response to selection (Falconer and Mackay, 1996), r_{PC} directly quantifies the proportion of genetic gain achieved in the purebred population that is transferred to the commercial level (Dekkers, 2007; Hidalgo et al., 2015; van Grevenhof and van der Werf, 2015).

Causes for r_{PC} to be less than 1 for a given trait were reviewed by Wientjes and Calus (2017). When a trait is affected not only by additive genetic effects but also by non-additive genetic effects (dominance, epistasis), the phenotype of a progeny in part depends on the effects of the combinations of alleles that come together in the progeny, i.e. genotype by genotype interactions (GxG). In this case, the value of a purebred animal as a parent, i.e. its breeding value, also depends on whether the parent's mates are from the same or a different breed or line (Fisher, 1918, 1930; Falconer and Mackay, 1996; Dekkers, 2007). In addition to GxG, r_{PC} can also be less than 1 as a result of differences in environment when the trait is affected by genotype by environment interactions (GxE) (Falconer, 1952). The impact of GxE is in particular important for disease resilience or related traits because the nucleus breeding herds of pig breeding programs must be maintained under high biosecurity to prevent transmission of disease to customers. Referring back to Fig. 1c, when EBV are based on nucleus performance, boar B would be selected over boar A because boar B has the best performance in a clean environment (with zero exposure). However, if most of the pigs that descend from these two boars are raised in a commercial environment with substantial disease exposure, then boar A would be preferred because his progeny perform better in such an environment. Finally, differences in how traits are measured can also reduce r_{PC} (Wientjes and Calus, 2017).

Wientjes and Calus (2017) summarized estimates of r_{PC} for pigs from 27 studies from 1964 to 2017, both by trait and across traits in six trait categories. The overall average estimate of r_{PC} across trait categories was about 0.6, with no average r_{PC} estimate exceeding 0.8 (Wientjes and Calus, 2017), indicating that a substantial portion of genetic gain achieved in purebred populations based on purebred phenotypes is not transferred to the commercial level. Two of these trait categories had less than ten estimates of $r_{PC,}$ with an average r_{PC} of 0.67 for feed traits (n = 9) and of 0.50 for the index trait category (n = 3) (Wientjes and Calus, 2017). The other four categories included more than 20 estimates, with an average r_{PC} of 0.54 for fertility traits (n = 62), of 0.67 for meat quality traits (n = 21), of 0.69 for meat amount traits (n = 63), and of 0.66 for growth traits (n = 43). Fertility traits had the second lowest average r_{PC} and the lowest average purebred and crossbred heritability estimates. This is consistent with the general finding that non-additive genetic effects tend to be more important for traits with lower heritability and that non-additive effects can result in lower r_{PC}. Although extensive data to assess the contribution of the different factors that contribute to r_{PC} being less than 1 are not available, a meta-analysis by Wientjes and Calus (2017) suggested that GxG may be more important than GxE, although this will obviously depend on the trait.

Strategies to select purebred animals for commercial crossbred performance were reviewed in Duenk et al. (2021). In their review, they considered:

- whether phenotypes should be collected in a commercial environment versus in the nucleus and on purebreds versus crossbred animals;
- whether and which animals should be genotyped; and
- what genetic evaluation model should be used.

They concluded that the standard strategy of selecting purebreds on EBV derived from purebred phenotypes collected in the nucleus is satisfactory for traits that have an r_{PC} greater than about 0.8, assuming the same number of phenotypes can be collected in the commercial environment as in the nucleus.

This conclusion applied even when the phenotyped animals are genotyped and genomic prediction is used. It is important to note, however, that the number of animals that must be genotyped will be greater when conducting genomic prediction on commercial animals because genomic selection still requires the purebred selection candidates to be genotyped.

For traits with r_{PC} less than about 0.8, alternate strategies should be considered, depending on a number of factors, including the contribution of GxG versus GxE to the low r_{PC}. If the low r_{PC} is primarily caused by GxG, rather than by GxE, the focus should be on predicting crossbred performance rather than on performance in a commercial environment. In this case, phenotypes can be collected on crossbreds within the most feasible environment or in an environment that results in higher trait heritabilities. When most of the GxG is caused by dominance, EBV of purebreds for crossbred performance can also be predicted directly from purebred phenotypes with a genomic evaluation model that accounts for dominance. In this case, there may not be a need to collect phenotypes on crossbreds to predict the EBV of purebreds for crossbred performance (Duenk et al., 2021). Whether the same holds when epistasis contributes to GxG is not clear.

In contrast, if the low r_{PC} is primarily caused by GxE, it will be crucial that phenotypes are collected in the commercial environment. If the contribution of GxG is low, this could involve testing purebreds in a commercial environment if that is more feasible than collecting commercial phenotypes on crossbreds. This also allows individuals with commercial phenotypes to be more closely related to the selection candidates, which increases accuracy (Wientjes et al., 2020). It should also be noted, however, that a phenotype on a crossbred animal can provide information on all purebred lines that contributed to it, while a phenotype on a purebred individual only provides information on that one purebred line (Duenk et al., 2021). In practice, given that the contribution of GxG versus GxE to r_{PC} is unknown, likely differs between traits, and the overall goal is to improve a combination of traits, phenotypes should be collected on both purebred individuals in the nucleus and in commercial environments, on either purebred or crossbred animals. Further systems analyses with knowledge of the contributions of GxG versus GxE to r_{PC} is needed to optimize breeding programs and this will need to be done on a case-by-case basis.

8 Phenotyping approaches for genetic improvement of disease resilience: collecting resilience-related phenotypes on animals outside the nucleus

Given the high biosecurity requirements for the nucleus herds where the purebred selection candidates are housed, an important question for breeding organizations is how to obtain information that can be used to select for disease resistance, tolerance, or resilience. Although mortality and veterinary treatment rates can be collected in the nucleus, given the low incidence of disease in the nucleus, the heritability of such binary phenotypes is expected to be low (Falconer and Mackay, 1996).

Possible strategies to obtain phenotypes that provide information to estimate breeding values of purebred selection candidates for disease resilience include:

- collecting phenotypes on animals that are exposed to disease outside the nucleus and
- collecting benign indicator trait phenotypes that are predictive of disease resilience on purebred animals in the nucleus.

This section discusses the first of these.

Phenotypes related to disease resilience can be collected in herds that are exposed to disease and used to predict breeding values of selection candidates in the nucleus for resilience, without having to expose the latter to disease. Here, the factors and issues discussed previously with regard to the use of phenotypes from commercial herds are relevant. Unfortunately, no clear estimates of r_{PC} are available for disease resilience traits, although r_{PC} is expected to be fairly low for such traits because of the large differences in disease pressure between the nucleus and commercial environments. In their review, Wientjes and Calus (2020) reported that fertility traits, which are related to fitness, similar to disease resilience, have the lowest estimates of r_{PC}. Given the large differences in environment, one might expect GxE to be a large contributor to r_{PC} also for disease resilience traits, although non-additive effects (i.e. GxG) have been hypothesized to be larger for fitness traits compared to other traits (Sellier, 1976; Nakavisut et al., 2005). However, empirical evidence and theory does not support this hypothesis (Hill et al., 2008). If the contribution of GxG is relatively small for disease resilience traits, they could be collected on purebred animals exposed to commercial-like environments. This approach is used in several poultry breeding programs (Kapell et al., 2012; Chu et al., 2019).

Additional factors to consider in the collection of disease resilience phenotypes outside the nucleus include the type(s) of herd(s) that these phenotypes should be collected on, including the diseases that are present in these herds, and what phenotypes should be collected on animals in these herds. These will be discussed in the following two sections.

8.1 Choice of herds

Recording of phenotypes related to disease resilience should be conducted in farms and facilities that enable expression of genetic differences in disease resilience and that enable the recording of related phenotypes. This requires farms or facilities in which animals are exposed to disease. These could be commercial herds or special disease challenge facilities. The latter option not only allows greater control over the disease challenge, over data recording, and over sample collection but is also more costly. Challenges with the use of commercial herds are the logistics of accurate phenotype recording on individual animals and the need for these animals to either be pedigreed or have DNA samples collected for genotyping. Pedigree recording is further complicated by the routine use of pooled semen in commercial herds.

Ideally, the diseases that animals in these herds or facilities are exposed to should represent the major infectious pathogens that are relevant in the industry. In commercial herds, the pathogens that are present cannot be controlled. However, if herds are routinely monitored for presence of pathogens or for indicators of outbreaks, data could be selected to represent specific diseases. For example, Mathur et al. (2014) showed how data from multiple sow herds could be used to identify PRRS outbreaks and to quantify the level of challenge, and Herrero-Medrano et al. (2015) showed how the resulting data could be used for genetic analysis of resilience to PRRS under different levels of challenge.

Disease resilience recording in specialized facilities could involve controlled experimental infection of pigs with specific diseases. An example is the nursery pigs PRRS infection studies described by Rowland et al. (2012), which led to the identification of a major gene for PRRS resistance (Boddicker et al., 2012). This series of experiments also included several in which pigs were experimentally infected with both PRRSV and PCV2, as a model for simultaneous exposure to multiple diseases, with or without prior vaccination to PRRSV, as summarized in Dekkers et al. (2017). In egg-type chickens, one breeding company has successfully used experimental infection of sibs of selection candidates with Marek's disease to reduce mortality to this disease (Fulton et al., 2013). Another example of the use of a specialized facility for the recording of phenotypes related to disease resilience is the natural disease challenge model described by Putz et al. (2019) and Cheng et al. (2020), and illustrated in Fig. 4. In this large-scale study, batches of crossbred nursery pigs were entered into a research facility that was initially seeded with multiple infectious pathogens by bringing in pigs from farms with known health problems. The disease challenge was maintained using continuous flow, where the entering batch was exposed to the previous batch that had itself been exposed by its preceeding batch of pigs. Under careful veterinary control, this allowed for a severe polymicrobial disease challenge.

The disease load under which resilience phenotypes are recorded, to the extent that it can be controlled, is ideally optimized to allow maximum genetic progress for traits related to disease resilience. This includes choice

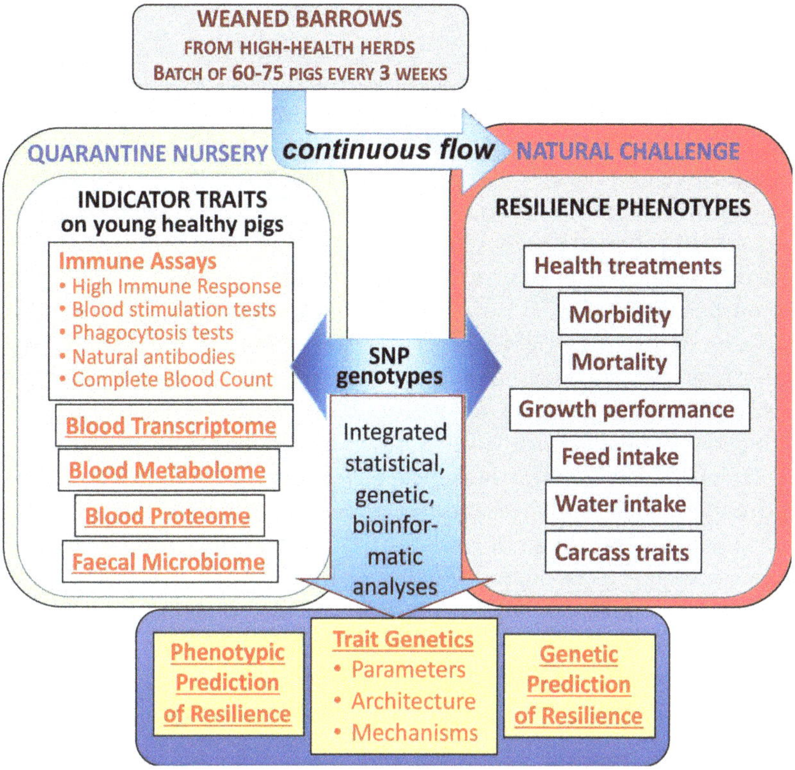

Figure 4 Illustration of the natural polymicrobial disease challenge model (Putz et al., 2019).

of the diseases that animals are exposed to and the level of exposure. Dekkers (2007) showed how, with knowledge of the (distribution of) disease loads in commercial herds and of genetic parameters of disease resilience traits, reaction norm models can be used to optimize the environment in which phenotypes should be collected in order to maximize their value for a breeding program that targets a specific or range of commercial environments.

8.2 Choice of traits

Choice of traits to record under a disease challenge should be driven by the cost and feasibility of recording in the herd(s) of choice, heritabilities of the traits, and their genetic correlations with traits in the breeding goal. Heritabilities should be high enough to provide information that is sufficiently reliable to provide accurate EBV for selection candidates in the nucleus. The low heritability of a trait can, however, be overcome by increasing the number of individuals that the trait is recorded on and by genomic prediction.

The recorded phenotype must, however, have a sufficiently high genetic correlation with at least one trait that is included in the breeding goal for it to be useful.

In commercial herds, the feasibility of recording phenotypes on individually identified animals that are either pedigreed or genotyped is a major challenge. This typically limits recording to the traits of litter size, pre-weaning mortality, weaning weight, and sow longevity in sow herds, and to growth rate (or age at slaughter) and mortality for grow-finish pigs (Knap, 2005). In special cases, individually recorded veterinary treatments may be available also. Such phenotypes were found to be moderately heritable by Henryon et al. (2001) and lowly heritable by Hermesch et al. (2017) and Guy et al. (2018). Apart from growth rate, traits recorded in commercial herds typically have fairly low heritabilities, in part because of data recording issues, and, therefore, require large numbers of individuals to be recorded on a routine basis.

Higher heritabilities for resilience-related phenotypes can be obtained when data recording is in specialized herds, as for example in the natural disease challenge model described by Putz et al. (2019) and Cheng et al. (2020), which was conducted at a research station (Fig. 4). Using trained staff, extensive and accurate phenotype recording was possible, including bi-weekly body weights, individual veterinary treatments, and subjective health scoring, resulting in reasonable heritabilities (Cheng et al., 2020). In addition, this research facility was equipped with electronic feeders and drinkers, allowing for the recording of feed and water intake and behavior on individual pigs housed in group pens.

Using phenotypes that are recorded on a longitudinal basis over time, such as egg production, body weight, milk production, and feed intake, resilience can also be quantified based on the distribution of an animal's phenotypes over time (Berghof et al., 2019). For these purposes, residuals from a statistical model that is fitted to the longitudinal data that is available on an animal should be used to account for systematic effects. Examples of such resilience measures include day-to-day variation of residuals (animals with larger variation are less resilient), skewness of residuals (animals with more negative residuals for, e.g. feed intake, are less resilient), the auto-correlation of residuals over time (animals with a negative autocorrelation of residuals are less resilient) (Berghof et al., 2019), or the proportion of residuals that are considered negative outliers, e.g. the proportion of off-feed days (Putz et al., 2019).

In the natural disease challenge model for grow-finish pigs (Fig. 4), Putz et al. (2019) and Cheng et al. (2020) found that day-to-day variation in feed intake or duration and the proportion of off-feed days based on feed intake or duration were low to moderately heritable and genetically positively correlated with disease incidence and veterinary treatment rates and negatively with growth rate under challenge. Laghouaouta et al. (2021) proposed deviations of body weight from an animal's expected growth curve, combined with the change in acute-phase protein haptoglobin levels in blood after applying a common vaccine, as a resilience indicator for growing pigs under standard growing conditions. Van der Zande et al. (2020) used ear tag accelerometers to evaluate several dynamic measures of activity in nursery pigs before and after an experimental challenge with PRRS virus. Some of these measures were found to be correlated with clinical signs and mortality. Both traits were shown to be moderately heritable but genetic correlations with resilience were not evaluated. As camera and sensor technology develops further for application in management and breeding (Perez-Enciso and Steibel, 2021), many other longitudinal measures that may relate to resilience will become available and should be investigated.

9 Phenotyping approaches for genetic improvement of disease resilience: collecting benign indicator traits of disease resilience on nucleus animals

Although nucleus breeding herds must remain free of major disease, animals in these herds are amenable to the recording of phenotypes that can serve as indicator traits for disease resilience. In order for an indicator trait to be useful to select for disease resilience, it must be heritable and genetically correlated to disease resilience. In addition, indicator traits are ideally recorded on selection candidates prior to their age of selection. Finally, to enable recording of these phenotypes on nucleus animals (ideally on selection candidates), they must not have negative effects on the health and performance of the animals, or negatively affect marketability of germ plasm from the nucleus.

There are numerous examples in the literature of potential indicator traits for disease resistance or resilience. These include phenotypes that quantify aspects of an animal's immune system, various -omics measures collected on blood, and responses to benign stressors. Many of these phenotypes can be collected on the live animal through in vivo measurements or tissue samples, e.g. blood.

9.1 Immune and hematological assays

The purpose of immune and hematological assays is to measure different aspects of an animal's immune response. Many immune and hematological assays can be measured on blood, including complete blood counts, natural antibodies, cytokine levels, and antibody levels to specific antigens, including vaccines. Multiple studies have found immune assays to be heritable, both when measured on healthy animals and on animals exposed to disease (Vissche et al., 2002; Henryon et al., 2006; Clapperton et al., 2008, 2009; Flori et al.,

2011; Ballester et al., 2020; Roth et al., 2022) and suggested their potential to select for disease resistance or resilience.

Several immunological and hematological assays were measured on blood of young healthy pigs prior to their in the natural polymicrobial disease challenge mentioned previously (Putz et al., 2019) and illustrated in Fig. 4, including

- complete blood counts (Bai et al., 2020);
- natural antibodies (Chen et al., 2020);
- mitogen-stimulated proliferation of peripheral blood mononuclear cells (Jeon et al., 2021); and
- antibody- and cell-mediated immune responses (Schmiedt et al., 2018) phagocytosis and immune stimulation assays (Bai et al., 2022).

9.2 Transcriptomics, proteomics, metabolomics, microbiomics, and other -omics data

Novel high-throughput molecular profiling technologies, such as transcriptomics, proteomics, metabolomics, and microbiomics, provide many opportunities to generate large numbers of additional phenotypes on individuals. While these phenotypes are highly context specific in terms of timing, conditions, and tissue, their collection on selection candidates typically requires non-invasive sampling procedures. For that purpose, blood is a convenient and highly informative tissue because it can be collected on live animals and provides a window into an organism's health (Yurkovich and Hood, 2019), as it is the animal's circulating 'connective' tissue and contains many immune cell types (Mohr and Liew, 2007).

The application of -omics technologies on samples that can be collected on live animals without exposure to disease provides extensive novel phenotypes that can be mined to identify potential indicator traits for disease resilience that can be measured on young healthy pigs, i.e. on selection candidates in the nucleus. These opportunities were explored in the previously described natural disease challenge model (Putz et al., 2019) (Fig. 4), using blood and fecal samples collected on over 900 nursery pigs prior to their entry in the disease challenge.

Analysis of phenotypic relationships of the expression of over 16 000 genes with subsequent disease resilience was described in Lim et al. (2021). Corresponding genetic analyses (heritabilities and genetic correlations with disease resilience) of over 3200 pigs with disease resilience data, of which nearly 1600 pigs had blood transcriptome data, were summarized by Lim et al. (2022). Phenotypic and genetic analyses of relationships of the abundance of 377 proteins and 44 metabolites in the blood of over 900 young healthy pigs with their subsequent disease resilience phenotypes are in Chen et al. (2023) and Dervisi et al. (2021), respectively.

Because of limited statistical power, these studies did not identify clear signals of individual features (i.e. the expression of a single gene or the abundance of a single protein or metabolite) that were both heritable and genetically correlated with resilience. However, when combining results across genes using enrichment analyses, there was evidence that the expression levels of genes that shared Gene Ontology terms associated with immune response were genetically correlated with disease resilience (Lim et al., 2022). Similar results were obtained for protein levels (Chen et al., 2023).

Another source of -omics data that can be collected on live animals is microbiome data from fecal samples. Recently, Ramayo-Caldas et al. (2021) demonstrated that gut microbiota composition is associated with immunity traits in pigs. They concluded that both host genome and microbial levels must be considered to accurately characterize immunocompetence in pigs.

9.3 Benign stressors

Although nucleus breeding programs don't enable selection candidates to be exposed to disease, they are exposed to multiple other stressors, including the stress of weaning, mixing, novel diets, human interaction, etc. Thus, some of the measures that were described in the previous section could also be recorded in the nucleus.

Colditz and Hines (2016) broadened the definition of resilience by combining resilience to infectious and non-infectious stressors as 'the capacity of an animal to be minimally affected by a disturbance or to rapidly return to the physiological, behavioral, cognitive, health, affective, and production states that pertained before exposure to a disturbance'. Their rational for considering general resilience rather than only disease resilience was that response to infectious and environmental stressors share a common defense strategy, starting with costly non-specific responses, followed by less costly but more specific responses and lower cost on re-exposure to the stressor. Indeed, in their review, Gimsa et al. (2018) documented the close connection between immune and stress response systems and the effect of psychosocial stress on innate and adaptive immune responses.

Another possible class of disease resilience indicator traits, therefore, are measures of response to non-infectious stressors. These could include measurements of stress hormones in blood or hair (Ballester et al., 2020), as well as standard animal behavior tests such as the back test (Bolhuis et al., 2005), the novel object test, the human approach test, or the novel environment test (Forkman et al., 2007). Colditz and Hine (2016) suggested using standard husbandry practices that provide physical and social stressors to animals, such as weaning, for characterization of resilience phenotypes. Hermesch et al. (2017, 2018) estimated genetic parameters for multiple immune and hematological traits recorded on weaner pigs for genetic improvement of disease resilience, along with multiple weight measurements post weaning. They found that the evaluated immune and hematological traits had moderate to high heritabilities and that multiple immune and hematological traits had significant genetic correlations with growth traits.

Breeding programs typically collect large amounts of individual feed intake data on nucleus animals in order to select for feed efficiency. As described previously, day-to-day variation in feed intake was found to be a potential genetic indicator trait for disease resilience (Putz et al., 2019; Cheng et al., 2020). This, however, was based on feed intake data collected under a severe polymicrobial disease challenge. Day-to-day variation in feed intake can also be computed based on the feed intake data that is routinely collected on nucleus animals, potentially reflecting resilience to minor pathogens and non-infectious stressors. Harzilius et al. (2020) found that boars with higher day-to-day variation in feed intake in a high-health nucleus herd had progeny that had higher death rates in a disease challenge. This suggests that, even without presence of major disease challenges, measures of resilience to disturbances such as social, metabolic, and environmental stressors, which is what boars are expected to experience in a nucleus environment, may be genetically correlated to resilience to disease and that an animal's resilience to such disturbances can be captured by day-to-day variation in feed intake. These findings, however, require further evaluation.

10 Conclusion

How pigs cope with infectious diseases has a substantial genetic component and can, therefore, be improved by genetic selection if pigs that have the right genetics can be identified. Therein, however, lies the problem. First of all, there are several mechanisms that can result in pigs to respond better in the face of disease, including resistance and tolerance. Although complete resistance may be the ultimate goal, it will likely be impossible to develop pigs that are completely resistant to the main infectious diseases that pigs are exposed to. In addition, pathogens can evolve much more quickly than we may be able to change the pig. Tolerance, resilience, and robustness refer to the ability to a pig to maintain performance in the face of disease and may be more beneficial targets for genetic improvement. The second challenge is how to obtain information that can be used to identify pigs that cope better with disease. Promising strategies in that regard are genetic markers or genomic prediction, as are indicator traits, such as immune response to vaccines or other inert compounds and data obtained from -omics technologies that can be obtained on young healthy pigs to identify genetic predictors of disease resilience. This is an active area of research, which is expected to result effective targets for selection to improve disease resilience.

11 References

Albers, G.A.A., Gray, G.D., Piper, L.R., Barker, J.S.F., Le Jambre, L.F. and Barger, I.A., 1987. The genetics of resistance and resilience to Haemonchus contortus infection in young Merino sheep. *International Journal for Parasitology*, 17(7), pp.1355–1363.

Bai, X., Cheng, J., Fortin, F., Harding, J.C.S., Dyck, M.K., PigGen Canada, Dekkers, J.C.M., Field, C.J., Rogel-Gaillard, C., Blanc, F. and Plastow, G.S., 2022. Indicators of disease resilience from complete blood count and in vitro immunoassays data from young-healthy pigs. *Proceedings WCGALP*. https://doi.org/10.3920/978-90-8686-940-4_161

Bai, X. and Plastow, G.S., 2022. Breeding for disease resilience: opportunities to manage polymicrobial challenge and improve commercial performance in the pig industry. *CABI Agriculture and Bioscience*, 3(1), pp.1–17.

Bai, X., Putz, A.M., Wang, Z., Fortin, F., Harding, J.C., Dyck, M.K., Dekkers, J.C., Field, C.J., Plastow, G.S. and Canada, P., 2020. Exploring phenotypes for disease resilience in pigs using complete blood count data from a natural disease challenge model. *Frontiers in Genetics*, 11, p.216.

Ballester, M., Ramayo-Caldas, Y., González-Rodríguez, O., Pascual, M., Reixach, J., Díaz, M., Blanc, F., López-Serrano, S., Tibau, J. and Quintanilla, R., 2020. Genetic parameters and associated genomic regions for global immunocompetence and other health-related traits in pigs. *Scientific Reports*, 10(1), pp.1–15.

Bastiaansen, J.W., Bovenhuis, H., Groenen, M.A., Megens, H.J. and Mulder, H.A., 2018. The impact of genome editing on the introduction of monogenic traits in livestock. *Genetics Selection Evolution*, 50(1), pp.1–14.

Benkaroun, J., Muir, K.F., Allshire, R., Tamer, C. and Weidmann, M., 2021. Isolation of a new infectious pancreatic necrosis virus (IPNV) variant from a fish farm in Scotland. *Viruses*, 13(3), p.385.

Berghof, T.V.L., Poppe, M. and Mulder, H.A., 2019. Opportunities to improve resilience in animal breeding programs. *Frontiers in Genetics*, 9, Article 692.

Bhatia, V., Stevens, T., Derks, M.F., Dunkelberger, J., Knol, E.F., Ross, J.W. and Dekkers, J., 2023. Identification of the genetic basis of sow pelvic organ prolapse. *Frontiers in Genetics*, 14, p.1154713.

Bidanel, J., 2011. Biology and genetics of reproduction. In M. F. Rothschild (Ed.), *Thegenetics of the pig*. CABI, pp.218–233.

Bishop S.C., 2004. Disease resistance: genetics. In W.G. Pond and A.W. Bell (Eds)., *Encyclopedia of animal science* (pp.288–290). Marcel Dekker, Inc., Ithaca, New York.

Bishop, S.C., Chesnais, J. and Stear, M.J., 2002, August. Breeding for disease resistance: issues and opportunities. In *Proceedings of the 7th World Congress on Genetics Applied to Livestock Production*, Montpellier, INRA, Castanet-Tolosan, France, pp.18–24.

Boddicker, N.J., Garrick, D.J., Rowland, R.R.R., Lunney, J.K., Reecy, J.M. and Dekkers, J.C.M., 2014. Validation and further characterization of a major quantitative trait locus associated with host response to experimental infection with porcine reproductive and respiratory syndrome virus. *Animal Genetics*, 45(1), pp.48–58.

Boddicker, N.J., Waide, E.H., Rowland, R.R.R., Lunney, J.K., Garrick, D.J., Reecy, J.M. and Dekkers, J.C.M., 2012. Evidence for a major QTL associated with host response to porcine reproductive and respiratory syndrome virus challenge. *Journal of Animal Science*, 90(6), pp.1733–1746.

Bolhuis, J.E., Schouten, W.G., Schrama, J.W. and Wiegant, V.M., 2005. Behavioural development of pigs with different coping characteristics in barren and substrate-enriched housing conditions. *Applied Animal Behaviour Science*, 93(3–4), pp.213–228.

Bondoc, O.L., Tec, J.D.P. and Santiago, C.A.T., 2001. Least-square analysis of published heterosis estimates in farm animals. *Philippine Journal of Veterinary and Animal Sciences*, 27(1), pp.1–1.

Burkard, C., Lillico, S.G., Reid, E., Jackson, B., Mileham, A.J., Ait-Ali, T., Whitelaw, C.B.A. and Archibald, A.L., 2017. Precision engineering for PRRSV resistance in pigs: macrophages from genome edited pigs lacking CD163 SRCR5 domain are fully resistant to both PRRSV genotypes while maintaining biological function. *PLoS Pathogens*, 13(2), p.e1006206.

Calvert, J.G., Slade, D.E., Shields, S.L., Jolie, R., Mannan, R.M., Ankenbauer, R.G. and Welch, S.K.W., 2007. CD163 expression confers susceptibility to porcine reproductive and respiratory syndrome viruses. *Journal of Virology*, 81(14), pp.7371–7379.

Chen, Y., Lonergan, S., Lim, K.S., Putz, A.M., Dyck, M.K., PigGen Canada, Fortin, F., Plastow, G.S., Harding, J.C. and Dekkers, J.C.M., 2023. Protein levels in the plasma of young, healthy pigs as potential indicators of disease resilience. *Journal of Animal Science*. https://doi.org/10.1093/jas/skad014

Chen, Y., Tibbs-Cortes, L.E., Ashley, C., Putz, A.M., Lim, K.S., Dyck, M.K., Fortin, F., Plastow, G.S., Dekkers, J. and Harding, J., 2020. The genetic basis of natural antibody titers of young healthy pigs and relationships with disease resilience. *BMC Genomics*, 21(1), pp.1–17.

Cheng, J., Putz, A.M., Harding, J.C., Dyck, M.K., Fortin, F., Plastow, G.S., Canada, P. and Dekkers, J.C., 2020. Genetic analysis of disease resilience in wean-to-finish pigs from a natural disease challenge model. *Journal of Animal Science*, 98(8), p.skaa244.

Chu, T.T., Bastiaansen, J.W., Berg, P., Romé, H., Marois, D., Henshall, J. and Jensen, J., 2019. Use of genomic information to exploit genotype-by-environment interactions for body weight of broiler chicken in bio-secure and production environments. *Genetics Selection Evolution*, 51(1), pp.1–13.

Clapperton, M., Diack, A.B., Matika, O., Glass, E.J., Gladney, C.D., Mellencamp, M.A., Hoste, A. and Bishop, S.C., 2009. Traits associated with innate and adaptive immunity in pigs: heritability and associations with performance under different health status conditions. *Genetics Selection Evolution*, 41(1), pp.1–11.

Clapperton, M., Glass, E.J. and Bishop, S.C., 2008. Pig peripheral blood mononuclear leucocyte subsets are heritable and genetically correlated with performance. *Animal*, 2(11), pp.1575–1584.

Colditz, I.G. and Hine, B.C. 2016. Resilience in farm animal: biology, management, breeding and implications for animal welfare. *Animal Production Science*, 56, 1961–1983.

Davies, G., Genini, S., Bishop, S.C. and Giuffra, E., 2009. An assessment of opportunities to dissect host genetic variation in resistance to infectious diseases in livestock. *Animal*, 3(3), pp.415–436.

Dekkers, J.C.M. 2007. Marker-assisted selection for commercial crossbred performance. *Journal of Animal Science*, 85(9), pp.2104–2114.

Dekkers, J.C.M. and Hospital, F., 2002. The use of molecular genetics in the improvement of agricultural populations. *Nature Reviews Genetics*, 3(1), pp.22–32.

Dekkers, J.C.M., Mathur, P.K. and Knol, E.F. 2010. Genetic improvement of the pig. Chapter 16. In A. Ruvinski and M.F. Rothschild (Eds.), *Genetics of the pig* (pp.390–425). CABI.

Dekkers, J.C.M., Rowland, R.R., Lunney, J.K. and Plastow, G., 2017. Host genetics of response to porcine reproductive and respiratory syndrome in nursery pigs. *Veterinary Microbiology*, 209, pp.107–113.

Dervishi, E., Yang, T., Dyck, M.K., Harding, J.C.S., Fortin, F., Cheng, J., Dekkers, J.C.M. and Plastow, G., 2021. Heritability and genetic correlations of plasma metabolites of pigs with production, resilience and carcass traits under natural polymicrobial disease challenge. *Scientific Reports*, 11(1), pp.1–13.

Doeschl-Wilson, A., Knap, P.W., Opriessnig, T. and More, S.J., 2021. Livestock disease resilience: from individual to herd level. *Animal*, 15, p.100286.

Doeschl-Wilson AB., Kyriazakis I., 2012 Should we aim for genetic improvement in host resistance or tolerance to infectious pathogens? Front Genet. 2012; 3:272.

Dong, Q., Dunkelberger, J., Lim, K.S., Lunney, J.K., Tuggle, C.K., Rowland, R.R. and Dekkers, J.C., 2021. Associations of natural variation in the CD163 and other candidate genes on host response of nursery pigs to porcine reproductive and respiratory syndrome virus infection. *Journal of Animal Science*, 99(10), p.skab274.

Duenk, P., Bijma, P., Wientjes, Y.C. and Calus, M.P., 2021. optimizing genomic selection for crossbred performance by model improvement and data collection. *Journal of Animal Science*, 99(8), p.skab205.

Falconer, D.S., 1952. The problem of environment and selection. *The American Naturalist*, 86(830), pp.293–298.

Falconer, D.S. and Mackay, T.F.C. 1996. *Introduction to quantitative genetics*. 4th ed. Longman Group Ltd, London.

Fisher, R.A. 1918. The correlation between relatives on the supposition of Mendelian inheritance. *Transactions of the Royal Society of Edinburgh*, 52, pp.399–433.

Fisher, R.A. 1930. *The genetical theory of natural selection*. Oxford University Press, Oxford, UK.

Flori, L., Gao, Y., Laloë, D., Lemonnier, G., Leplat, J.J., Teillaud, A., Cossalter, A.M., Laffitte, J., Pinton, P., de Vaureix, C. and Bouffaud, M., 2011. Immunity traits in pigs: substantial genetic variation and limited covariation. *PloS One*, 6(7), p.e22717.

Forkman, B., Boissy, A., Meunier-Salaün, M.C., Canali, E. and Jones, R.B., 2007. A critical review of fear tests used on cattle, pigs, sheep, poultry and horses. *Physiology & Behavior*, 92(3), pp.340–374.

Fulton, J.E., Arango, J., Arthur, J.A., Settar, P., Kreager, K.S. and O'Sullivan, N.P., 2013. Improving the outcome of a Marek's disease challenge in multiple lines of egg type chickens. *Avian Diseases*, 57(2s1), pp.519–522.

Gimsa, U., Tuchscherer, M. and Kanitz, E., 2018. Psychosocial stress and immunity–what can we learn from pig studies? *Frontiers in Behavioral Neuroscience*, 12, p.64.

Glass, E.J., 2012. The molecular pathways underlying host resistance and tolerance to pathogens. *Frontiers in Genetics*, 3, p.263.

Godoy, M., Kibenge, M.J., Montes de Oca, M., Pontigo, J.P., Coca, Y., Caro, D., Kusch, K., Suarez, R., Burbulis, I. and Kibenge, F.S., 2022. Isolation of a New Infectious

Pancreatic Necrosis Virus (IPNV) Variant from Genetically Resistant Farmed Atlantic Salmon (Salmo salar) during 2021–2022. *Pathogens*, 11(11), p.1368.

Guy, S.Z.Y., Li, L., Thomson, P.C. and Hermesch, S., 2018. Genetic parameters for health of the growing pig using medication records. In *Proceedings of the 11th World Congress on Genetics Applied to Livestock Production*, Auckland.

Guy, S.Z.Y., Thomson, P.C. and Hermesch, S., 2012. Selection of pigs for improved coping with health and environmental challenges: breeding for resistance or tolerance? *Frontiers in Genetics*, 3, p.281.

Harzilius, B, P. Mathur, and E.F. Knol (2020) Breeding for resilience: new opportunities in a modern pig breeding program. *Journal of Animal Science* 98(Supp1): 150–554.

Henderson, C.R., 1988. Theoretical basis and computational methods for a number of different animal models. *Journal of Dairy Science*, 71, pp.1–16.

Henryon, M., Berg, P., Jensen, J. and Andersen, S. 2001. Genetic variation for resistance to clinical and subclinical disease exists in growing pigs. *Animal Science, 73*, 375–387.

Henryon, M., Heegaard, P.M.H., Nielsen, J., Berg, P., Juul-Madsen, H.R., 2006. Immunological traits have the potential to improve selection of pigs for resistance to clinical and subclinical disease. *Animal Science,* 82, pp.597–606.

Hermesch, S., Guy, S., Sales, N., McKenna, T. and Bauer, M.M., 2017. 2B-105: *Genetic parameters for health, survival, immune competence, post-weaning growth and disease resilience of pigs*. Report prepared for the Co-operative Research Centre for High Integrity Australian Pork. Retrieved December 20, 2022 from http://porkcrc.com.au/wp-content/uploads/2017/10/2B-105-Final-Research-Report-final.pdf

Hermesch, S. and Luxford, B., 2018, February. Genetic parameters for white blood cells, haemoglobin and growth in weaner pigs for genetic improvement of disease resilience. In *Proceedings of the 11th world congress on genetics applied to livestock production,* pp. 11–16, Massey University, New Zeeland.

Herrero-Medrano, J.M., Mathur, P.K., Napel, J.T., Rashidi, H., Alexandri, P., Knol, E.F. and Mulder, H.A., 2015. Estimation of genetic parameters and breeding values across challenged environments to select for robust pigs. *Journal of Animal Science*, 93(4), pp.1494–1502.

Hidalgo, A.M., Bastiaansen, J.W., Lopes, M.S., Harlizius, B., Groenen, M.A. and de Koning, D.J. 2015a. Accuracy of predicted genomic breeding values in purebred and crossbred pigs. *G3,* 5(8), pp.1575–1583.

Hill, W.G., Goddard, M.E. and Visscher, P.M., 2008. Data and theory point to mainly additive genetic variance for complex traits. *PLoS Genetics*, 4(2), p.e1000008.

Hillestad, B., Johannessen, S., Melingen, G.O. and Moghadam, H.K., 2021. Identification of a new infectious pancreatic necrosis virus (IPNV) variant in atlantic salmon (Salmo salar L.) that can cause high mortality even in genetically resistant fish. *Frontiers in Genetics*, 12, p.635185.

Hine, B.C., Mallard, B.A., Ingham, A. and Colditz, I.G., 2014. Immune competence in livestock. In S. Hermesch and S. Dominik (Eds.), *Breeding focus 2014 - Improving resilience*. Animal Genetics and Breeding Unit, Armidale.

Houston, R.D., Haley, C.S., Hamilton, A., Guy, D.R., Mota-Velasco, J.C., Gheyas, A.A., Tinch, A.E., Taggart, J.B., Bron, J.E., Starkey, W.G. and McAndrew, B.J., 2010. The susceptibility of Atlantic salmon fry to freshwater infectious pancreatic necrosis is largely explained by a major QTL. *Heredity*, 105(3), pp.318–327.

Houston, R.D., Haley, C.S., Hamilton, A., Guy, D.R., Tinch, A.E., Taggart, J.B., McAndrew, B.J. and Bishop, S.C., 2008. Major quantitative trait loci affect resistance to infectious pancreatic necrosis in Atlantic salmon (Salmo salar). *Genetics*, 178(2), pp.1109–1115.

Jeon, R.L., Gilbert, C., Cheng, J., Putz, A.M., Dyck, M.K., Plastow, G.S., Fortin, F., PigGen Canada, Dekkers, J.C. and Harding, J.C., 2021. Proliferation of peripheral blood mononuclear cells from healthy piglets after mitogen stimulation as indicators of disease resilience. *Journal of Animal Science*, 99(8), p.skab084.

Johnsson, M., Ros-Freixedes, R., Gorjanc, G., Campbell, M.A., Naswa, S., Kelly, K., Lightner, J., Rounsley, S. and Hickey, J.M., 2018. Sequence variation, evolutionary constraint, and selection at the CD163 gene in pigs. *Genetics Selection Evolution*, 50(1), pp.1–9.

Kapell, D.N.R.G., Hill, W.G., Neeteson, A.M., McAdam, J., Koerhuis, A.N.M. and Avendano, S., 2012. Genetic parameters of foot-pad dermatitis and body weight in purebred broiler lines in 2 contrasting environments. *Poultry Science*, 91(3), pp.565–574.

Knap, P.W., 2005. Breeding robust pigs. *Australian Journal of Experimental Agriculture*, 45(8), pp.763–773.

Knap, P.W. and Doeschl-Wilson, A., 2020. Why breed disease-resilient livestock, and how?. *Genetics Selection Evolution*, 52(1), pp.1–18.

Koltes, J.E., Fritz-Waters, E., Eisley, C.J., Choi, I., Bao, H., Kommadath, A., Serão, N.V., Boddicker, N.J., Abrams, S.M., Schroyen, M., Loyd, H., Tuggle, C.K., Plastow, G.S., Guan, L., Stothard, P., Lunney, J.K., Liu, P., Carpenter, S., Rowland, R.R.R., Dekkers, J.C.M, and Reecy, J.M. 2015. Identification of a putative quantitative trait nucleotide in guanylate binding protein 5 for host response to PRRS virus infection. *BMC Genomics*, 16(1), pp.1–13.

Kristiansen, M., Graversen, J.H., Jacobsen, C., Sonne, O., Hoffman, H.J., Law, S.K. and Moestrup, S.K., 2001. Identification of the haemoglobin scavenger receptor. *Nature*, 409(6817), pp.198–201.

Laghouaouta, H., Pena, R.N., Ros-Freixedes, R., Reixach, J., Díaz, M., Estany, J., Armengol, R., Bassols, A. and Fraile, L., 2021. A methodology to quantify resilience in growing pigs. *Animals*, 11(10), p.2970.

Lim, B., Khatun, A., Kim, S.W., Nazki, S., Jeong, C.G., Gu, S., Lee, J., Lee, K.T., Park, C.K., Lee, S.M. and Kim, W.I., 2018. Polymorphisms in the porcine CD163 associated with response to PRRSV infection. *Animal Genetics*, 49, pp.98–99.

Lim, K.S., Cheng, J., Putz, A., Dong, Q., Bai, X., Beiki, H., Tuggle, C.K., Dyck, M.K., Fortin, F., Harding, J. and Plastow, G.S., 2021. Quantitative analysis of the blood transcriptome of young healthy pigs and its relationship with subsequent disease resilience. *BMC Genomics*, 22(1), pp.1–18.

Lim, K.S., Cheng, J., Tuggle, C.K., Dyck, M., Canada, P., Fortin, F., Harding, J., Plastow, G. and Dekkers, J.C., 2022. PSXIII-B-14 investigation of the blood transcriptome of young healthy pigs to identify genetic indicators for disease resilience. *Journal of Animal Science*, 100(Supplement_3), pp.325–325.

Lipschutz-Powell, D., Woolliams, J.A., Bijma, P. and Doeschl-Wilson, A.B., 2012. Indirect genetic effects and the spread of infectious disease: are we capturing the full heritable variation underlying disease prevalence?. *PloS One*, 7(6), p.e39551.

Mathur, P.K., Herrero-Medrano, J.M., Alexandri, P., Knol, E.F., Napel, J.T., Rashidi, H. and Mulder, H.A., 2014. Estimating challenge load due to disease outbreaks and other challenges using reproduction records of sows. *Journal of Animal Science*, 92(12), pp.5374–5381.

Meijerink, E., Fries, R., Vögeli, P., Masabanda, J., Wigger, G., Stricker, C., Neuenschwander, S., Bertschinger, H.U. and Stranzinger, G., 1997. Two α (1, 2) fucosyltransferase genes on porcine chromosome 6q11 are closely linked to the blood group inhibitor (S) and Escherichia coli F18 receptor (ECF18R) loci. *Mammalian Genome*, 8(10), pp.736–741.

Meijerink, E., Neuenschwander, S., Fries, R., Dinter, A., Bertschinger, H.U., Stranzinger, G. and Vögeli, P., 2000. A DNA polymorphism influencing α (1, 2) fucosyltransferase activity of the pig FUT1 enzyme determines susceptibility of small intestinal epithelium to Escherichia coli F18 adhesion. *Immunogenetics*, 52(1), pp.129–136.

Mellencamp, M.A., Galina-Pantoja, L., Gladney, C.D. and Torremorell, M., 2008. Improving pig health through genomics: a view from the industry. In *Animal genomics for animal health* (Vol. 132, pp.35–41). Karger Publishers.

Meuwissen, T.H.E., Hayes, B.J. and Goddard, M.E., 2001. Prediction of total genetic value using genome-wide dense marker maps. *Genetics*, 157, pp.1819–1829.

Misztal, I., Legarra, A. and Aguilar, I., 2009. Computing procedures for genetic evaluation including phenotypic, full pedigree, and genomic information. *Journal of Dairy Science*, 92(9), pp.4648–4655.

Moen, T., Baranski, M., Sonesson, A.K. and Kjøglum, S., 2009. Confirmation and fine-mapping of a major QTL for resistance to infectious pancreatic necrosis in Atlantic salmon (Salmo salar): population-level associations between markers and trait. *BMC Genomics*, 10(1), pp.1–14.

Moen, T., Torgersen, J., Santi, N., Davidson, W.S., Baranski, M., Ødegård, J., Kjøglum, S., Velle, B., Kent, M., Lubieniecki, K.P. and Isdal, E., 2015. Epithelial cadherin determines resistance to infectious pancreatic necrosis virus in Atlantic salmon. *Genetics*, 200(4), pp.1313–1326.

Mohr, S. and Liew, C.C., 2007. The peripheral-blood transcriptome: new insights into disease and risk assessment. *Trends in Molecular Medicine*, 13(10), pp.422–432.

Mulder, H.A. and Rashidi, H., 2017. Selection on resilience improves disease resistance and tolerance to infections. *Journal of Animal Science*, 95(8), pp.3346–3358.

Nakavisut, S., Crump, R., Suarez, M. and Graser, H.U., 2005. Genetic correlations between the performance of purebred and crossbred pigs. In *Proceedings of the association for the advancement of animal breeding and genetics* (Vol. 16, pp.99–102).

Nakov, D., Hristov, S., Stankovic, B., Pol, F., Dimitrov, I., Ilieski, V., Mormede, P., Hervé, J., Terenina, E., Lieubeau, B. and Papanastasiou, D.K., 2019. Methodologies for assessing disease tolerance in pigs. *Frontiers in Veterinary Science*, 5, p.329.

Pérez-Enciso, M. and Steibel, J.P., 2021. Phenomes: the current frontier in animal breeding. *Genetics Selection Evolution*, 53(1), pp.1–10.

Putz, A.M., Harding, J.C., Dyck, M.K., Fortin, F., Plastow, G.S., Dekkers, J.C. and Canada, P., 2019. Novel resilience phenotypes using feed intake data from a natural disease challenge model in wean-to-finish pigs. *Frontiers in Genetics*, 9. https://doi.org/10.3389/fgene.2018.00660

Raberg, L., Sim, D. and Read, A.F. 2007. Disentangling genetic variation for resistance and tolerance to infectious diseases in animals. *Science*, 318, pp.812–814.

Ramayo-Caldas, Y., Zingaretti, L.M., Pérez-Pascual, D., Alexandre, P.A., Reverter, A., Dalmau, A., Quintanilla, R. and Ballester, M., 2021. Leveraging host-genetics and gut microbiota to determine immunocompetence in pigs. *Animal Microbiome*, 3(1), pp.1–11.

Ren, J., Yan, X., Ai, H., Zhang, Z., Huang, X., Ouyang, J., Yang, M., Yang, H., Han, P., Zeng, W. and Chen, Y., 2012a. Susceptibility towards enterotoxigenic Escherichia coli F4ac diarrhea is governed by the MUC13 gene in pigs. *PLoS One,* 7(9), p.e44573.

Ren, Y.W., Zhang, Y.Y., Affara, N.A., Sargent, C.A., Yang, L.G., Zhao, J.L., Fang, L.R., Wu, J.J., Fang, R., Tong, Q. and Xiao, J., 2012b. The polymorphism analysis of CD169 and CD163 related with the risk of porcine reproductive and respiratory syndrome virus (PRRSV) infection. *Molecular Biology Reports*, 39(11), pp.9903–9909.

Roth, K., Pröll-Cornelissen, M.J., Heuß, E.M., Dauben, C.M., Henne, H., Appel, A.K., Schellander, K., Tholen, E. and Große-Brinkhaus, C., 2022. Genetic parameters of immune traits for Landrace and Large White pig breeds. *Journal of Animal Breeding and Genetics*, 139(6), pp.695–709.

Rowland, R.R., Lunney, J. and Dekkers, J., 2012. Control of porcine reproductive and respiratory syndrome (PRRS) through genetic improvements in disease resistance and tolerance. *Frontiers in Genetics*, 3, p.260.

Schmied, J.D., Putz, A., Dekkers, J., Dyck, M. and Fortin, F., 2018. Mortality rate and survival of pigs classified by immune response phenotype using the high immune response (HIR™) technology. In *Proceedings of the world congress on genetics applied to livestock production* (p.653).

Sellier, P., 1976. The basis of crossbreeding in pigs; a review. *Livestock Production Science*, 3(3), pp.203–226.

Strandén, I. and Garrick, D.J., 2009. Derivation of equivalent computing algorithms for genomic predictions and reliabilities of animal merit. *Journal of Dairy Science*, 92(6), pp.2971–2975.

Van der Zande, L.E., Dunkelberger, J.R., Rodenburg, T.B., Bolhuis, J.E., Mathur, P.K., Cairns, W.J., Keyes, M.C., Eggert, J.M., Little, E.A., Dee, S.A. and Knol, E.F., 2020. Quantifying individual response to PRRSV using dynamic indicators of resilience based on activity. *Frontiers in Veterinary Science*, 7, p.325.

van Grevenhof, I.E. and van der Werf, J.H. 2015. Design of reference populations for genomic selection in crossbreeding programs. *Genetics, Selection, Evolution: GSE,* 47, p.14.

VanRaden, P.M., 2008. Efficient methods to compute genomic predictions. *Journal of Dairy Science*, 91(11), pp.4414–4423.

Vissche, A.H., Janss, L.L.G., Niewold, T.A. and De Greef, K.H., 2002. Disease incidence and immunological traits for the selection of healthy pigs: a review. *Veterinary Quarterly*, 24(1), pp.29–34.

Visscher, P., Pong-Wong, R., Whittemore, C. and Haley, C., 2000. Impact of biotechnology on (cross) breeding programmes in pigs. *Livestock Production Science*, 65(1–2), pp.57–70.

Walker, L.R., Engle, T.B., Vu, H., Tosky, E.R., Nonneman, D.J., Smith, T.P., Borza, T., Burkey, T.E., Plastow, G.S., Kachman, S.D. and Ciobanu, D.C., 2018. Synaptogyrin-2 influences replication of Porcine circovirus 2. *PLoS Genetics*, 14(10), p.e1007750.

Wei, M. and van der Steen, H.A.M. 1991. Comparison of reciprocal recurrent selection with pure-line selection systems in animal breeding (a review). *Animal Breeding Abstracts,* 59, pp.281–298.

Wells, K.D., Bardot, R., Whitworth, K.M., Trible, B.R., Fang, Y., Mileham, A., Kerrigan, M.A., Samuel, M.S., Prather, R.S. and Rowland, R.R., 2017. Replacement of porcine CD163 scavenger receptor cysteine-rich domain 5 with a CD163-like homolog confers resistance of pigs to genotype 1 but not genotype 2 porcine reproductive and respiratory syndrome virus. *Journal of Virology*, 91(2), pp.e01521–16.

Whitworth, K.M., Rowland, R.R., Ewen, C.L., Trible, B.R., Kerrigan, M.A., Cino-Ozuna, A.G., Samuel, M.S., Lightner, J.E., McLaren, D.G., Mileham, A.J. and Wells, K.D., 2016. Gene-edited pigs are protected from porcine reproductive and respiratory syndrome virus. *Nature Biotechnology*, 34(1), pp.20–22.

Wientjes, Y.C.Y., Bijma, P. and Calus, M.P., 2020. Optimizing genomic reference populations to improve crossbred performance. *Genetics Selection Evolution*, 52(1), pp.1–18.

Wientjes, Y.C.J. and Calus, M.P.L., 2017. Board invited review: the purebred-crossbred correlation in pigs: a review of theory, estimates, and implications. *Journal of Animal Science*, 95(8), pp.3467–3478.

Wilkie, B.N. and Mallard, B.A., 1999. Selection for high immune response: an alternative approach to animal health maintenance? *Veterinary Immunology and Immunopathololog*y, 72, pp.231–235.

Yurkovich, J.T. and Hood, L., 2019. Blood is a window into health and disease. *Clinical Chemistry*, 65(10), pp.1204–1206.

Zhang, B., Ren, J., Yan, X., Huang, X., Ji, H., Peng, Q., Zhang, Z. and Huang, L., 2008. Investigation of the porcine MUC13 gene: isolation, expression, polymorphisms and strong association with susceptibility to enterotoxigenic Escherichia coli F4ab/ac. *Animal Genetics*, 39(3), pp.258–266.

Chapter 5

Advances in understanding the genetics of sow prolificacy: reproductive performance, trade-offs, and management solutions

Glen W. Almond, North Carolina State University, USA; and Diego F. Leal, University of São Paulo, Brazil

1 Introduction

The domestication of the wild boar (*Sus scrofa*), about 10 000 years ago, marked the start of genetic improvement in pigs. This was followed by the selection of pigs (*Sus scrofa domesticus*) for specific traits, giving rise to a host of domestic breeds with different physical, behavioral, physiological, and reproductive characteristics (Dekkers et al., 2011).

During the early part of the last century, the focus of selection was on breed characteristics and physical appearance. From the 1950s onward, breeding goals were diverted toward production traits, resulting in increased growth rate and carcass leanness. Despite the genetic progress in lean pork production and growth rate, reproduction traits showed little gain, and litter size remained relatively unaltered during those decades (Larson et al., 2011). In the early 1990s, however, selection targets shifted toward litter size, driven by improved knowledge of genetics, such as family selection and the application

http://dx.doi.org/10.19103/AS.2024.0137.10

of Best Linear Unbiased Prediction, led to clear genetic progress for litter size in the dam lines (Merks, 2000).

In the last four decades, there has been a steady increase in litter size. Genetic selection for higher ovulation rates, in parallel with advances in nutrition and reproductive management, has increased the number of piglets born per litter (Koketsu et al., 2017). The litter size of hyperprolific sow lines increased by 0.60 piglets/year, from an average of 11.7 live piglets in 2000 to 17.5 in 2019 (Sell-Kubiak, 2021). In comparison, the litter size of European wild pigs averages four to five with a range of 1–12 piglets (Andersen et al., 2005). However, some trade-offs arose as unintended consequence of increased prolificacy (Fig. 1).

Increased litter size is negatively associated with piglet birth weight and positively associated with variation in piglet birth weight within a litter (Feldpausch et al., 2019), which is associated with low piglet vitality and survivability (Quiniou et al., 2002; Fix et al., 2010). Larger litter size is positively correlated with the duration of farrowing (Van Dijk et al., 2005). Protracted farrowing duration increases the risk of stillbirths (Oliviero et al., 2010). Genetic selection for highly prolific sows was accompanied by an increased proportion of intrauterine growth-restricted (IUGR) piglets. The IUGR piglets represent a phenotype associated with increased rates of pre-weaning morbidity and mortality, resulting from physiological immaturity and dysfunction of organs and tissues, which are important to digestion, nutrient absorption, and metabolism (Wang et al., 2017).

Although contemporary sow genotypes give birth to a greater number of piglets, the number of teats has not increased in step with litter size, leading to

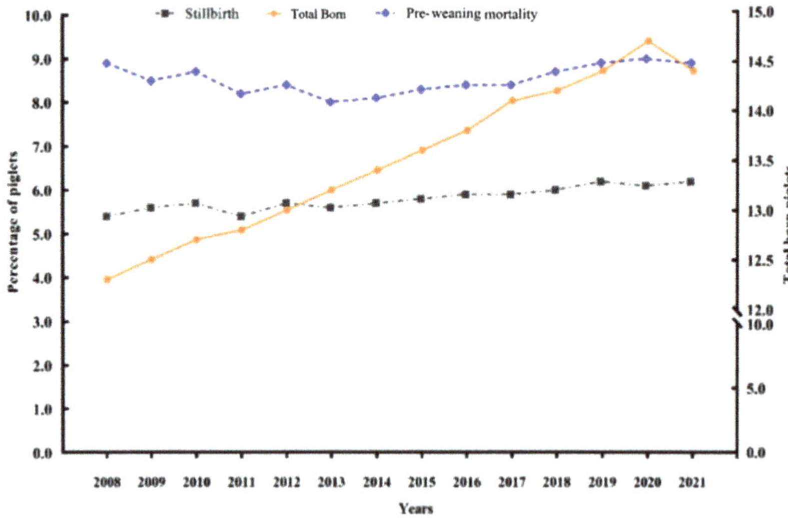

Figure 1 Litter size, percentage of stillbirths, and pre-weaning mortality over a period of 14 years in Brazil (adapted from Bortolozzo et al., 2023).

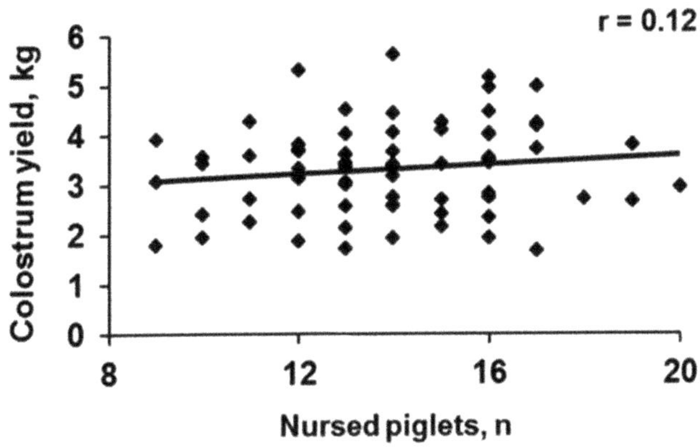

Figure 2 Relations between colostrum yield during the first 24 h postpartum and number of piglets (Quesnel et al., 2012).

greater competition among siblings for teats (Andersen et al., 2011). Even when the number of functional teats equals the number of offspring, colostrum intake may not be adequate since colostrum yield does not seem to be correlated with litter size (Quesnel et al., 2012; Fig. 2). It was estimated that 35–55% of sows from a hyperprolific herd do not produce enough colostrum to meet the demands of their litters (Le Dividich et al., 2005; Quesnel et al., 2012).

Collectively, these phenotypes represent more than a mere economic problem for the pig industry; they also raise ethical and welfare concerns for the entire chain. Therefore, genetic selection efforts should be directed toward increasing litter quality and reducing postnatal piglet mortality. This can be accomplished by selecting for uterine capacity, milk yield/lactation performance, and maternal behavior.

The current chapter is primarily a review of the advances and trade-offs of genetic selection for improved reproductive performance in modern sows. Management strategies to increase piglet survival are discussed elsewhere; thus, here we will only pinpoint some of those, with a special focus on sow maternal behavior.

2 Breeding for improved conception rate

Sow reproductive performance relies on a complex of physiological pathways and is influenced by external factors such as ambient temperature, nutritional and health status, farm management, and housing conditions. From the sow side, this includes estrus, ovulation, and farrowing. The boar contributes to this complex by providing the semen that enables fertilization (Merks et al., 2000). Conception rates have increased substantially over the years as a result of a combination of genetic selection for ovulation rate, application of advanced reproductive biotechnologies, and nutritional management. Nowadays, high conception rate (+95%) is achievable but requires superior management practices (Koketsu et al., 2017).

The direct boar effect on fertilization appears to be small, probably due to the superabundance of sperm cells available in each dose of semen. However, since semen dose can limit total born numbers by 0.3 pigs, the use of boars with high-quality sperm parameters and the adequate storage of semen doses (i.e. temperature and time) play central roles in achieving high conception rates (Merks et al., 2000).

To achieve high conception rates, it is important to breed the sow at an optimum time relative to ovulation. Considering that sows ovulate 2/3 through estrus, the fertilization rate is optimal in sows when insemination occurs between 0 and 24 h before ovulation (Soede et al., 1994). Appropriate detection of estrus is pivotal for inseminating at an optimum time relative to ovulation (Soede et al., 2011). Identifying signs of estrus in gilts can be challenging because some gilts do not show strong signs of estrus when compared to multiparous sows. In this context, Rydhmer et al. (1994) reported that gilts can be selected for strong signs of estrus, as these parameters have genetic variation and heritabilities around 0.2–0.3, respectively.

The quality of the ovulated oocytes is another important component of the conception rate. The quality of the oocytes can be affected by nutritional management (Ferguson et al., 2007). Indeed, nutritional status alters the secretion of IGF-I, and a positive association exists among nutrient intake, circulating IGF-I, and oocyte quality in sows (van Wettere et al., 2011).

Body condition during gilt development can influence fertility. Body weight and backfat thickness of gilts at the first observed standing estrus significantly influenced total number of piglets born per litter and the number of piglets born alive per litter up to parity three (Tummaruk et al., 2007). Tummaruk et al. (2001) observed that gilts with higher growth rates (500–550 g/d) until reaching 100 kg of body weight gave birth to larger litter sizes and had greater farrowing rates than gilts with lower growth rates (350–450 g/d).

A large litter poses a challenge to the sow's metabolism, which may compromise ovarian cyclicity after weaning (Peltoniemi et al., 2019), thereby impairing the conception rate. This is especially true for first-parity sows. Sows with excessive weight loss during lactation have reduced pregnancy rates and embryonic survivals (Thaker and Bilkei, 2005). Interestingly, before the 1990s, weight loss during lactation often resulted in an increased wean-to-estrus interval; however, as a result of selection for shorter wean-to-estrus intervals, contemporary sow genotypes do not seem to be affected in the same way. Nevertheless, this shorter wean-to-estrus interval increases the chances of selecting follicles of poor quality at weaning, thereby affecting conception rate and embryo survival (Soede and Kemp, 2019). Therefore, to support optimal fertility, management strategies should be implemented to avoid excessive weight loss (approximately 10–12% body weight loss) during lactation (Thaker and Bilkei, 2005). The feed composition and feeding management during lactation must be addressed to maximize feed intake. Maintaining adequate supplies of fresh feed and water is crucial, as well as stimulating the sow to stand at least six times per day. It is recommended to feed sows at least four daily meals during lactation. Furthermore, the inclusion of 2–10% fat sources in the lactation diet is suggested to increase the energy density and palatability, especially when the females are housed at high temperature (>24°C) (Muro et al., 2022).

3 Breeding for improved litter size

The pig production landscape has completely changed over the years, driven by scientific knowledge and the massive application of cutting-edge technologies. Modern sows are arguably the best representation of this progress. Indeed, sow productivity has dramatically increased during the last four decades due to continuous genetic selection for ovulation rate, leading to the high prolificacy of sows and the production of high lean pigs. It is not uncommon for contemporary highly prolific sows to ovulate more than 20 oocytes per estrus, with more than 90% conception rate.

Genetic regulation of litter size is complex, and the main component traits are ovulation rate, embryonic/fetal survival, uterine capacity, and pre-weaning mortality (Distl, 2007). The heritability for traits associated with litter size was well described by Zak et al. (2017) as shown in Table 1. Most of the references are more than 20 years old and one wonders if these heritabilities have changed over time and selection, as modern sows are genetically distinct from their predecessors given the genetic progress realized in the past two decades.

With the emergence of genomic tools, more focus was given to specific genes that control litter size. One of the first studies to demonstrate a major gene associated with litter size was conducted by Rothschild et al. (1996). The result of the study showed that a specific allele of the estrogen receptor (ER) locus was associated with increased litter size. The improved litter size, associated with a Meishan background and found in Large White ancestry, persisted in all parities of sows. This was confirmed by Short et al. (1997) in four synthetic lines of European breeds. Owing to its important role in conceptus physiology during the time of pregnancy establishment in pigs, providing the conceptus with appropriate amounts of retinoic acid, the retinol-binding protein 4 (RBP4) on *Sus scrofa* chromosomes (SSCs) 14 is another candidate gene for litter size (Marantidis et al., 2016; Rothschild et al., 2000; Spötter and Distl, 2006). Mucin 4 plays a protective role during the implantation period in pigs (Bazer and Johnson, 2014). Balcells et al. (2011) evaluated the MUC4 gene

Table 1 Estimates of heritability for traits associated with litter size

Trait	Heritability (h^2)[a]	References
Ovulation rate	0.24–0.42	Johnson et al. (1999); Ruíz-Flores and Johnson (2001)
Embryonal survival (%)	0.14	Johnson et al. (1999)
Number of fetuses at day 50	0.18	Johnson et al. (1999)
Number born alive per litter	0.10–0.23	Holm et al. (2004); Rosendo et al. (2007); Ruiz-Florez and Johnson (2001)
Number born alive at first litter	0.10 ± 0.01	Holm et al. (2004)
Number born alive at second litter	0.1 ± 0.01	Holm et al. (2004)
Number still born per litter	0.17–0.29	Johnson et al. (1999); Ruiz-Flores and Johnson (2001)
Proportion of still births within a litter (%)	0.08 ± 0.01	Rosendo et al. (2007)
Total litter birth weight (kg)	0.32	Johnson et al. (1999)

[a]Heritability traits presented as mean ± standard error (Zack et al., 2017).

as a candidate gene for litter size in Iberian × Meishan F_2 pigs. The authors concluded that MUC4 is associated with litter size in the Iberian × Meishan F_2 population and is differentially expressed regarding the number of embryos in uterus at 30 days of gestation.

Using GWAS data from a systematic review combined with sequencing data, Martins et al. (2022) used two candidate genes (*GRID2* and *PALB2*) for litter size and six candidate genes (*GHR*, *IFT80*, *FSTL3*, *SKOR1*, *SMURF1*, and *AKT3*) for the number of teats. Transcription factors (TFs) associated with candidate genes were also identified for litter size (*PALB2* and *GRID2*) and the number of teats (RIN, LTBP2, and COL6A6). Sell-Kubiak et al. (2015) identified a possible candidate gene for number of total born piglets is *ENOX1*, which is involved in cell growth and survival. Moreover, Tao et al. (2013) identified *TCF12*, *CTNNAL1*, and *WNT10B* as candidate genes for litter size. The direct application this knowledge of specific genes to modern sow reproductive management remains to be elucidated.

3.1 The tradeoffs of prolificacy

3.1.1 Increased piglet mortality

Despite the increases in litter sizes, the pig industry has been confronted with ethical concerns, economic losses, and animal welfare issues due to higher-than-acceptable mortality rates among young piglets. Genetic selection schemes targeting only few traits (e.g. litter size, growth rate, and carcass leanness) have had the untoward effect of decreasing piglet survival due to undesirable genetic correlations among these traits. For instance, the Danish pig industry was selected only for litter size until 2003. This resulted in a significant increase in prolificacy but was accompanied by greater piglet mortality (Zak et al., 2017). Nowadays, the genetic correlation of litter size and pre-weaning survival is negative in most pig breeds (Spötter and Distl, 2006). Stillborn and mummified fetuses represent 9.8% of total born pigs, while the mean piglet pre-weaning mortality rate in commercial pig operations can be as high as 20%; thus, more than 25% of total born pigs do not survive until weaning (Gourley et al., 2020). Of these, 50–80% of all piglet deaths occur in the first week after birth (Muns et al., 2016).

Physiological alterations in embryo/fetal development are considered as the main factors underpinning the high rate of perinatal mortality. This occurs because selection for higher ovulation rates has created a discrepancy between the number of ovulations and the number of fully formed fetuses that the uterus can maintain to term. This has led to more competition among fetuses for nutrients and uterine space, increasing the proportion of piglets born underweight (<1.0 kg) and in the variation of birth weight within the litter (Kemp et al., 2018). Piglet birth weight is the most influential factor affecting piglet survival (Baxter et al., 2008). Low-birth weight piglets have a higher probability of dying as a result of crushing, chilling, or starvation than heavier littermates (Feldpausch et al., 2019).

Colostrum uptake by piglets born underweight is often delayed and inadequate. Inadequate intake of colostrum by newborn piglets is considered a major underlying cause of piglet deaths in the first 24 h after birth (Quesnel et al., 2012). It has been estimated that a piglet should consume at least 250 g of colostrum in the first 24 h after birth to stay alive until weaning (Devillers et al., 2004). However, the chance of ingesting an adequate amount of colostrum decreases as litter size increases (Quesnel et al., 2012). Likewise, Peltoniemi et al. (2021) suggested that protracted farrowing duration, a trait associated with larger litters, can shorten the optimum window for suckling colostrum, reducing the chances of high-quality colostrum intake.

Although the average individual birth weight is reduced in larger litters, there are risks associated with selecting to increase piglet birth weight. Large birth weight piglets are genetically (additive and maternal) correlated with an increased incidence of stillbirth (Knol et al., 2022); thus, while selection for heavier piglets at birth might decrease piglet deaths, especially by crushing, it may not improve the overall survival rate (Hellbrügge et al., 2008; Knol et al., 2022). Selection for the number of pigs alive on day 5 post-farrowing would be more effective to improve piglet survival and litter size at weaning (Su et al., 2007). However, it should be noted that the survival rate has low heritability (Grandinson et al., 2003; Hellbrügge, 2008) and is a trait affected by several environmental factors; additionally, it is challenging to separate the effects of cross-fostering (Gäde et al., 2008). Furthermore, piglet survival is affected by both the genes of the piglet (direct genetic effects) and the genes of the sow (maternal genetic effects). Maternal genetic effects are the main limiting factor for improving piglet survival (Nguyen et al., 2021)

The number of sow teats has not increased in step with litter size. A larger litter size in pigs requires a sufficient number of functional teats for piglets to thrive (Zhuang et al., 2020). Teat number can be efficiently improved by selection (Earnhardt and Knauer, 2019). According to Chalkias et al. (2013), the estimated heritability of the total number of teats and the number of functional teats, both at 3 weeks of age and at 100 kg live weight, ranged from 0.36 to 0.42. Nonetheless, the heritability of the number of non-functional teats at 3 weeks of age was low, 0.02. In the study by Earnhardt and Knauer (2019), the heritability estimates for total teat number, functional teat number, and non-functional teat number ranged from 0.19 to 0.30, 0.17 to 0.24, and 0.12 to 0.16, respectively. The authors concluded that sufficient genetic variation exists for total teat number and functional teat number, meaning that genetic gains from selecting for these traits are possible.

3.1.2 Increase in the incidence of intra-uterine growth restriction

As genetic selection increased the litter size, it is evident that small 'runt' pigs have also become more common. It is assumed that these small pigs are exposed to IUGR. IUGR piglets are characterized by low birth weight and a dolphin-like head conformation due to the brain-sparing response to fetal undernutrition (Amdi et al., 2013). IUGR presumably takes place due to placental insufficiency (Almeida and Dias, 2022; Felicioni et al., 2020; Foxcroft et al., 2006). For pig producers, these small IUGR piglets become a challenge since they often die or are considerably smaller than their littermates at weaning.

Several factors apparently contribute to the diminished growth and survival of the IUGR piglets. The skeletal muscle of IUGR piglets differs from that of normal-weight piglets. The semitendinosus muscle weight, muscle fiber diameter, and muscle cross-sectional area were smaller in IUGR pigs than in normal birth-weight pigs at birth (Felicioni et al., 2020). IUGR also affects the myogenic factors in postnatal myogenesis and development. Evidently, immune function in the IUGR pigs is impacted, with an increase in neutrophil percentages and fibrinogen levels, but decreased CD4+ T cells (Amdi et al., 2020). These authors suggested that IUGR piglets have an immunological hypo-responsiveness and altered bone marrow hematopoiesis at 24 days of age. Thus, it would not be surprising that the IUGR piglets are more susceptible to diseases after weaning. The altered immune function also was noted in preterm IUGR piglets (Bæk et al., 2020). These piglets had diminished *ex vivo* cord blood responses to LPS for genes related to innate and adaptive immunity: spontaneous infections were more common in the IUGR piglets at 9 days of age. Furthermore, reduced levels of TNF-α and IFN-γ, and decreased gene expression of cytokines were observed in the small intestine (ileum) of IUGR piglets (Dong et al., 2014). Recently, investigators concluded that IUGR impairs nutrient digestion and absorption in the small intestine and altered microbial diversity in the cecum (Li et al., 2021).

Collectively, it is evident that the selection of female pigs for increased litter size also created an issue for pork production efficiency. The IUGR piglets are more common and one of the first challenges is related to the number of functional mammary glands. With the large litters (+15), many sows do not have enough mammary glands to support the piglets. The IUGR piglets often are the first to be compromised since they fail to effectively compete for a functional mammary gland. Based on the aforementioned studies, the IUGR piglets are disadvantaged at birth. Producers may cross-foster these IUGR piglets to sows with sufficient functional mammary glands or reduce the number of nursing piglets, via cross-fostering, to allow the IUGR piglets an opportunity to nurse. Cross-fostering, albeit well-intentioned, also creates additional challenges for the piglets, and may increase mortality (Calderón Díaz et al., 2018). It should be pointed out that due to compromised development of the gastrointestinal tract, the intake and absorption of colostrum by IUGR piglets is hindered (Amdi et al., 2013). It was demonstrated that IUGR piglets have a smaller stomach capacity compared with normal piglets (Lynegaard et al., 2020). Moreover, since the IUGR piglets are much lighter at weaning, with a number of distinct problems, these pigs typically do not grow or compete in the nursery. Of particular note, in the study of Moreira et al. (2017) low-birth-weight piglets provided with supplemental colostrum (200 mL), with the aid of an orogastric tube in the first 24 h of life, had satisfactory pre-weaning survival rate (≥87%); notwithstanding this, the weight gain until weaning of the supplemented piglets was not affected. Some producers actually provide a specialized housing system to provide the disadvantaged IUGR piglets an opportunity to gain weight prior to transfer to nursery facilities. Despite sustained efforts, there are no management strategies to date that can fully counteract the deleterious consequences of IUGR restriction.

According to Matheson et al. (2018), selection against IUGR is possible at the maternal level by using the within-litter proportion of IUGR-affected piglets trait; the inclusion of litter size, in addition to the within-litter proportion of IUGR-affected piglets, in the breeding program should maintain current litter sizes. In addition, according to those authors, the proportion of IUGR-affected piglets in a litter is a suitable indirect measure of uterine capacity for inclusion in breeding programs that aim at reducing IUGR in piglets and improving piglet survival.

4 Selecting sows for uterine capacity

Uterine capacity is defined as the number of conceptuses (embryos/fetuses and associated membranes) that the pig uterus can successfully carry to term. Evidence suggests that limitations of uterine capacity are set early in gestation, possibly around the time of conceptus elongation on days 11 through 13 (Ford et al., 2002). Uterine capacity and placental function are component traits of litter size and are regarded as the ultimate limitation on litter size. For instance, it is not uncommon for contemporary highly prolific sows to ovulate >20 oocytes per estrus, with more than 90% conception rate. However, 30–50% of embryos fail to survive due to a lack of uterine space and blood supply. Furthermore, the surviving embryos must compete to acquire adequate placental area for blood flow and the delivery of vital nutrients (Rutherford et al., 2013). Thus, selection for uterine capacity and placental efficiency may be the most productive approach in genetic selection programs (Wilson et al., 1998). Selection of sows with increased uterine capacity and placental efficiency would offer the best opportunity for increasing the number of pigs born per litter without compromising birth weight and, consequently, post-natal survival (Foxcroft, 2012).

As discussed by Freking et al. (2016), direct selection for uterine capacity is difficult, and most improvement programs have utilized indirect measures to improve component traits, such as the number of live piglets on day 5 after birth. In light of this, direct selection for uterine capacity for 11 generations was associated with a slight decrease in ovulation but resulted in 1.6 more live pigs of similar birth weight at term compared to an unselected control line (Freking et al., 2007). Likewise, selection for uterine capacity improved sow retention and significantly increased total piglet weaning weight per sow through four parities (Freking et al., 2016). Moreover, evaluation of the placental transcriptome of these selected lines identified differentially expressed genes predicted to have roles in cell migration, suggesting participation in placental folded bilayer formation.

5 Breeding for improved maternal behavior

In modern pig husbandry, the ability of the sow to raise large litters and to meet the piglets' needs, especially related to colostrum/milk provision, is an important component of piglet survival (Wischner et al., 2009). Genetic selection for increased litter size does not necessarily translate into a direct increase in the number of weaned piglets if the capacity of the sow to care for her litter is a limiting factor (Johnson and McGlone, 2011). Litters of contemporary sow genotypes can consist of up to 18–20 piglets (Oliviero et al., 2019). Considering that most piglet mortality occurs during the first 72 h after birth (Muns et al., 2016), with crushing by the sow as one of the main causes, good mothering ability (i.e. proper bonding between mother and offspring, nursing behavior, responsiveness, and attentiveness toward the offspring) during the first days of life plays a very important role in piglet survival (Canario et al., 2014). Therefore, the long-term aim of selection for better maternal behavior is to increase the piglet survival rate until weaning. As suggested by Knol et al. (2022), there is a high correlation between the observed survival of piglets until weaning and the estimated breeding value for mothering ability. Due to welfare concerns, there is increased demand for the use of loose-farrowing facilities in substitution of conventional farrowing crates. Based on this trend, Rydhmer (2021) stated that 'Genetic predisposition for larger litters in combination with loose-housing are motives for selection for maternal abilities'.

In general, few studies have assessed the genetic variation in the sows' maternal behavior. One study showed that the sows' reaction to screaming piglets when handled (h^2 = 0.14), fear of humans (h^2 = 0.14), and aggressive behavior toward humans (h^2 = 0.11) were possible traits for selection (Vangen et al., 2005). Unfortunately, the three traits were the only ones identified out of the 11 traits examined. Similar heritabilities (h^2 = 0.03 ± 0.02 to h^2 = 0.19 ± 0.03), albeit somewhat lower, were identified for farrowing behavior, rearing performance, usability of the sow, sow aggression, and udder quality in the farrowing unit (Appel et al., 2016). Setting aside the heritabilities and genetic influences, it was demonstrated that sow communication (i.e. sniffing, grunting, and nudging) was a good predictor of maternal care, piglet survival, and litter quality in three different breeds of domestic pigs (Ocepek and Andersen, 2018), making it a possible mothering ability trait for selection. It is important that sows respond promptly to their piglet's distress calls and behave carefully when lying down to minimize the risk of crushing (Grandinson, 2005). Using the piglet scream test, Grandinson et al. (2003) noted that a strong sow response is genetically associated with a higher piglet survival rate during the first days of lactation. Therefore, selection of sows showing a strong response on the piglet scream test may result in an improved early piglet survival rate.

Maternal infanticide (savaging) is the extreme manifestation arising from the lack of maternal bond between the sow and her newborn piglets (Chen et al., 2008). Maternal infanticide is more frequently observed in gilts; it has a high repeatability (~0.4), meaning that a sow that has savaged her piglets once is likely to repeat this abnormal behavior in subsequent parities (Chen et al., 2008). The etiology of maternal infanticide is still not well understood; however, environment, hormonal patterns, parturition experience, and genetic factors are all associated (Chen et al., 2011). The heritability of maternal infanticide is estimated to be 0.02 (Gäde et al., 2008). Chen et al. (2011) confirmed the existence of QTL for maternal infanticide behavior on several pig chromosomes. Thus, genomic selection against maternal infanticide may be possible.

Farrowing is a critical time in pig production, as events occurring around this stage can influence the health of the sow and piglet survival (Björkman et al., 2018; Ison et al., 2018). According to Stratz et al. (2016), good farrowing behavior includes no postural changes during farrowing, no aggression toward piglets or caretakers, and short farrowing without the need of human intervention. Domestication and breeding have not given rise to major changes in sow behavior around farrowing; this means that contemporary sows still have the same behavioral needs as their undomesticated ancestor (Wischner et al., 2009). Therefore, the application of management practices for sows to fulfill their behavioral needs around farrowing may improve the farrowing process, reducing piglet mortality.

In large litters, the fights for teats during nursing are increased. As a result, the sow may shortens nursing bouts or even terminate colostrum letdown (Quesnel et al., 2012). Udder morphology (i.e. teat pair distance, teat length, and diameter of functional teats) can also influence sibling competition for teats. Vasdal and Andersen (2012) demonstrated that only 46% of functional teats in primiparous sows were suckled, with the lower teat row being the least used. Ocepek et al. (2016) showed that a greater distance between teat pairs in the middle and posterior positions was associated with a higher proportion of not-used functional teats. Moreover, the proportion of not-used teats on day 1 post farrowing influenced teats becoming nonfunctional during lactation. It has been demonstrated that genetic selection for number of functional teats is possible (Earnhardt and Knauer, 2019). In the study of Balzani et al. (2016), the heritabilities of udder morphology traits varied from high (h^2 = 0.46 for teat length and h^2 = 0.56 for teat diameter) to moderate (h^2 = 0.37 for inter-teat distance within the same row and h^2 = 0.22 for teat distance from the abdominal midline). Based on the above observations, breeding sows for these traits can be another strategy to improve nursing behavior.

5.1 Management strategies to promote maternal behavior

Baxter (2011) stated that 'agricultural practices in livestock farming systems should take into consideration the biological needs of the animals involved'. However, this is the exception rather than the rule during farrowing and lactation. Farrowing crates are widely used in modern pig husbandry, mainly to reduce piglet mortality due to crushing (Yun et al., 2013). Nevertheless, they have created other issues, most notably for sow behavior.

In spite of domestication and artificial selection, a number of studies demonstrate that prepartum sows are still highly motivated to build a nest before the onset of farrowing (Jensen et al., 1997; Plush et al., 2021; Yun et al.,

2013, 2014). The lack of opportunity to express appropriate nest-building behavior is associated with increased stress levels (Jarvis et al., 2006). Research efforts indicate that provision of nesting materials to prepartum sows is a means to satisfy the behavioral needs of nest-building, and is associated with several measures of sow welfare (Baxter et al., 2011; Wischner et al., 2009; Yun et al., 2013). Results from a recent meta-analysis by our research group strengthened this notion. It was observed that provision of nesting material to sows prior to farrowing increased sow maternal care traits and piglet performance. Further, sows provided with nesting material had increased response to piglet distress calls and nursing bout duration and frequency (Monteiro et al., 2023). However, due to waste management constraints, the provision of nesting substrates (e.g. straw or hay) may not be possible. As an alternative, farmers should provide sows with at least man-made materials such as burlap sheets or hessian sacks.

Over 20 years ago, English and McPherson (1995) remarked that 'the belief of some industry leaders that the lack of quality stockpeople is now becoming the "Achilles Heel" of many swine production enterprises and a constraint to achieving further improvements in production efficiency and product quality'. Although genetic selection and technological advances have improved the reproductive performance in sow herds, stockmanship remains an obstacle. Various studies and reports in the 1980s and 1990s pointed out the critical importance of training and retention of farm workers (Hemsworth et al., Hansen, 1986; Hemsworth et al., 1994; English and McPhersont, 1995). Indeed, routine, positive interactions with farm personnel improve piglet survival in loose-housed sows (Rosvold et al., 2017). More recently, Losada-Espinosa and coworkers (2020) referred to stockpeople as the cornerstone of any system of livestock production since they are responsible for the welfare and productivity of the animals they work with. The authors continued: 'in spite of its importance the scientific literature on stockpeople is limited, dispersed, and specially focused on animals'. Although it is an important topic to be discussed, the role of stockpeople in modern pig husbandry is beyond the scope of this chapter. More information on this subject can be found in the excellent review of Losada-Espinosa et al. (2020).

6 Conclusion and future trends

Sow prolificacy has increased to such a high level that piglet quality, sow longevity, and welfare are being compromised under current management practices. A paradigm shift in genetic selection, coupled with enhanced management practices, is needed to resolve these issues. Genetic selection efforts should be directed toward increasing litter quality and reducing post-natal piglet mortality. This can be accomplished by selecting uterine capacity, milk yield/lactation performance, and maternal behavior. Selecting for some of these traits may not be easy, especially regarding sow maternal behavior. Therefore, management strategies to mitigate the effects of large litter sizes should be implemented.

7 Where to look for more information

The following articles provide a good overview of the subject discussed in this book chapter:

- Farmer, C. and Edwards, S.A. (2022), 'Review: Improving the performance of neonatal piglets', Animal, 16:100350.
- Gourley, K. M., Calderon, H. I., Woodworth, J. C., DeRouchey, J. M., Tokach, M. D., Dritz, S. S., and Goodband, R. D. (2020), 'Sow and piglet traits associated with piglet survival at birth and to weaning', J Anim Sci, 98:1-9.
- Kemp, B., Da Silva, C.L.A and Soede, N.M. (2018), 'Recent advances in pig reproduction: Focus on impact of genetic selection for female fertility', Reprod Domest Anim, 53:28-36.
- Rydhmer, L. (2021), 'Advances in understanding the genetics of pig behaviour', in S. Edwards (Ed), 'Understanding the behaviour and improving the welfare of pigs', Burleigh Dodds Science Publishing.
- Rothschild, M.F and Ruvinsky, A. (2011), 'The genetics of the pig', CABI, Wallingford, UK.

8 References

Almeida, F. R. C. L. & Dias, A. L. N. A. (2022). Pregnancy in pigs: the journey of an early life. *Domestic Animal Endocrinology*, *78*, 106656. https://doi.org/10.1016/j.domaniend.2021.106656

Amdi, C., Krogh, U., Flummer, C., Oksbjerg, N., Hansen, C. F. & Thei, P. K. (2013). Intrauterine growth restricted piglets defined by their head shape ingest insufficient amounts of colostrum. *Journal of Animal Science*, *91*(12), 5605–5613. https://doi.org/10.2527/jas.2013-6824

Amdi, C., Lynegaard, J., Thymann, T. & Williams, A. (2020). Intrauterine growth restriction in piglets alters blood cell counts and impairs cytokine responses in peripheral mononuclear cells 24 days post-partum. *Scientific Reports*, *10*(1), 4683. https://doi.org/10.1038/s41598-020-61623-w

Andersen, I. L., Berg, S. & Bøe, K. E. (2005). Crushing of piglets by the mother sow (Sus scrofa) - Purely accidental or a poor mother? *Applied Animal Behaviour Science*, *93*(3–4), 229–243. https://doi.org/10.1016/j.applanim.2004.11.015

Andersen, I. L., Nævdal, E. & Bøe, K. E. (2011). Maternal investment, sibling competition, and offspring survival with increasing litter size and parity in pigs (Sus scrofa). *Behavioral Ecology and Sociobiology*, *65*(6), 1159–1167. https://doi.org/10.1007/s00265-010-1128-4

Appel, A. K., Vo, B., Tönepöhl, B., König Von Borstel, U. & Gauly, M. (2016). Genetic associations between maternal traits and aggressive behaviour in Large White sows. *Animal*, *10*(7), 1234–1242. https://doi.org/10.1017/S1751731116000045

Bæk, O., Ren, S., Brunse, A., Sangild, P. T. & Nguyen, D. N. (2020). Impaired neonatal immunity and infection resistance following fetal growth restriction in preterm pigs. *Frontiers in Immunology*, *11*(August), 1–13. https://doi.org/10.3389/fimmu.2020.01808

Balcells, I., Castelló, A., Mercadé, A., Noguera, J. L., Fernández-Rodríguez, A., Sànchez, A. & Tomàs, A. (2011). Analysis of porcine MUC4 gene as a candidate gene for prolificacy QTL on SSC13 in an Iberian × Meishan F2population. *BMC Genetics*, *12*, 2–7. https://doi.org/10.1186/1471-2156-12-93

Balzani, A., Cordell, H. J., Sutcliffe, E. & Edwards, S. A. (2016). Heritability of udder morphology and colostrum quality traits in swine. *Journal of Animal Science*, *94*(9), 3636–3644. https://doi.org/10.2527/jas.2016-0458

Baxter, E. M., Jarvis, S., D'Eath, R. B., Ross, D. W., Robson, S. K., Farish, M., Nevison, I. M., Lawrence, A. B. & Edwards, S. A. (2008). Investigating the behavioural and physiological indicators of neonatal survival in pigs. *Theriogenology*, *69*(6), 773–783. https://doi.org/10.1016/j.theriogenology.2007.12.007

Baxter, E. M., Lawrence, A. B. & Edwards, S. A. (2011). Alternative farrowing systems: Design criteria for farrowing systems based on the biological needs of sows and piglets. *Animal*, *5*(4), 580–600. https://doi.org/10.1017/S1751731110002272

Baxter, E. M., Jarvis, S., Sherwood, L., Farish, M., Roehe, R., Lawrence, A. B. & Edwards, S. A. (2011). Genetic and environmental effects on piglet survival and maternal behaviour of the farrowing sow. *Applied Animal Behaviour Science*, *130*(1–2), 28–41. https://doi.org/10.1016/j.applanim.2010.11.020

Bazer, F. W. & Johnson, G. A. (2014). Pig blastocyst-uterine interactions. *Differentiation*, *87*(1–2), 52–65. https://doi.org/10.1016/j.diff.2013.11.005

Björkman, S., Oliviero, C., Kauffold, J., Soede, N. M. & Peltoniemi, O. A. T. (2018). Prolonged parturition and impaired placenta expulsion increase the risk of postpartum metritis and delay uterine involution in sows. *Theriogenology*, *106*, 87–92. https://doi.org/10.1016/j.theriogenology.2017.10.003

Bortolozzo, F. P., Zanin, G. P., Ulguim, R. da R. & Mellagi, A. P. G. (2023). Managing reproduction in hyperprolific sow herds. *Animals*, *13*(11), 1842. https://doi.org/10.3390/ani13111842

Calderón Díaz, J. A., García Manzanilla, E., Diana, A. & Boyle, L. A. (2018). Cross-fostering implications for pig mortality, welfare and performance. *Frontiers in Veterinary Science*, *5*(Jun), 1–10. https://doi.org/10.3389/fvets.2018.00123

Canario, L., Bidanel, J. P. & Rydhmer, L. (2014). Genetic trends in maternal and neonatal behaviors and their association with perinatal survival in french large white swine. *Frontiers in Genetics*, *5*(Dec), 1–16. https://doi.org/10.3389/fgene.2014.00410

Chalkias, H., Rydhmer, L. & Lundeheim, N. (2013). Genetic analysis of functional and non-functional teats in a population of Yorkshire pigs. *Livestock Science*, *152*(2–3), 127–134. https://doi.org/10.1016/j.livsci.2013.01.003

Chen, C., Gilbert, C. L., Yang, G., Guo, Y., Segonds-Pichon, A., Ma, J., Evans, G., Brenig, B., Sargent, C., Affara, N. & Huang, L. (2008). Maternal infanticide in sows: incidence and behavioural comparisons between savaging and non-savaging sows at parturition. *Applied Animal Behaviour Science*, *109*(2–4), 238–248. https://doi.org/10.1016/j.applanim.2007.02.008

Chen, C., Yang, Z., Li, Y., Wei, N., Li, P., Guo, Y., Ren, J., Ding, N. & Huang, L. (2011). Association and haplotype analysis of candidate genes in five genomic regions linked to sow maternal infanticide in a white Duroc × Erhualian resource population. *BMC Genetics*, *12*. https://doi.org/10.1186/1471-2156-12-24

Dekkers, J. C. M., Mathur, P. K. & Knol, E. F. (2011). Genetic improvement of the pig. In M. F. Rothschild and A. Ruvinsky (Eds.), *The genetics of the pig* (pp. 390–425). CABI. https://doi.org/10.1079/9781845937560.0390

Devillers, N., van Milgen, J., Prunier, A. & Le Dividich, J. (2004). Estimation of colostrum intake in the neonatal pig. *Animal Science*, *78*(2), 305–313. https://doi.org/10.1017/S1357729800054096

Distl, O. (2007). Mechanisms of regulation of litter size in pigs on the genome level. In *Reproduction in domestic animals* (Vol. 42, Issue Suppl. 2, pp. 10–16). https://doi.org/10.1111/j.1439-0531.2007.00887.x

Dong, L., Zhong, X., Ahmad, H., Li, W., Wang, Y. & Zhang, L. (2014). *Intrauterine Growth Restriction Impairs Small Intestinal Mucosal Immunity in Neonatal Piglets, 62*(1), 510–518. https://doi.org/10.1369/0022155414532655

Earnhardt, A. L. & Knauer, M. (2019). PSII-30 the genetics of functional teats in swine. *Journal of Animal Science*, *97*(Supplement_3), 215–216. https://doi.org/10.1093/jas/skz258.441

English, P. R. & McPherson, O. (1995). Stockmanship: The 'Achilles Heel' of the pig industry and the role of training, education and motivational procedures in enhancing pig care and performance. Proceedings Iowa State University Conference on 'Swine Breeding Herd Management'. Des Moines, Iowa, USA.

Feldpausch, J. A., Jourquin, J., Bergstrom, J. R., Bargen, J. L., Bokenkroger, C. D., Davis, D. L., Gonzalez, J. M., Nelssen, J. L., Puls, C. L., Trout, W. E. & Ritter, M. J. (2019). Birth weight threshold for identifying piglets at risk for preweaning mortality. *Translational Animal Science*, *3*(2), 633–640. https://doi.org/10.1093/tas/txz076

Felicioni, F., Pereira, A. D., Caldeira-Brant, A. L., Santos, T. G., Paula, T. M. D., Magnabosco, D., Bortolozzo, F. P., Tsoi, S., Dyck, M. K., Dixon, W., Martinelli, P. M., Jorge, E. C., Chiarini-Garcia, H. & Almeida, F. R. C. L. (2020). Postnatal development of skeletal muscle in pigs with intrauterine growth restriction: morphofunctional phenotype and molecular mechanisms. *Journal of Anatomy*, *236*(5), 840–853. https://doi.org/10.1111/joa.13152

Ferguson, E. M., Slevin, J., Hunter, M. G., Edwards, S. A. & Ashworth, C. J. (2007). Beneficial effects of a high fibre diet on oocyte maturity and embryo survival in gilts. *Reproduction*, *133*(2), 433–439. https://doi.org/10.1530/REP-06-0018

Fix, J. S., Cassady, J. P., Holl, J. W., Herring, W. O., Culbertson, M. S. & See, M. T. (2010). Effect of piglet birth weight on survival and quality of commercial market swine. *Livestock Science*, *132*(1–3), 98–106. https://doi.org/10.1016/j.livsci.2010.05.007

Ford, S. P., Vonnahme, K. A. & Wilson, M. E. (2002). Uterine capacity in the pig reflects a combination of uterine environment and conceptus genotype effects. *Journal of Animal Science*, *80*(E. Suppl. 1), E66–E73. http://dx.doi.org/10.1016/j.anireprosci.2016.01.018

Foxcroft, G. R. (2012). Reproduction in farm animals in an era of rapid genetic change: Will genetic change outpace our knowledge of physiology? *Reproduction in Domestic Animals*, *47*(Suppl. 4), 313–319. https://doi.org/10.1111/j.1439-0531.2012.02091.x

Foxcroft, G. R., Dixon, W. T., Novak, S., Putman, C. T., Town, S. C. & Vinsky, M. D. (2006). The biological basis for prenatal programming of postnatal performance in pigs.

Journal of Animal Science, *84*, 105–112. https://doi.org/10.2527/2006.8413_supplE105x

Freking, B. A., Lents, C. A. & Vallet, J. L. (2016). Selection for uterine capacity improves lifetime productivity of sows. *Animal Reproduction Science*, *167*, 16–21. https://doi.org/10.1016/j.anireprosci.2016.01.018

Freking, B. A., Leymaster, K. A., Vallet, J. L. & Christenson, R. K. (2007). Number of fetuses and conceptus growth throughout gestation in lines of pigs selected for ovulation rate or uterine capacity. *Journal of Animal Science*, *85*(9), 2093–2103. https://doi.org/10.2527/jas.2006-766

Gäde, S., Bennewitz, J., Kirchner, K., Looft, H., Knap, P. W., Thaller, G. & Kalm, E. (2008). Genetic parameters for maternal behaviour traits in sows. *Livestock Science*, *114*(1), 31–41. https://doi.org/10.1016/j.livsci.2007.04.006

Gourley, K. M., Calderon, H. I., Woodworth, J. C., DeRouchey, J. M., Tokach, M. D., Dritz, S. S. & Goodband, R. D. (2020). Sow and piglet traits associated with piglet survival at birth and to weaning. *Journal of Animal Science*, *98*(6), 1–9. https://doi.org/10.1093/JAS/SKAA187

Grandinson, K. (2005). Genetic background of maternal behaviour and its relation to offspring survival. *Livestock Production Science*, *93*(1), 43–50. https://doi.org/10.1016/j.livprodsci.2004.11.005

Grandinson, K., Rydhmer, L., Strandberg, E. & Thodberg, K. (2003). Genetic analysis of on-farm tests of maternal behaviour in sows. *Livestock Production Science*, *83*(2–3), 141–151. https://doi.org/10.1016/S0301-6226(03)00095-2

Hellbrügge, B., Tölle, K. H., Bennewitz, J., Henze, C., Presuhn, U. & Krieter, J. (2008). Genetic aspects regarding piglet losses and the maternal behaviour of sows. Part 1. Genetic analysis of piglet mortality and fertility traits in pigs. *Animal*, *2*(9), 1273–1280. https://doi.org/10.1017/S1751731108002504

Hemsworth, P. H., Barnett, J. L. & Hansen, C. (1986). The influence of handling by humans on the behaviour, reproduction and corticosteroids of male and female pigs. *Applied Animal Behaviour Science*, *15*(4), 303–314. https://doi.org/10.1016/0168-1591(86)90123-1

Hemsworth, P. H., Coleman, G. J. & Barnett, J. L. (1994). Improving the attitude and behaviour of stockpersons towards pigs and the consequences on the behaviour and reproductive performance of commercial pigs. *Applied Animal Behaviour Science*, *39*(3–4), 349–362. https://doi.org/10.1016/0168-1591(94)90168-6

Holm, B., Bakken, M., Klemetsdal, G. & Vangen, O. (2004). Genetic correlations between reproduction and production traits in swine. *Journal of Animal Science*, *82*(12), 3458–3464. https://doi.org/10.2527/2004.82123458x

Ison, S. H., Jarvis, S., Hall, S. A., Ashworth, C. J. & Rutherford, K. M. D. (2018). Periparturient behavior and physiology: further insight into the farrowing process for primiparous and multiparous sows. *Frontiers in Veterinary Science*, *5*(Jun), 1–15. https://doi.org/10.3389/fvets.2018.00122

Jarvis, S., D'Eath, R. B., Robson, S. K. & Lawrence, A. B. (2006). The effect of confinement during lactation on the hypothalamic-pituitary- adrenal axis and behaviour of primiparous sows. *Physiology and Behavior*, *87*(2), 345–352. https://doi.org/10.1016/j.physbeh.2005.10.004

Jensen, M. S., Jensen, S. K. & Jakobsen, K. (1997). Development of digestive enzymes in pigs with emphasis on lipolytic activity in the stomach and pancreas. *Journal of Animal Science*, *75*(2), 437–445. https://doi.org/10.2527/1997.752437x

Johnson, A. K. & McGlone, J. J. (2011). Behaviour genetics of the domestic pig. In A. R. M. F. Rothschild & V. Information (Eds.), *The genetics of the pig* (pp. 200–217). CABI. https://doi.org/10.1079/9781845937560.0200

Johnson, R. K., Nielsen, M. K. & Casey, D. S. (1999). Responses in ovulation rate, embryonal survival, and litter traits in swine to 14 generations of selection to increase litter size. *Journal of Animal Science*, *77*(3), 541. https://doi.org/10.2527/1999.773541x

Kemp, B., Da Silva, C. L. A. & Soede, N. M. (2018). Recent advances in pig reproduction: Focus on impact of genetic selection for female fertility. *Reproduction in Domestic Animals*, *53*(May), 28–36. https://doi.org/10.1111/rda.13264

Knol, E. F., Van Der Spek, D. & Zak, L. J. (2022). Genetic aspects of piglet survival and related traits: a review. *Journal of Animal Science*, *100*(6), 1–9. https://doi.org/10.1093/jas/skac190

Koketsu, Y., Tani, S. & Iida, R. (2017). Factors for improving reproductive performance of sows and herd productivity in commercial breeding herds. *Porcine Health Management*, *3*(1), 1. https://doi.org/10.1186/s40813-016-0049-7

Larson, G., Cucchi, T. & Dobney, K. (2011). Genetic aspects of pig domestication. In R. M. and R. A. (Eds.), *The genetics of the pig* (pp. 14–37). CABI. https://doi.org/10.1079/9781845937560.0014

Le Dividich, J., Rooke, J. A. & Herpin, P. (2005). Nutritional and immunological importance of colostrum for the new-born pig. *Journal of Agricultural Science*, *143*(6), 469–485. https://doi.org/10.1017/S0021859605005642

Li, T., Huang, S., Lei, L., Tao, S., Xiong, Y., Wu, G., Hu, J., Yuan, X., Zhao, S., Zuo, B., Yang, H., Xiao, Y., Lin, G. & Wang, J. (2021). Intrauterine growth restriction alters nutrient metabolism in the intestine of porcine offspring. *Journal of Animal Science and Biotechnology*, *12*(1), 1–12. https://doi.org/10.1186/s40104-020-00538-y

Losada-Espinosa, N., Miranda-De la Lama, G. C. & Estévez-Moreno, L. X. (2020). Stockpeople and Animal Welfare: Compatibilities, Contradictions, and Unresolved Ethical Dilemmas. *Journal of Agricultural and Environmental Ethics*, *33*(1), 71–92. https://doi.org/10.1007/s10806-019-09813-z

Lynegaard, J. C., Hales, J., Nielsen, M. N., Hansen, C. F. & Amdi, C. (2020). The stomach capacity is reduced in intrauterine growth restricted piglets compared to normal piglets. *Animals*, *10*(8), 1291. https://doi.org/10.3390/ani10081291

Marantidis, A., Laliotis, G. P. & Avdi, M. (2016). Association of RBP4 genotype with phenotypic reproductive traits of sows. *Genetics Research International*, *2016*, 0–5. https://doi.org/10.1155/2016/4940532

Martins, T. F., Braga Magalhães, A. F., Verardo, L. L., Santos, G. C., Silva Fernandes, A. A., Gomes Vieira, J. I., Irano, N. & dos Santos, D. B. (2022). Functional analysis of litter size and number of teats in pigs: from GWAS to post-GWAS. *Theriogenology*, *193*, 157–166. https://doi.org/10.1016/j.theriogenology.2022.09.005

Matheson, S. M., Walling, G. A. & Edwards, S. A. (2018). Genetic selection against intrauterine growth retardation in piglets: a problem at the piglet level with a solution at the sow level. *Genetics Selection Evolution*, *50*(1), 1–11. https://doi.org/10.1186/s12711-018-0417-7

Merks, J., Ducro-Steverink, D. & Feitsma, H. (2000). Management and genetic factors affecting fertility in sows. *Reproduction in Domestic Animals*, *35*(6), 261–266. https://doi.org/10.1046/j.1439-0531.2000.00269.x

Merks, J. W. M. (2000). One century of genetic changes in pigs and the future needs. *BSAP Occasional Publication*, *27*, 8–19. https://doi.org/10.1017/s1463981500040498

Monteiro, M. S., Muro, B. B. D., Carnevale, R. F., Poor, A. P., Araújo, K. M., Viana, C. H. C., Almond, G. W., Moreno, A. M., Garbossa, C. A. P. & Leal, D. F. (2023). The beneficial effects of providing prepartum sows with nesting materials on farrowing traits, piglet performance and maternal behavior: a systematic review and meta-analysis. *Applied Animal Behaviour Science*, *259*, 105795. https://doi.org/10.1016/j.applanim.2022.105795

Moreira, L. P., Menegat, M. B., Barros, G. P., Bernardi, M. L., Wentz, I. & Bortolozzo, F. P. (2017). Effects of colostrum, and protein and energy supplementation on survival and performance of low-birth-weight piglets. *Livestock Science*, *202*, 188–193. https://doi.org/10.1016/j.livsci.2017.06.006

Muns, R., Nuntapaitoon, M. & Tummaruk, P. (2016). Non-infectious causes of pre-weaning mortality in piglets. *Livestock Science*, *184*, 46–57. https://doi.org/10.1016/j.livsci.2015.11.025

Muro, B. B. D., Carnevale, R. F., Leal, D. F., Almond, G. W., Monteiro, M. S., Poor, A. P., Schinckel, A. P. & Garbossa, C. A. P. (2022). The importance of optimal body condition to maximize reproductive health and perinatal outcomes in pigs. *Nutrition Research Reviews*, *6*. https://doi.org/10.1017/S0954422422000129

Nguyen, T. Q., Knap, P. W., Simm, G., Edwards, S. A. & Roehe, R. (2021). Evaluation of direct and maternal responses in reproduction traits based on different selection strategies for postnatal piglet survival in a selection experiment. *Genetics Selection Evolution*, *53*(1), 1–16. https://doi.org/10.1186/s12711-021-00612-7

Ocepek, M., Andersen-Ranberg, I., Edwards, S. A. & Andersen, I. L. (2016). Udder characteristics of importance for teat use in purebred and crossbred pigs. *Journal of Animal Science*, *94*(2), 780–788. https://doi.org/10.2527/jas.2015-9420

Ocepek, M. & Andersen, I. L. (2018). Sow communication with piglets while being active is a good predictor of maternal skills, piglet survival and litter quality in three different breeds of domestic pigs (Sus scrofa domesticus). *PLoS One*, *13*(11), 1–18. https://doi.org/10.1371/journal.pone.0206128

Oliviero, C., Heinonen, M., Valros, A. & Peltoniemi, O. (2010). Environmental and sow-related factors affecting the duration of farrowing. *Animal Reproduction Science*, *119*(1–2), 85–91. https://doi.org/10.1016/j.anireprosci.2009.12.009

Oliviero, C., Junnikkala, S. & Peltoniemi, O. (2019). The challenge of large litters on the immune system of the sow and the piglets. *Reproduction in Domestic Animals*, *54*, 12–21. https://doi.org/10.1111/rda.13463

Peltoniemi, O., Björkman, S., Oropeza-Moe, M. & Oliviero, C. (2019). Developments of reproductive management and biotechnology in the pig. *Animal Reproduction*, *16*(3), 524–538. https://doi.org/10.21451/1984-3143-AR2019-0055

Peltoniemi, O., Han, T. & Yun, J. (2021). Coping with large litters: Management effects on welfare and nursing capacity of the sow. *Journal of Animal Science and Technology*, *63*(2), 199–210. https://doi.org/10.5187/JAST.2021.E46

Plush, K. J., McKenny, L. A., Nowland, T. L. & van Wettere, W. H. E. J. (2021). The effect of hessian and straw as nesting materials on sow behaviour and piglet survival and growth to weaning. *Animal*, *15*(7), 100273. https://doi.org/10.1016/j.animal.2021.100273

Quesnel, H., Farmer, C. & Devillers, N. (2012). Colostrum intake: influence on piglet performance and factors of variation. *Livestock Science*, *146*, 105–114. https://doi.org/10.1016/j.livsci.2012.03.010

Quiniou, N., Dagorn, J. & Gaudré, D. (2002). Variation of piglets' birth weight and consequences on subsequent performance. *Livestock Production Science*, *78*(1), 63–70. https://doi.org/10.1016/S0301-6226(02)00181-1

Rosendo, A., Druet, T., Gogué, J. & Bidanel, J. P. (2007). Direct responses to six generations of selection for ovulation rate or prenatal survival in Large White pigs. *Journal of Animal Science*, *85*(2), 356–364. https://doi.org/10.2527/jas.2006-507

Rosvold, E. M., Kielland, C., Ocepek, M., Framstad, T., Fredriksen, B., Andersen-Ranberg, I., Næss, G. & Andersen, I. L. (2017). Management routines influencing piglet survival in loose-housed sow herds. *Livestock Science*, *196*, 1–6. https://doi.org/10.1016/j.livsci.2016.12.001

Rothschild, M., Jacobson, C., Vaske, D., Tuggle, C., Wang, L., Short, T., Eckardt, G., Sasaki, S., Vincent, A., Mclaren, D., Southwood, O., Van Der Steen, H., Mileham, A. & Plastow, G. (1996). The estrogen receptor locus is associated with a major gene influencing litter size in pigs. *Proceedings of the National Academy of Sciences of the United States of America*, *93*(1), 201–205. https://doi.org/10.1073/pnas.93.1.201

Rothschild, M. F., Messer, L., Day, A., Wales, R., Short, T., Southwood, O. & Plastow, G. (2000). Investigation of the retinol-binding protein 4 (RBP4) gene as a candidate gene for increased litter size in pigs. *Mammalian Genome*, *11*(1), 75–77. https://doi.org/10.1007/s003350010015

Ruíz-Flores, A. & Johnson, R. K. (2001). Direct and correlated responses to two-stage selection for ovulation rate and number of fully formed pigs at birth in swine. *Journal of Animal Science*, *79*(9), 2286–2297. https://doi.org/10.2527/2001.7992286x

Rutherford, K., Baxter, E., D'Eath, R., Turner, S., Arnott, G., Roehe, R., Ask, B., SandØe, P., Moustsen, V., Thorup, F., Edwards, S., Berg, P. & Lawrence, A. (2013). The welfare implications of large litter size in the domestic pig I: biological factors. *Animal Welfare*, *22*(2), 199–218. https://doi.org/10.7120/09627286.22.2.199

Rydhmer, L. (2021). Advances in understanding the genetics of pig behaviour. In *Understanding the behaviour and improving the welfare of pigs* (pp. 3–26). https://doi.org/10.19103/AS.2020.0081.01

Rydhmer, L., Eliasson-Selling, L., Johansson, K., Stern, S. & Andersson, K. (1994). A genetic study of estrus symptoms at puberty and their relationship to growth and leanness in gilts. *Journal of Animal Science*, *72*(8), 1964–1970. https://doi.org/10.2527/1994.7281964x

Sell-Kubiak, E. (2021). Selection for litter size and litter birthweight in Large White pigs: maximum, mean and variability of reproduction traits. *Animal*, *15*(10), 100352. https://doi.org/10.1016/j.animal.2021.100352

Sell-Kubiak, E., Duijvesteijn, N., Lopes, M. S., Janss, L. L. G., Knol, E. F., Bijma, P. & Mulder, H. A. (2015). Genome-wide association study reveals novel loci for litter size and its variability in a large white pig population. *BMC Genomics*, *16*(1), 1–13. https://doi.org/10.1186/s12864-015-2273-y

Short, T. H., Rothschild, M. F., Southwood, O. I., McLaren, D. G., De Vries, A., Van Der Steen, H., Eckardt, G. R., Tuggle, C. K., Helm, J., Vaske, D. A., Mileham, A. J. & Plastow, G. S. (1997). Effect of the estrogen receptor locus on reproduction and production traits in four commercial pig lines. *Journal of Animal Science*, *75*(12), 3138–3142. https://doi.org/10.2527/1997.75123138x

Soede, N. M., Helmond, F. A. & Kemp, B. (1994). Periovulatory profiles of oestradiol, LH and progesterone in relation to oestrus and embryo mortality in multiparous sows using transrectal ultrasonography to detect ovulation. *Reproduction*, *101*(3), 633–641. https://doi.org/10.1530/jrf.0.1010633

Soede, N. M., Langendijk, P. & Kemp, B. (2011). Reproductive cycles in pigs. *Animal Reproduction Science*, *124*(3–4), 251–258. https://doi.org/10.1016/j.anireprosci.2011.02.025

Soede, Nicoline M, & Kemp, B. (2019). Recent advances in pig reproduction: 1. Shorter weaning-to-oestrus interval. *Anais Do XXIII Congresso Brasileiro de Reproducilo Animal 2019*, *2009*, 79–83.

Spötter, A. & Distl, O. (2006). Genetic approaches to the improvement of fertility traits in the pig. *Veterinary Journal*, *172*(2), 234–247. https://doi.org/10.1016/j.tvjl.2005.11.013

Stratz, P., Just, A., Faber, H. & Bennewitz, J. (2016). Genetic analyses of mothering ability in sows using field-recorded observations. *Livestock Science*, *191*, 1–5. https://doi.org/10.1016/j.livsci.2016.06.014

Su, G., Lund, M. S. & Sorensen, D. (2007). Selection for litter size at day five to improve litter size at weaning and piglet survival rate1. *Journal of Animal Science*, *85*(6), 1385–1392. https://doi.org/10.2527/jas.2006-631

Tao, H., Mei, S., Sun, X., Peng, X., Zhang, X., Ma, C., Wang, L., Hua, L. & Li, F. (2013). Associations of TCF12, CTNNAL1 and WNT10B gene polymorphisms with litter size in pigs. *Animal Reproduction Science*, *140*(3–4), 189–194. https://doi.org/10.1016/j.anireprosci.2013.05.013

Thaker, M. Y. C. & Bilkei, G. (2005). Lactation weight loss influences subsequent reproductive performance of sows. *Animal Reproduction Science*, *88*(3–4), 309–318. https://doi.org/10.1016/j.anireprosci.2004.10.001

Tummaruk, P., Lundeheim, N., Einarsson, S. & Dalin, A. M. (2001). Effect of birth litter size, birth parity number, growth rate, backfat thickness and age at first mating of gilts on their reproductive performance as sows. *Animal Reproduction Science*, *66*(3–4), 225–237. https://doi.org/10.1016/S0378-4320(01)00095-1

Tummaruk, P., Tantasuparuk, W., Techakumphu, M. & Kunavongkrit, A. (2007). Age, body weight and backfat thickness at first observed oestrus in crossbred Landrace × Yorkshire gilts, seasonal variations and their influence on subsequence reproductive performance. *Animal Reproduction Science*, *99*(1–2), 167–181. https://doi.org/10.1016/j.anireprosci.2006.05.004

Van Dijk, A. J., Van Rens, B. T. T. M., Van Der Lende, T. & Taverne, M. A. M. (2005). Factors affecting duration of the expulsive stage of parturition and piglet birth intervals in sows with uncomplicated, spontaneous farrowings. *Theriogenology*, *64*(7), 1573–1590. https://doi.org/10.1016/j.theriogenology.2005.03.017

van Wettere, W. H. E. J., Mitchell, M., Revell, D. K. & Hughes, P. E. (2011). Nutritional restriction of pre-pubertal liveweight gain impairs ovarian follicle growth and oocyte developmental competence of replacement gilts. *Theriogenology*, *75*(7), 1301–1310. https://doi.org/10.1016/j.theriogenology.2010.12.001

Vangen, O., Holm, B., Valros, A., Lund, M. S. & Rydhmer, L. (2005). Genetic variation in sows' maternal behaviour, recorded under field conditions. *Livestock Production Science*, *93*(1), 63–71. https://doi.org/10.1016/j.livprodsci.2004.11.007

Vasdal, G. & Andersen, I. L. (2012). A note on teat accessibility and sow parity - consequences for newborn piglets. *Livestock Science*, *146*(1), 91–94. https://doi.org/10.1016/j.livsci.2012.02.005

Wang, J., Feng, C., Liu, T., Shi, M., Wu, G. & Bazer, F. W. (2017). Physiological alterations associated with intrauterine growth restriction in fetal pigs: Causes and insights for nutritional optimization. In *Molecular Reproduction and Development* (Vol. 84, Issue 9, pp. 897–904). https://doi.org/10.1002/mrd.22842

Wilson, M. E., Biensen, N. J. & Ford, S. P. (1998). *Novel Insight into the Control of Litter Size in Pigs, Using Placental Efficiency as a Selection Tool 1, 2 ABSTRACT* : 1654–1658.

Wischner, D., Kemper, N. & Krieter, J. (2009). Nest-building behaviour in sows and consequences for pig husbandry. In *Livestock science* (Vol. 124, Issues 1–3, pp. 1–8). Elsevier B.V. https://doi.org/10.1016/j.livsci.2009.01.015

Yun, J., Swan, K., Farmer, C., Oliviero, C., Peltoniemi, O. & Valros, A. (2014). Prepartum nest-building has an impact on postpartum nursing performance and maternal behaviour in early lactating sows. *Applied Animal Behaviour Science*, *160*, 31–37. https://doi.org/10.1016/j.applanim.2014.08.011

Yun, J., Swan, K., Vienola, K., Farmer, C., Oliviero, C., Peltoniemi, O. & Valros, A. (2013). Nest-building in sows : Effects of farrowing housing on hormonal modulation of maternal characteristics. *Applied Animal Behaviour Science*, *148*(1–2), 77–84. https://doi.org/10.1016/j.applanim.2013.07.010

Zak, L. J., Gaustad, A. H., Bolarin, A., Broekhuijse, M. L. W. J., Walling, G. A. & Knol, E. F. (2017). Genetic control of complex traits, with a focus on reproduction in pigs. *Molecular Reproduction and Development*, *84*(9), 1004–1011. https://doi.org/10.1002/mrd.22875

Zhuang, Z., Ding, R., Peng, L., Wu, J., Ye, Y., Zhou, S., Wang, X., Quan, J., Zheng, E., Cai, G., Huang, W., Yang, J. & Wu, Z. (2020). Genome-wide association analyses identify known and novel loci for teat number in Duroc pigs using single-locus and multi-locus models. *BMC Genomics*, *21*(1), 1DUMM. https://doi.org/10.1186/s12864-020-6742-6

Chapter 6

Advances in the genetic improvement of sow longevity

Luiz Brito, Purdue University, USA; and Susanne Hermesch, AGBU a joint venture of NSW Department of Primary Industries and University of New England, Australia

1 Introduction

Sow longevity plays a major role in the long-term sustainability and profitability of swine production systems. The number and viability of piglets produced during a sow's productive life is directly related to the number of pigs available on farms to produce pork and to the length of the sow's productive life and reproductive efficiency. From an economic perspective, a sow must remain in the herd for at least three-to-four parities to produce enough piglets to reach a positive net return and pay for its rearing costs (Stalder et al., 2003; Lacy and Stalder, 2004). Sow longevity indirectly affects performance of her progeny. For example, mortality rates are higher for gilt progeny when compared to sow progeny. Daily weight gain in gilt progeny is also lower than that observed in progeny of older sows (Carney-Hinkle et al., 2013; Craig et al., 2019; Wijesiriwardana et al., 2022). Early replacements of sows attributed to death have also had a great impact in the destabilization of the immune status of the farm (Niemi et al., 2017). Reduced longevity and involuntary culling have ethical and animal welfare implications. For instance, recent data from North America has reported an increase in sow mortality from 8% in 2012 to nearly 14% in 2021 (Eckberg, 2022). Another study

http://dx.doi.org/10.19103/AS.2024.0137.19

evaluating sow mortality in midwestern USA reported that sow death occurred at a median of parity two and the main reasons for death were locomotion (27%) and reproduction (24%) (Kikuti et al., 2023). In this context, there is an urgent need for developing efficient management and breeding strategies to minimize involuntary culling of sows and improve sow longevity in commercial herds.

Sows are culled or die from multiple reasons, including infectious, metabolic, and structural diseases; fertility and reproductive disorders; conformation characteristics (e.g. feet and leg disorders); and poor performance (Stalder et al., 2004; Kikuti et al., 2023). For instance, Monteiro et al. (2022) investigated the risk factors of sow mortality based on post-mortem examination and microbiological investigation on 123 sows that had died on a breeding herd in Brazil with 15 000 sows in total. The mortality of spontaneous death in sows occurred mostly in the peripartum period (53%). According to the authors, sow mortality involved multiple risk factors and several bacterial agents. The spontaneous deaths were mainly associated with heart failures, hemorrhagic and perforating gastric ulcers, and liver torsion, while in the euthanized sows, the post-mortem findings were primarily associated with locomotor disorders (Monteiro et al., 2022). Therefore, it is clear that sow longevity is a complex trait influenced by multiple biological mechanisms. While nutritional strategies, management practices, and housing conditions directly impact sow longevity, they need to be combined with effective selection strategies for long-term genetic improvements.

Multiple descriptors of sow longevity have been proposed reflecting the complex biological nature of this trait. Genetic improvement of sow longevity requires a clear definition of the breeding objective, the identification of indicator traits that are heritable and can be measured in many individuals as early as possible, preferably at a low cost and in a non-invasive manner. Well-designed recording schemes need to be available for comprehensively capturing most of the phenotypic variability underlying sow longevity across environments and multiple types of production systems. The wide adoption of genomic selection in pig breeding programs and the investigation of novel traits and biomarkers provide substantial opportunities for obtaining accurate genomic breeding values for longevity indicators in young animals. In this chapter, we cover these main opportunities and progress obtained in the genetic improvement of sow longevity with a focus on sow longevity in intensive production systems. We also highlight the key challenges faced by pig breeders to improve sow longevity and outline potential avenues to overcome them.

2 Definition of sow longevity and indicator traits

For swine breeding purposes, sow longevity can be defined based on the ability of a commercial sow to remain healthy and productive in the herd for a long

Published by Burleigh Dodds Science Publishing Limited, 2025.

duration. However, there are many nuances to this definition of sow longevity and over time, various indicators of sow longevity have been proposed for genetic improvement, including length of productive life in days, total number of parities before culling, age at culling, stayability and lifetime piglet production (e.g. Tholen et al., 1996; Serenius and Stalder, 2006; Kerssen et al., 2019). Some of these indicators of sow longevity have not been defined consistently across studies. For instance, length of productive life has been defined as the number of days between the date when a sow entered the breeding herd to culling (Serenius et al., 2006) or as the number of days between first farrowing and culling (Meszaros et al., 2010). These slight differences in the definition of each indicator of sow longevity need to be considered when interpreting survival distribution functions for different genetic lines. For example, Serenius et al. (2006) found a significant drop in survival of sows at 160–180 days of length of productive life which represented the time around the first parity. This decline in survival of sows around the first parity was also observed by Meszaros et al. (2010). However, the dip in sow survival occurred at about 30 days of length of productive days in the study by Meszaros et al. (2010) because the definition of sow longevity, which was length of productive days, started with the date of the first farrowing instead of the date when a sow entered the breeding herd. These studies illustrate that the risk of sows leaving a herd is largest around each lactation event and sow longevity is closely related to the number of parities.

Stayability describes whether a sow stays in the herd until a certain time point and is defined as a binary trait for each parity (e.g. Tholen et al., 1996). Stayability from the first to the second, third, or higher parity is defined as 0 if a sow has not reached the relevant parity and 1 otherwise. Stayability has the advantage that it is defined after each parity rather than at the desired number of parities. As such, information for this trait is available earlier in the life of the sows and is less affected by data censoring, which occurs when a proportion of sows are still alive and therefore have a missing observation for their lifetime performance. Furthermore, sows are most at risk of leaving the herd around each lactation and stayability corresponds closely to the period when sows are at the highest risk of leaving the herd. Another important trait group is represented by functional longevity, in which the lifetime performance of sows are considered. Functional longevity depends on lifespan, prolificacy, fertility rate, reproductive efficiency, and maternal ability.

Additional longevity traits have been proposed for other livestock species and could be adapted to genetically evaluate sow longevity. For instance, Oliveira et al. (2020) proposed novel indicators of functional longevity in North American Angus cattle based on how long the cow stayed in the herd while also calving, considering different culling reasons (natural death, structural problems, disease, fertility, performance, and miscellaneous) based on random regression models. The authors found that functional longevity based on different culling reasons have a different genetic background and suggested that a longevity selection sub-index should be developed (Oliveira et al., 2020). A major challenge for adapting this trait to pig breeding programs might be the minimal documentation and low accuracy of culling reasons recorded in commercial herds (less variables can be included in the evaluation models), which is an important point to be addressed for the development of a sow longevity sub-index. In addition to direct selection for sow longevity, genetically improving traits related to sow culling reasons (e.g. fertility, reproduction, health, conformation, resilience) is another key route for increasing sow longevity in commercial herds. In summary, given the complexity of factors involved in sow longevity, it is no surprise that multiple definitions have been proposed over time. Moving forward, pig breeding companies should identify a consistent definition and description of sow longevity; firstly as a well-defined breeding objective trait and secondly as effective selection criteria (indicator traits), which reflect the sow's lifetime productive efficiency and her ability to live a long and healthy life.

3 Sow longevity in the breeding objective

Early studies conducted in the 1980s developed extensive bioeconomic models which described the life cycle of sows for the derivation of economic values of traits (e.g. Tess et al. 1983, de Vries 1989). These bioeconomic models considered changes in body weight, body composition, and reproductive performance of sows over their lifetime in different parities. An economic value for sow longevity *per se* was not presented in these studies. Furthermore, the review by Quinton et al. (2006) of economic values for sow productivity traits presented in studies mainly conducted in the 1980s and 1990s did not include sow longevity. De Vries (1989) included culling percentage of sows at three different time points after farrowing. These culling categories were linked to specific days after farrowing and indirectly reflect an improvement in the rebreeding success of sows within each parity thereby reducing the number of unproductive days of sows and the costs of keeping sows in the breeding herd.

The length of productive life has been considered as an indicator trait of sow longevity in most studies (Serenius and Stalder, 2006) and has been commonly defined based on the number of days from first farrowing until the sow is removed from the herd. Reducing the length of productive life while maintaining the same productivity improves the profitability of pig enterprises resulting from the reduction in the costs of keeping a sow in the herd. This approach was used by Serenius et al. (2007) to derive an economic value for each day of length of productive life. The model was similar to the bioeconomic model of de Vries (1989). Lifetime production was based on 10 parities assuming a specific culling percentage of sows for each parity.

Other traits of sows considered in the breeding objective and also expressed per day were age at first farrowing and extra days between farrowings. The economic values for age at first farrowing, extra days between farrowings and length of productive life were −0.05, −0.187, and 0.014 Euro per day, respectively. These economic values are not directly comparable because they have different genetic expressions. Using the discounted genetic expressions presented by Amer et al. (2014) and Hermesch et al. (2014) for traits expressed at equivalent time points and number of genetic expressions showed that length of productive life accounted for 1% of selection emphasis of the overall breeding objective based on underlying genetic standard deviations of traits presented by Serenius et al. (2007). In comparison, age at first farrowing and extra days between farrowings accounted for 2% and 8% of selection emphasis in the overall breeding objective. The low selection emphasis on length of productive life was based on the underlying bio-economic model used by Serenius et al. (2007) which was not outlined in detail. Wolfova and Wolf (2013) point out that the economic value for length of productive life differs substantially among authors and depends predominantly on which component traits were changed in the model. Wolfova and Wolf (2013) state that "the definition of 'longevity' of breeding females is perhaps the most controversial among traits included in the breeding goals of livestock species."

Intuitively, the cost of keeping a sow in the herd for a day represents the economic value for length of productive life. However, this economic value depends on the assumptions made in regard to other component traits of this definition of sow longevity that are also associated with daily cost of keeping a sow in the herd and how they are considered in the breeding objective. Differences in days between farrowings that reflect different cost components of each parity can be accommodated by considering individual traits such as weaning to conception interval, farrowing rate, and possibly gestation length. Gestation length and weaning to conception interval are usually not considered in the breeding objectives, however, their economic value can be easily derived. Further, the economic value of farrowing rate was derived by Hermesch (2021). Farrowing interval is usually not considered in pig breeding programs because it is influenced by management decisions, including lactation length (litter weaning age) or the decision to use a sow as a nurse sow. Furthermore, farrowing interval encompasses gestation length and weaning to conception interval which are both heritable traits (e.g. Tholen et al., 1996; ten Napel et al., 1995) and may be considered in the breeding objective as separate traits. Farrowing interval has been considered a factor for length of productive life (e.g. Heusing et al., 2005). However, the majority of studies ignored explicit factors to account for differences in management decisions that could influence the length of productive life of sows. This overview illustrates the importance of clearly defining the underlying assumptions for considering length of productive life and its components in a breeding objective.

These issues when deriving an economic value for length of productive life taking components of this trait into account may be less when the number of parities is included in the breeding objective to describe sow longevity. This approach was used by Amer et al. (2014) who derived the economic value for average number of parities per sow based on the effect of a 1% increase in the probabilities of sows surviving between parities on profit. The resulting increase in profit was then divided by the average change in the number of parities resulting from this unit change in inter-parity survival of sows for the derivation of the economic value for the number of parities per sow as a trait describing sow longevity in the breeding objective. The model considered energy requirements per sow, revenue, and costs of sows and piglets for each parity to derive the profit per parity taking into account replacement rate and replacement costs of sows. This definition of sow longevity represented 6% of the selection emphasis in a breeding objective that considered both traits of sows and piglets and 12% of selection emphasis in a breeding objective that considered only sow traits. The effects of sow longevity on progeny performance and hence profitability of pork production were not considered in the underlying calculations which should be extended to quantify the effects of improved sow longevity on profitability through improved progeny performance.

Sow longevity influences progeny performance through the parity effect of dams on the performance of their progeny (e.g. Schinckel et al., 2010). Gilt offspring have reduced growth rate (e.g. Hermesch et al., 2017; Pineiro et al., 2019) as well as greater mortality and poorer feed efficiency in the nursery and grow-finish phases of production in comparison to progeny of multiparous sows (e.g. Calderón-Diaz et al., 2014). Hermesch and Li (2013) showed that the effect of dam parity on growth performance differs between farms. Gilt progeny had a reduced growth rate of 5 to 20 g per day for different farms in comparison to progeny from third-parity sows. A similar reduction in growth rate was also observed for progeny of older sows that had at least six parities illustrating that progeny performance is maximized for second- to fifth-parity sows. In this context, a literature review by Wijesiriwardana et al. (2022) highlighted that progeny born to primiparous sows farrowing their first litter had poorer overall production performance than progeny from multiparous sows. Based on multiple studies, the authors concluded that gilt progeny consistently grow slower, are born and weaned lighter, and have higher postweaning illness and mortality rates than progeny from multiparous sows (Wijesiriwardana et al., 2022). Therefore, high replacement rates of sows (i.e. greater proportion of first-parity sows in the herd) can result in significant losses for the pig industry. The economic implications of this parity effect of dams can be easily incorporated in the economic model outlined by Amer et al. (2014) for calculating the economic value for the number of parities. For this extension of the model, the difference in performance of progeny from each parity relative to the average performance of progeny across parity is required. This difference in performance of progeny from different dam parities is then multiplied with the corresponding economic value of progeny performance (e.g. growth rate, feed efficiency) to quantify the effect of dam parity on profit taking into account the proportion of progeny from each dam parity. This is possible, because economic values can be used to quantify the effects of non-genetic factors on profitability. Incorporating the economic values for traits describing progeny performance such as growth rate, feed efficiency, or post-weaning survival (e.g. Hermesch et al., 2014) quantify the indirect effects of sow longevity on profitability of pork production thereby extending existing economic models to derive the economic value for sow longevity. The parity effects of dams on progeny performance are additional cost components of sow longevity and considering these in bio-economic models to derive the economic value for number of parities per sow will ultimately lead to higher economic values for sow longevity.

In addition to the economic and welfare benefits of improving sow longevity, the environmental footprints of pig production will be minimized as less replacement sows would be needed on commercial farms and therefore, reducing the need for feed inputs and minimizing nutrient losses. Furthermore, improved progeny performance of multiparous sows is expected to reduce the environmental footprint of producing pork. The existing framework and bio-economic models to derive economic values for sow longevity and other traits should be extended to include the environmental impact of producing pork and its implication for the economic value of sow longevity.

4 Genetic evaluation of sow longevity traits

4.1 Genetic analyses of sow longevity traits

The rate of genetic progress in a breeding program relies on the identification of heritable indicator traits with substantial additive genetic variability. In general, most traditional indicators of sow longevity are heritable but still largely influenced by environmental factors, with heritability estimates ranging from 0.03 (Engblom et al., 2009) to 0.39 (Heusing et al., 2005) (Table 1). Although variance components are population specific, the range of heritability estimates for each indicator trait are consistent across breeds and populations. A key factor influencing the heritability estimates relates to the statistical models used for the genetic or genomic analyses. There is no consensus on the best statistical methods for genetically evaluating sow longevity (Serenius and Stalder, 2006), as all approaches proposed have advantages and disadvantages.

Published by Burleigh Dodds Science Publishing Limited, 2025.

Table 1 Average and range of heritability estimates for indicators of sow longevity

Sow longevity indicators	Average	Range	Studies[a]
Length of productive life	0.12	0.04–0.39	2, 3, 4, 5, 6, 8, 10, 11, 12, 13
Culling age	0.10	0.05–0.14	7, 20
Number of parities	0.13	0.08–0.19	7, 9, 11, 19
Number of total piglets produced per sow lifetime	0.20	0.08–0.39	3, 4, 7, 8, 11
Number of alive piglets produced per sow lifetime	0.13	0.09–0.17	7, 8, 10
Stayability to second parity	0.09	0.01–0.29	1, 6, 9, 14
Stayability to third parity	0.07	0.05–0.09	1, 6
Stayability to fourth parity	0.08	0.08–0.09	1, 11

[a]Studies:
1: Tholen et al. (1996a),
2: Yazdi et al. (2000b),
3: Serenius and Stalder (2004),
4: Heusing *et al.* (2005),
5: Serenius *et al.* (2008),
6: Engblom *et al.* (2009),
7: Nikkilä *et al.* (2013),
8: Sevón-Aimonen and Uimari (2013),
9: Aasmundstad *et al.* (2014),
10: Noppibool *et al.* (2016),
11: Kerssen *et al.* (2019),
12: Paixão *et al.* (2019),
13: Meszaros *et al.* (2019),
14: Poulsen *et al.* (2020).

Overall, longevity in pigs and other livestock species (e.g. dairy cattle) has been evaluated based on survival analysis models (e.g. Ducrocq and Sölkner, 1998; Engblom et al., 2009), linear models (Aasmundstad et al., 2014; Abell et al., 2016), or random regression models (Veerkamp et al., 1999).

Since many indicators of sow longevity are categorically defined with a non-Gaussian distribution, standard linear methods might underperform in the genetic evaluation of these traits in breeding programs. Therefore, the most-used methods are based on survival analysis or modifications of linear models. The survival analysis method is based on proportional hazard models, in which baseline hazard is usually assumed to follow a Weibull distribution (Ducrocq and Sölkner, 1998; Yazdi et al., 2000a, 2000b). The resulting estimated breeding values (EBVs) are presented with a hazard rate indicating proportional risk of sows being culled at any given time. Survival analysis properly accounts for censored observations (animals that are still alive and thus do not have their own longevity phenotypic records), non-normal distribution, and model time-dependent effects, and the most commonly used software for performing survival analysis is the Survival Kit (Ducrocq and Sölkner, 1998). Some of the limitations of this method are the computational demands of the analyses and the fact that only one trait (i.e. longevity) can be evaluated at a time. Therefore, it is difficult to estimate their genetic relationships with other traits and one cannot take advantage of information provided by genetically correlated traits for the calculation of sow longevity EBVs.

Other studies have ignored the underlying assumptions of linear models in the genetic evaluation of categorical sow longevity indicators but still captured genetic variability for sow longevity. An alternative to the linear animal model when genetically evaluating categorical traits is to use threshold models, which fit the categorical data as if there was an underlying normally distributed random variable that determines in which category an observation will exist. These methods have been widely compared in the literature. For instance, Serenius and Stalder (2004) compared sow longevity EBVs from survival analysis and multi-trait linear models. The heritability estimates for sow longevity obtained from linear model analyses were lower (0.05–0.10) than those from survival analyses (0.16–0.19). When only animals with no-censored records were included, they observed relatively low correlations (0.40–0.72) between the EBVs using the survival and linear models. Although the survival analysis might be considered superior due to the higher heritability estimates obtained, one should not make the same conclusion based on the low-to-moderate correlations because the true breeding values are not known when working with real (instead of simulated) datasets. Sobczynska and Blicharski (2015) also compared survival and linear models based on heritabilities and ranking of EBVs of sires using the same dataset and including censored observations. The heritability estimates from the survival analysis were higher based on sire models (0.21 and 0.38) as compared to animal models (0.09 and 0.16). The use of animal or sire models in the linear analyses gave similar heritability estimates (0.12 and 0.10). The authors reported stronger correlations between EBVs within methodologies (0.83–0.99) than within models with different methodologies (0.51–0.63).

As shown in Table 1, all indicators of sow longevity are heritable although some capture greater genetic variability. Overall, there is a relatively high genetic correlation among the different sow longevity definitions (e.g. Engblom et al., 2009) suggesting that they are influenced by similar sets of genes. The most commonly reported indicator trait has been the length of productive life, in which a single EBV is obtained per animal. Using random regression models based on different covariance functions (Legendre orthogonal polynomial and linear spline), Plaengkaeo et al. (2020a) reported heritability estimates for sow survival ranging over time from 0.12 to 0.15 and from 0.14 to 0.18 for Landrace and Large White sows, respectively. Most of the studies investigating the genetic background of sow longevity have focused on maternal-line pig populations, especially Landrace, Large White, and their crosses. Although genetic selection is usually done at the purebred level (nucleus herds), where sow longevity in the herd is of a lesser concern, sows in commercial breeding herds are crossbred, usually a two-breed cross between maternal lines such as Landrace and Large White, and therefore tend to outperform their purebred parents for traits related to reproduction and fitness due to heterosis effects. In addition to the narrow-sense heritability estimates, few studies have also estimated the impact of non-additive genetic effects on sow longevity indicating the importance of considering dominance genetic effects when designing matings plans. For instance, Serenius et al. (2006) estimated variance components for length of productive life and sow stayability until the third and fifth parity in Finnish pig populations, including purebred Landrace and Large White and crossbred animals of these two breeds. The authors reported estimates for the proportion of dominance variance of phenotypic variance ranging from 0.01 to 0.17, with the highest values for the crossbred population. The genetic correlations of the same traits in purebred and crossbred were all high (>0.75) and the authors recommended accounting for dominance effects when estimating breeding values of sow longevity incorporating crossbred data in the analyses. Additionally, dominance genetic variation could also be exploited when planning matings for commercial production.

Another key point when genetically evaluating sow longevity is the fact that this trait has a substantial economic value in commercial farms, but it is usually not measured in nucleus herds. This is because in the nucleus herds, sows produce fewer parities to reduce the generation interval and increase the rate of genetic progress for other economically relevant traits. However, early voluntary culling of breeding sows does not enable them to express their full longevity potential (Abell et al., 2016). One potentially feasible alternative could be to move sows from breeding herds to facilities with similar conditions of commercial farms and keep them for many generations to develop a reference population for genomic selection to improve sow longevity. In addition, breeding programs have searched for early indicators of sow longevity such as age at first farrowing, leg conformation, and reproductive performance in the first few parities.

Another challenge is that the environmental conditions in the nucleus herds tend to be different from those in commercial farms and there could be significant genotype-by-environment interactions (GxE). This can be evaluated by assessing the genetic correlation between sow longevity measured in genetically related nucleus and commercial (crossbred) populations. Abell et al. (2016) evaluated the genetic relationship between purebred (Landrace) and crossbred (F1 Landrace and Large White) sow longevity and reported little to no genetic correlations between crossbred and purebred performance. As the authors concluded, this indicates that selection for longevity or lifetime performance at the nucleus level may not result in improved longevity and lifetime performance at the crossbred level. Poulsen et al. (2020) also reported

Published by Burleigh Dodds Science Publishing Limited, 2025.

genetic correlations between crossbred and purebred sow longevity indicators, including data from 470 824 sows and 300 herds of purebred Landrace or Yorkshire and crossbreds between the two breeds. Commercial longevity was defined as age in days when culled, and stayability to service after first parity was defined for both commercial sows and multiplier sows. For Landrace and Yorkshire, the genetic correlations between commercial longevity and commercial stayability was 0.86 ± 0.02 and 0.81 ± 0.03, respectively. However, lower genetic correlations were observed between the traits measured in the commercial and multiplier herds. For Landrace and Yorkshire, the genetic correlations between commercial longevity and multiplier stayability were 0.24 ± 0.05 and 0.17 ± 0.05, respectively. The genetic correlations between commercial and multiplier stayability were also low, i.e. 0.34 ± 0.06 for Landrace and 0.18 ± 0.07 for Yorkshire. In summary, sow longevity in commercial herds is more strongly correlated with stayability when stayability is measured in commercial herds rather than multiplier herds (Poulsen et al., 2020). Iversen et al. (2020) also investigated whether the use of crossbred phenotypes could improve prediction accuracy of sow longevity for purebred populations using data from two purebred dam lines and their F1 cross. The traits evaluated were (1) whether or not the sow was inseminated for a second litter within 85 days of first farrowing, (2) how many litters the sow can produce within 570 d of first farrowing, and (3) a repeatability trait that indicated whether or not the sow survived until the next parity. Again, the authors observed low genetic correlations between the same trait in purebred and crossbred populations, indicating that longevity is not expressed as the same trait in purebred and crossbred animals. Overall, these studies provide a clear message that for the genetic evaluation of commercial sow longevity, phenotypic records from crossbred sows should be added to the genetic and genomic evaluations of purebred animals. Thus, increasing the amount of sow longevity records from crossbred and multiplier females is an essential aspect of enabling genetic companies to more accurately select animals for improved performance in commercial herds.

Although sow longevity is under genetic control, there are challenges with the incorporation of the trait into a swine breeding program. Sow longevity is a complex trait as there are multiple reasons for sows being involuntarily removed from the herd, including reproductive performance, health issues, and locomotion problems. Therefore, sows might not get the opportunity to express their full longevity potential. In this context, direct selection for these traits that will reduce the culling rates is a promising route for increasing sow longevity. Some of the traits that are directly linked with sow longevity include lameness, sow prolapses, and heat tolerance, which are all heritable traits (e.g. Bhatia et al., 2023; Freitas et al., 2023; Nikkilä et al., 2013). For instance, Anil et al. (2009) reported that lame sows were almost two times more likely to be removed within 350 days after lameness diagnosis compared to non-lame sows. Aasmundstad et al. (2014) evaluated the genetic relationship between conformation and longevity indicators (the ability to stay in production for two parities and the maximum parity number before culling) in Landrace pigs. Interestingly, the authors considered two models – with and without correcting for EBV as a covariate in the model to account for the fact that in nucleus herds sows are culled due to low EBVs. The authors reported higher heritability and genetic correlation estimates when adjusting for EBVs of sows at culling in the models. However, in this specific population, the conformation traits evaluated were not highly correlated with sow longevity, indicating that direct selection for sow longevity would still be needed.

4.2 Genetic associations between sow longevity and other relevant traits

Sow longevity can be genetically improved through direct selection for longevity indicators (e.g. length of productive life) or by selecting for correlated traits such as feet and leg conformation, disease resistance, and reproductive performance. Overall, there is large variability on the level of genetic relationship among sow longevity indicators and other relevant traits. First, sow longevity and reproductive performance are favorably correlated at the genetic level, which is expected as poor reproductive performance is a primary culling reason. For instance, Kerssen et al. (2019) reported genetic correlation estimates between sow longevity and age at first farrowing ranging from −0.13 to −0.22 and first litter size with sow longevity traits ranging from 0.49 to 0.65. Other studies in the literature have reported similar results with genetic correlations between age at first farrowing and sow longevity indicators ranging from −0.29 to 0.17 (although the positive estimates had high standard errors) (Serenius and Stalder, 2004; Kerssen et al., 2019). Sobczynska et al. (2013) reported a positive genetic correlation between the number of live-born piglets in the first litter with lifetime pig production and lifetime pig efficiency in Landrace and Large White pigs. These results indicate that first litter size and age at first farrowing could be suitable auxiliary genetic indicators for sow longevity (Kerssen et al., 2019; Knauer et al., 2011; Serenius et al., 2008). The genetic correlations between sow longevity and growth and body composition tend to be small or non-significant. For example, Sobczynska et al. (2013) reported antagonistic genetic correlations (0.21–0.26) between sow longevity indicators and lean meat percentage in Large White sows. The same authors reported small or non-significant genetic correlations of sow longevity with average daily gain, backfat thickness, loin depth, and exterior traits. Other studies have also reported unfavorable genetic correlations with average daily gain and backfat thickness (Serenius and Stalder, 2004; Knauer et al., 2011; López-Serrano et al., 2000). Feet and leg conformation are also favorably correlated with sow longevity (Serenius et al., 2004; Le et al., 2015). Heritability estimates between leg conformation scores and sow longevity indicators are usually positive and of low-to-moderate magnitude (e.g. 0.04–0.36; Serenius and Stalder, 2004). Sevilla et al. (2008) investigated the effect of leg confirmation on sow longevity in Duroc, Landrace, and Large White populations. The authors reported that leg conformation had a substantial effect on sow longevity, where an accurate removal of candidate gilts with poor conformation of legs before first mating could improve sow survival and reduce culling costs. Furthermore, selection for traits that would reduce involuntary culling (e.g. lameness, reproduction performance) would allow sows to better express their productive lifetime potential in commercial herds.

4.3 Genomic avenues to improve sow longevity

The use of genomic information has revolutionized pig breeding. Genomic selection enables the prediction of more accurate genomic breeding values for non-phenotyped individuals at an early age; increase in selection intensity, as more selection candidates can be screened, including within-litter selection; reduction in generation interval; and, more accurate genetic evaluation of traits that are sex-limited, difficult or expensive to measure, or measured late in the life of the animals. Therefore, sow longevity is a great example of the traits that can greatly benefit from genomic selection, as it is a sex-limited trait (phenotype is measured only on sows) that is measured late in the life of the animals. Furthermore, sow longevity is not adequately measured in nucleus herd animals because breeding companies aim at reducing generation interval for greater genetic progress per year. Therefore, collecting phenotypic data in commercial and multiplier herds and integrating the data with nucleus herds animals through genomic information could greatly improve the rates of genetic progress for sow longevity at the commercial level. The success of genomic selection depends on the accuracy, bias, and dispersion of the genomic EBVs (GEBVs). There is a limited number of studies reporting the performance of genomic predictions for sow longevity indicators. Iversen et al. (2020) investigated the value of incorporating crossbred phenotypes on the genomic analyses to improve prediction accuracy of sow longevity in two purebred maternal-line pigs. Traits were evaluated both as the same trait across breeds and as different traits between breeds. The authors reported a low genetic correlation between sow longevity in purebred and crossbred animals, indicating that it is not the same trait in genetically related purebred and crossbred populations. Although reasonable accuracies were obtained for the purebred populations, the prediction accuracies for the crossbreds was generally poor. The authors also emphasized the need for large-scale datasets for obtaining more accurate estimates of breeding values for sow longevity and that crossbred phenotypes should be used in the genetic evaluations. As most crossbred herds do not record pedigree, genomic information is a feasible alternative for integrating crossbred and purebred datasets for genomic evaluations and additional genomic analyses. Another important avenue for further improving sow longevity is to use precision technologies (e.g. cameras, wearable sensors) for collecting high-throughput phenotypes in a subset of the breeding population, generate reference populations (individuals with both genomic and phenotypic datasets), and then perform genomic predictions for these novel traits in a large number of animals, which would indirectly contribute to improving sow longevity. Examples of traits that could be derived are sow feet and leg conformation based on digital imagery (e.g. Stock et al., 2017).

Various studies in different pig populations have indicated that sow longevity is a highly polygenic trait and no major genes have been identified so far, with most of the significant genomic regions explaining less than 1% of the total additive genetic variance for sow longevity indicators. There is also low-to-no overlap among the genomic regions identified across studies (even within the same breed). For instance, Kang et al. (2017) performed a genome-wide association study and identified no significant genomic markers associated with sow longevity indicators. Onteru et al. (2011) reported genomic regions explaining up to 0.15 of the phenotypic variance for sow longevity indicators and overlap with key genes such as *SLC22A18*, *FUT9*, *P2RY6*, *KCNC2*, *TRAFD1*, and *FGF2*. Mote et al. (2009) investigated genetic polymorphisms in 20 selected genes and reported that the genes *CCR7*, *IGFBP1*, and *CPT1A* are key candidate genes related to sow longevity in Large White and Landrace populations. Plaengkaeo et al. (2020b), working with Landrace animals, reported 27 candidate genes associated with sow longevity, including the *TMLHE*, *PDHX*, and *KCNJ6* genes. As a large number of biological processes influence sow longevity (e.g. growth rate, fertility, conformation, immune response), and consequently numerous genes involved in these biological processes, it is very unlikely that genetic tests based on few genomic markers could be effective. For example, in the PigQTL database (www.animalgenome.org/cgi-bin/QTLdb/SS/index) there are more than 16 000 quantitative trait loci (QTL) identified to be associated with traits directly linked with sow longevity. Therefore, although genome-wide association studies can provide key insights on the biological processes underlying sow longevity, in our view, the best path moving forward is to develop large reference populations including phenotypes from both purebred and crossbred animals to calculate GEBVs based on methods such as the single-step Genomic Best Linear Unbiased Prediction (ssGBLUP).

In addition to genomics, other multi-omic datasets might also be useful for improving the accuracy of genomic predictions for sow longevity, such as transcriptomics, metabolomics, epigenomics, and metagenomics. These types of multi-omic datasets could be added to the prediction models or used for identifying key genetic variants to be prioritized in the genomic models. Closer-to-biology biomarkers could also be developed. For example, telomere length and attrition rate have been linked to longevity in various mammal species (Zhang et al., 2023). Telomeres are repeated sequences of nucleotides at the end of chromosomes, which can deteriorate across mitotic divisions of a cell. As reviewed by Zhang et al. (2023), telomere length in dairy cattle is heritable with heritability estimates ranging from 0.12 to 0.46. This is a promising area of research in sow longevity.

5 Future trends

Although genetic selection in pigs is usually performed at the nucleus herd level, sow longevity is a trait expressed in the multiplier and crossbred herds. Furthermore, sow longevity is a sex-limited trait measured late in the life of the animals. Therefore, it is a trait with potential to greatly benefit from genomic selection. There are limited genomic studies evaluating sow longevity in purebred and crossbred populations. However, there is evidence of GxE interactions for sow longevity, highlighting the need for adding phenotypes from crossbred and multiplier herds in the genomic evaluations of purebred animals. There is also evidence of non-additive genetic effects influencing sow longevity in crossbred populations, indicating the potential for taking advantage of that when designing mating plans. Additional research in different populations using large-scale datasets are needed to assess the value of each data source (e.g. data from nucleus sows, sows at multiplication herds, and crossbred commercial sows) for more accurate genomic predictions of sow longevity in nucleus herds.

The integration of multi-omic datasets and high-throughput phenotypes for novel traits related to sow longevity also hold great potential to increase the accuracy of genomic predictions for sow longevity. Novel traits related to disease susceptibility, resistance, or resilience could also be developed based on variability in longitudinal records (e.g. feed intake, body condition score, activity level). For instance, various indicators of animal resilience based on longitudinal datasets have been proposed in dairy cattle (e.g. Poppe et al., 2020; Chen et al., 2023), pigs (e.g. Cheng et al., 2020; Gorssen et al., 2023; Putz et al., 2019), and other species.

6 Conclusion

In summary, multiple indicators of sow longevity have been proposed and all of them are heritable and of polygenic inheritance. There is substantial GxE between sow longevity in purebred and crossbred herds, indicating the need for incorporating crossbred phenotypes in the purebred genetic evaluations. As sow culling and longevity are influenced by various factors, breeding programs should consider genetically evaluating novel traits, potentially generated by precision technologies, that might be more heritable and could fastener genetic progress for sow longevity. More productive, robust, and resilient sows are likely to stay longer in the herds and increase the long-term sustainability of the pig industry.

7 Acknowledgements

We would like to thank Dr. Allan P. Schinckel (Purdue University, West Lafayette, IN, USA) and Sharif-Islam (AGBU, University of New England, Armidale, Australia) for providing valuable comments and suggestions during the preparation of this book chapter. We would also like to express our gratitude to Dr. Kenneth Stalder (Iowa State University, Ames, IA, USA) who devoted his career researching strategies and approaches for improving sow longevity and conducting extension activities and training of the next generation of scientists before his passing in 2022. We greatly benefitted from his published work in the preparation of this chapter.

8 Where to look for further information

- Moeller, G.A. and Stalder, K.J. 2023. Sow longevity. Advances in Pig Welfare – Second Edition. 163–184. https://doi.org/10.1016/C2020-0-02489-8
- Mote, B.E., Serenius, T.V., Supakorn, C. and Stalder K.J. 2019. Genetic improvement of sow longevity and its economic impact on commercial pork production. South African Journal of Animal Science, 49 (6), 1036–1046.

9 References

Aasmundstad, T., Olsen, D., Sehested, E. and Vangen, O., 2014. The genetic relationships between conformation assessment of gilts and sow production and longevity. *Livestock Science*, 167, pp.33–40.

Abell, C.E., Fernando, R.L., Serenius, T.V., Rothschild, M.F., Gray, K.A. and Stalder, K.J., 2016. Genetic relationship between purebred and crossbred sow longevity. *Journal of Animal Science and Biotechnology*, 7, pp.1–6.

Amer, P.R., Ludemann, C.I. and Hermesch, S., 2014. Economic weights for maternal traits of sows, including sow longevity. *Journal of Animal Science*, 92(12), pp.5345–5357.

Anil, S.S., Anil, L., Deen, J., Baidoo, S.K., Wilson, M.E. and Ward, T.L., 2009. Comparison of claw lesion scores in sows housed in gestation stalls and in group pens with electronic sow feeders (ESF) during gestation. In: *Proceedings of the Leman Conference*, Minnesota, USA.

Bhatia, V., Stevens, T., Derks, M.F., Dunkelberger, J., Knol, E.F., Ross, J.W. and Dekkers, J., 2023. Identification of the genetic basis of sow pelvic organ prolapse. *Frontiers in Genetics*, 14, p.1154713.

Calderón-Díaz, J.A.C., Nikkilä, M.T. and Stalder, K. 2014. Chapter 19: sow longevity. In *The gestating and lactating sow*. pp.423–452. https://doi.org/10.3920/978-90-8686-803-2_19.

Carney-Hinkle, E.E., Tran, H., Bundy, J.W., Moreno, R., Miller, P.S. and Burkey, T.E., 2013. Effect of dam parity on litter performance, transfer of passive immunity, and progeny microbial ecology. *Journal of Animal Science*, 91(6), pp.2885–2893.

Chen, S.Y., Boerman, J.P., Gloria, L.S., Pedrosa, V.B., Doucette, J. and Brito, L.F., 2023. Genomic-based genetic parameters for resilience across lactations in North American Holstein cattle based on variability in daily milk yield records. *Journal of Dairy Science*, 106(6), pp.4133–4146.

Cheng, J., Putz, A.M., Harding, J.C., Dyck, M.K., Fortin, F., Plastow, G.S., Canada, P. and Dekkers, J.C., 2020. Genetic analysis of disease resilience in wean-to-finish pigs from a natural disease challenge model. *Journal of Animal Science*, 98(8), p.skaa244.

Craig, J.R., Dunshea, F.R., Cottrell, J.J., Wijesiriwardana, U.A. and Pluske, J.R., 2019. Primiparous and multiparous sows have largely similar colostrum and milk composition profiles throughout lactation. *Animals*, 9(2), p.35.

de Vries, 1989. A model to estimate economic values of traits in pig breeding. *Livestock Production Science*, 21, pp.49–66.

Ducrocq, V. and Sölkner, J., 1998, January. Implementation of a routine breeding value evaluation for longevity of dairy cows using survival analysis techniques. In *Proceedings of the 6th World Congress on Genetics Applied to Livestock Production*, Armidale, Australia, Vol. 23, pp.359–362.

Eckberg, B., 2022. *2021 Sow mortality analysis*. National Hog Farmer. https://www.nationalhogfarmer.com/hog-health/2021-sow-mortality-analysis?_sp=5dfb5702-6f15-449e-9e47-8d515c591513.1645194115506.

Engblom, L., Calderón Díaz, J.A., Nikkilä, M., Gray, K., Harms, P., Fix, J., Tsuruta, S., Mabry, J. and Stalder, K., 2016. Genetic analysis of sow longevity and sow lifetime reproductive traits using censored data. *Journal of Animal Breeding and Genetics*, 133, 1–7. https://doi.org/10.1111/jbg.12177.

Engblom, L., Lundeheim, N., Schneider, M.D.P., Dalin, A.M. and Andersson, K., 2009. Genetics of crossbred sow longevity. *Animal*, 3, 783–790.

Freitas, P.H., Johnson, J.S., Wen, H., Maskal, J.M., Tiezzi, F., Maltecca, C., Huang, Y., DeDecker, A.E., Schinckel, A.P. and Brito, L.F., 2023. Genetic parameters for automatically-measured vaginal temperature, respiration efficiency, and other thermotolerance indicators measured on lactating sows under heat stress conditions. *Genetics Selection Evolution*, 55(1), p.65.

Gorssen, W., Winters, C., Meyermans, R., Chapard, L., Hooyberghs, K., Janssens, S., Huisman, A., Peeters, K., Mulder, H. and Buys, N., 2023. A promising resilience parameter for breeding: the use of weight and feed trajectories in growing pigs. *Journal of Animal Science and Biotechnology*, 14(1), p.101.

Hermesch, S., 2021. Economic values for farrowing rate to improve seasonal fertility. In *Proceedings of the 24th Conference of the Association for the Advancement of Animal Breeding and Genetics*. 2–4 November, pp.159–162.

Hermesch, S. and Li, L., 2013. The effect of dam parity on growth of pigs differs between herds. In *Proceedings of the 14th Biennial Conference of the Australasian Pig Science Association (APSA)*. Australasian Pig Science Association Inc., p.53.

Hermesch, S., Ludemann, C. I. and Amer, P.R., 2014. Economic weights for performance and survival traits of growing pigs. *Journal of Animal Science*, 92, 5358–5366. https://doi.org/10.2527/JAS2014-7944.

Hermesch, S., McKenna, T., Bauer, M.M. and Sales, N., 2017. The effect of dam parity on growth, white blood cell count, haemoglobin and immunoglobulin levels of weaner pigs. In *Proceedings of the Sixtheenth Biennial Conference of the Australasian Pig Science Association (APSA)*. Australasian Pig Science Association Inc.

Heusing, M., Hamann, H. and O. Distl, 2005. Genetische analyse von Lebensleistungs - und Fruchbarkeitsmerkmalen bei sauen der rassen deutsches Edelschwein, Deutsche Landrasse und Pietrain. *Züchtungskunde,* 77(1), pp.15–34.

Hoge, M.D. and Bates, R.O., 2011. Developmental factors that influence sow longevity. *Journal of Animal Science*, 89(4), pp.1238–1245.

Iversen, M.W., Nordbø, Ø., Gjerlaug-Enger, E., Grindflek, E. and Meuwissen, T.H., 2020. Predicting survival and longevity of sows using purebred and crossbred data. *Translational Animal Science*, 4(2), pp.993–1005.

Kang, J.H., Lee, E.A., Lee, S.H., Kim, S.H., Lee, D.H., Hong, K.C. and Park, H.B., 2017. Genome-wide association study for sow lifetime productivity related traits in a Landrace purebred population. *Livestock Science*, 202, 21–24.

Kerssen, N., Ducro, B.J. and Hermesch, S., 2019, October. Genetic analyses of sow longevity traits, age at first farrowing and first-litter characteristics. In Proceedings of the the 23rd Association for the Advancement of Animal Breeding and Genetics Conference, Armidale, NSW, Australia, Vol. 23, pp.167–170.

Kikuti, M., Preis, G.M., Deen, J., Pinilla, J.C. and Corzo, C.A., 2023. Sow mortality in a pig production system in the midwestern USA: reasons for removal and factors associated with increased mortality. *Veterinary Record*, 192(7), p.e2539.

Knauer, M.T., Cassady, J.P., Newcom, D.W. and See, M.T., 2011. Phenotypic and genetic correlations between gilt estrus, puberty, growth, composition, and structural conformation traits with first-litter reproductive measures. *Journal of Animal Science,* 89, pp.935–942.

Lacy, R.C. and Stalder, K.J., 2004. Sow Longevity Calculator Ver. 2.0: Farrow-to-Finish. Metric and English unit measures. http://www.extension.iastate.edu/ipic/information/FFSLv2.html.

Lopez-Serrano, M., Reinsch, N., Looft, H. and Kalm, E., 2000. Genetic correlations of growth, backfat thickness and exterior with stayability in large white and landrace sows. *Livestock Production Science,* 64, pp.121–131.

Meszaros, G., Palos, J., Ducrocq, V. and J. Sölkner, 2010. Heritability of longevity in Large White and Landrace sows using continuous time and grouped data models. *Genetics Selection Evolution,* 42. https://doi.org/10.1186/1297-9686-42-13.

Monteiro, M.S., Matias, D.N., Poor, A.P., Dutra, M.C., Moreno, L.Z., Parra, B.M., Silva, A.P.S., Matajira, C.E.C., de Moura Gomes, V.T., Barbosa, M.R.F. and Sato, M.I.Z., 2022. Causes of sow mortality and risks to post-mortem findings in a Brazilian intensive swine production system. *Animals*, 12(14), p.1804.

Mote, B.E., Koehler, K.J., Mabry, J.W., Stalder, K.J. and Rothschild, M.F., 2009. Identification of genetic markers for productive life in commercial sows. *Journal of Animal Science,* 87(7), pp.2187–2195.

Niemi, J.K., Bergman, P., Ovaska, S., Sevón-Aimonen, M.L. and Heinonen, M., 2017. Modeling the costs of postpartum dysgalactia syndrome and locomotory disorders on sow productivity and replacement. *Frontiers in Veterinary Science*, 4, p.181.

Nikkilä, M.T., Stalder, K.J., Mote, B.E., Rothschild, M.F., Gunsett, F.C., Johnson, A.K., Karriker, L.A. Boggess, M.V. and T.V. Serenius, 2013. Genetic associations for gilt growth, compositional, and structural soundness traits with sow longevity and lifetime reproductive performance. *Journal of Animal Science,* 91(4), pp.1570–1579. https://doi.org/10.2527/jas.2012-5723.

Noppibool, U., Elzo, M.A., Koonawootrittriron, S. and Suwanasopee, T., 2016. Estimation of genetic parameters and trends for length of productive life and lifetime production traits in a commercial Landrace and Yorkshire Swine population in Northern Thailand. *Asian-Australasian journal of Animal Sciences,* 29(9), p.1222.

Oliveira, H. R., Brito, L.F., Miller, S.P. and Schenkel, F.S. 2020. Using random regression models to genetically evaluate functional longevity traits in North American Angus cattle. *Animals,* 10(12), p.2410. https://doi.org/10.3390/ani10122410

Onteru, S.K., Fan, B., Nikkilä, M.T., Garrick, D.J., Stalder, K.J. and Rothschild, M.F., 2011. Whole-genome association analyses for lifetime reproductive traits in the pig. *Journal of Animal Science*, 89(4), pp.988–995.

Paixão, G., Martins, Â., Esteves, A., Payan-Carreira, R. and Carolino, N., 2019. Genetic parameters for reproductive, longevity and lifetime production traits in Bísaro pigs. *Livestock Science*, 225, pp.129–134.

Piñeiro, C., Manso, A., Manzanilla, E.G. and Morales, J. 2019. Influence of sows' parity on performance and humoral immune response of the offspring. *Porcine Health Management,* 5, p.1. https://doi.org/10.1186/s40813-018-0111-8

Plaengkaeo, S., Duangjinda, M., Boonkum, W., Stalder, K.J. and Mabry, J.W., 2020a. Genetic evaluation of sow survival in Thailand commercial farms using random regression models. *Livestock Science*, 233, p.103970.

Plaengkaeo, S., Duangjinda, M. and Stalder, K., 2020b. Identification of potential genetic variants associated with longevity and lifetime production traits in a Thai Landrace pig population using weighted single-step genome-wide association methods. *Genetics and Molecular Research,* 19(3), p.gmr18465.

Poppe, M., Veerkamp, R.F., Van Pelt, M.L. and Mulder, H.A., 2020. Exploration of variance, autocorrelation, and skewness of deviations from lactation curves as resilience indicators for breeding. *Journal of Dairy Science*, 103(2), pp.1667–1684.

Poulsen, B.G., Nielsen, B., Ostersen, T. and Christensen, O.F., 2020. Genetic associations between stayability and longevity in commercial crossbred sows, and stayability in multiplier sows. *Journal of Animal Science*, 98(6), p.skaa183.

Putz, A.M., Harding, J.C., Dyck, M.K., Fortin, F., Plastow, G.S., Dekkers, J.C. and PigGen Canada, 2019. Novel resilience phenotypes using feed intake data from a natural disease challenge model in wean-to-finish pigs. *Frontiers in Genetics*, 9, p.660.

Quinton, V.M., Wilton, J.W., Robinson, J.A. and Mathur, K.A., 2006. Economic weights for sow productivity traits in nucleus pig populations. *Livestock Science*, 99, pp.69–77.

Schinckel, A.P., Einstein, M.E., Stewart, T.S., Schwab, C. and Olynkf, N.J., 2010. Use of a stochastic model to evaluate the growth performance and profitability of pigs from different litter sizes and parities of dams. *The Professional Animal Scientist*, 26(5), pp.547–560.

Serenius, T., Muhonen, P. and Stalder, K., 2007. Economic values of pork production related traits in Finland. *Agricultural and Food Science*, 16, pp.79–88.

Serenius, T. and Stalder, K.J., 2004. Genetics of length of productive life and lifetime prolificacy in the Finnish Landrace and Large White pig populations. *Journal of Animal Science,* 83, pp.3111–3117.

Serenius, T. and Stalder, K.J. 2006. Selection for sow longevity. *Journal of Animal Science,* 84, pp.E166–E171.

Serenius, T., Stalder, K.J. and Puonti, M. 2006. Impact of dominance effects on sow longevity. *Journal of Animal Breeding and Genetics*, 123(6), pp.355–361.

Serenius, T., Stalder, K.J. and Fernando, R.L. 2008. Genetic associations of sow longevity with age at first farrowing, number of piglets weaned, and wean to insemination interval in the Finnish Landrace swine population. *Journal of Animal Science,* 86(12), pp.3324–3329.

Sevilla, X.F., Fàbrega, E., Tibau, J. and Casellas, J., 2008. Effect of leg conformation on survivability of Duroc, Landrace, and Large White sows. *Journal of Animal Science,* 86(9), pp.2392–2400.

Sevón-Aimonen, M.L. and Uimari, P., 2013. Heritability of sow longevity and lifetime prolificacy in Finnish Yorkshire and Landrace pigs. *Agricultural and Food Science,* 22(3), pp.325–330. https://doi.org/10.23986/afsci.799.

Sobczyńska, M. and Blicharski, T., 2015. Phenotypic and genetic variation in longevity of Polish Landrace sows. *Journal of Animal Breeding and Genetics*, 132(4), pp.318–327.

Sobczyńska, M., Blicharski, T. and Tyra, M., 2013. Relationships between longevity, lifetime productivity, carcass traits and conformation in Polish maternal pig breeds. *Journal of Animal Breeding and Genetics*, 130(5), pp.361–371.

Sölkner, J. and Ducrocq, V., 1999. The Survival Kit: a tool for analysis of survival data. *Interbull Bulletin*, 21, pp.11–11.

Stalder, K.J., Knauer, M., Baas, T.J., Rothschild, M.F. and Mabry, J.W. 2004. Sow longevity. *Pig News and Information,* 25(2), pp.53N–74N.

Stalder, K.J., Lacy, R.C, Cross, T.L. and Conatser, G.E., 2003. Financial impact of average parity of culled females in a breed-to-wean swine operation using replacement gilt net present value analysis. *Journal of Swine Health and Production*, 11(2), pp.69–74.

Stock, J.D., Calderón Díaz, J.A., Mote, B., Rothschild, M.F. and Stalder, K.J., 2017. Development of an objective feet and leg conformation evaluation method using digital imagery in swine. *Journal of Animal Sciences and Livestock Production,* 1(2), p.006.

ten Napel, J., Kemp, B., Luiting, P. and de Vries, A.G., 1995. A biological approach to examine genetic variation in weaning-to-oestrus interval in first-litter sows. *Livestock Production Science,* 41, pp.81–93.

Tess, M.W., Bennett, G.L. and Dickerson, G.E., 1983. Simulation of genetic changes in life cycle efficiency of pork production. I. A bioeconomic model. *Journal of Animal Science*, 56(2), pp.336–363.

Tholen, E., Bunter, K.L., Hermesch, S. and Graser, H.-U., 1996. The genetic foundation of fitness and reproduction traits in Australian pig populations: I. Genetic parameters for weaning to conception interval, farrowing interval and stayability. *Australian Journal of Agricultural Research,* 47, pp.1261–1274.

Veerkamp, R.F., Brotherstone, S. and Meuwissen, T., 1999. Survival analysis using random regression models. *Interbull Bulletin*, 21, pp.36–36.

Wijesiriwardana, U.A., Craig, J.R., Cottrell, J.J., Dunshea, F.R. and Pluske, J.R., 2022. Animal board invited review: factors affecting the early growth and development of gilt progeny compared to sow progeny. *Animal,* 16(8), p.100596.

Wolfova, M. and Wolf, J. 2013. Strategies for defining traits when calculating economic values for livestock breeding: a review. *Animal*, 7(9), pp.1401–1413.

Yazdi, M.H., Lundeheim, N., Rydhmer, L., Ringmar-Cederberg, E. and Johansson, K., 2000a. Survival of Swedish Landrace and Yorkshire sows in relation to osteochondrosis: a genetic study. *Animal Science*, 71, pp.1–9.

Yazdi, M., Rydhmer, L. Ringmar-Cederberg, E., Lundeheim, N. and Johansson, K., 2000b. Genetic study of longevity in Swedish Landrace sows. *Livestock Production Science*, 63, pp.255–264.

Zhang, N., Baker, E.C., Welsh, T.H. Jr. and Riley, D.G., 2023. Telomere dynamics in livestock. *Biology*, 12(11), p.1389. https://doi.org/10.3390/biology12111389.

Part 2

Reproduction

Chapter 7

Improving the reproductive efficiency of boars: past, present, and future

Ian Shofner and Karl Kerns, Iowa State University, USA

1 Introduction

Antonie Van Leeuwenhoek, often recognized as the father of optic microscopy, was the pioneer in studying spermatozoa (Karamanou et al., 2010). In 1677, when Leeuwenhoek first reported his observations on spermatozoa, many of his peers surmised that the sample had simply decayed. Contrarily, Leeuwenhoek perceived them as an intrinsic component of semen, coining the term 'animalcules spermatiques' or the seeds of life. This marked the initiation of the andrology field of study (Howards, 1997; Lane, 2015). While Leeuwenhoek is not credited for inventing the light microscope, he was instrumental in popularizing and refining its use. This technology not only facilitated the discovery of sperm cells but also fostered the growth of microbiology.

Following the identification of spermatozoa, there was a prevalent belief, understood by Leeuwenhoek and his contemporaries, that these sperm enclosed a 'homunculus' or an unformed embryo. It was thought that this homunculus initiated the embryo's development post-coitus. However, the intricacies of this mechanism remained enigmatic (Baker, 1739; Wollman et al.,

http://dx.doi.org/10.19103/AS.2024.0137.11

2015). A profound understanding of male gametes did not emerge until the advent of the cell theory. Proposed by Schleiden (1838) and Schwann (1839), this theory posited that all organisms are cellular, cells arise from existing cells, and cells are life's foundational units (Mazzarello, 1999; Müller-Wille, 2010). The origins of life and the purpose of the male contribution continued to develop, change, and be highly contested until the 1950s when, in separate studies, the development of *in vitro* fertilization occurred, proving the necessity of semen in fertilization (Austin, 1951; Chang, 1951). This provides a backdrop that shows our understanding of andrology is still evolving.

This chapter provides a critical review of key developments in understanding and improving boar reproductive efficiency, building on the seminal discoveries outlined above to contemporary advances in artificial insemination (AI) and sperm analysis technologies. Key developments in identifying fertile boars are discussed, including sperm kinematics, proteomics, lipidomics, and flow cytometry (FC), along with computer-assisted sperm analysis (CASA). Environmental influences, such as heat stress, sperm handling, and neonatal development, on semen quality and fertility are explored. The chapter also discusses future advancements such as sex-sorted sperm, nanoparticle separation, and colloid centrifugation hold promise for improving semen quality and AI efficiency. As the chapter shows, technologies like image-based FC, microfluidics, and advanced artificial intelligence (AI^2) are paving the way for a new era in boar semen andrology testing.

2 Developments in reproductive techniques: artificial insemination

Technological advancements in reproductive techniques and insights into reproductive functions have historically had a considerable influence on livestock production. Research has continued to propel evolution to more efficient livestock farming practices. Boars exhibit significant variability in their sperm count per ejaculate, typically oscillating between 50 billion and 100 billion. This can vary for several reasons, including, but not limited to, season, breed, and boar genetics (Sonderman et al., 2008). Breed plays a significant role in semen amount and quality, with genetic lines having a crucial role where crossbreds tend to have the highest overall semen quality. Multiple studies have shown that breed can have effects on sperm concentration, total sperm count, ejaculate volume, and even post-freeze viability (Smital et al., 2004; Smital, 2009; Sonderman et al., 2008; Wolf, 2009).

Leveraging AI techniques, a single boar's ejaculate can impregnate multiple sows by fractioning an ejaculate into several doses. Each dose comprises of 1.5–2.5 billion sperm cells in a 35–80 mL diluted dose. Post-cervical artificial insemination (PCAI), a technique where semen is deposited past the cervix, has significantly improved reproductive efficiency. This method enhances semen retention, increases the likelihood of fertilization, and reduces retrograde transport, thereby allowing the use of smaller semen doses while maintaining high fertility rates. AI has emerged as a paramount technique to optimize genetic utilization and manage sow reproductive performance (Waberski et al., 2019a).

By the early 2000s, the adoption of AI skyrocketed, with its application spanning >95% of sows in the United States, enhancing litter size and pregnancy rates. A common practice, known as pooling, entails combining multiple boar ejaculates per dose. This approach serves as a biological insurance, ensuring fertilization success when individual boar fertility is unknown, a strategy predominant in current boar studs (Foxcroft et al., 2008). However, pooling eliminates the ability to track individual boar fertility metrics and progeny data, potentially diminishing the pool's overall genetic quality and the proportion of high-quality spermatozoa. Advanced sperm analysis techniques, optimizing both accuracy, precision, and automation, facilitate optimal semen usage by understanding sample metrics. The swine industry remains at the forefront, deploying novel technologies to boost productivity. Ongoing studies in fertility assessment and bacterial management are vital for advancing the agricultural sector. Simultaneously, efforts to increase fertilization rates with fewer sperm and to advance sperm-freezing techniques are crucial in meeting the nutritional needs of the world's expanding population. Throughout this chapter, we will discuss boar development, identifying fertile boars, factors affecting semen quality, the development of AI, and the future of AI.

3 Understanding sperm fertility

The evolution of sperm fertility dogma reflects a fascinating journey from the seventeenth century to the present day, as illustrated in Figure 1. Antonie van Leeuwenhoek's discovery of spermatozoa in 1677 laid the foundation for modern andrology (Karamanou et al., 2010; Howards, 1997). At the time, prevailing beliefs held that all motile, morphologically normal sperm cells were fertile, a perspective referred to as Dogma #1. This assumption persisted for centuries, bolstered by early cell theory and the limitations of microscopy. Compensable sperm traits, such as poor motility or abnormal morphology, fit neatly within this framework, as their impact on fertilization could often be mitigated by increasing the number of sperm in an insemination dose (Saacke et al., 2008). However, Dogma #1 fails to account for more nuanced defects, especially those that cannot be overcome by increasing sperm numbers, highlighting the need for a deeper understanding of sperm function.

The shift to Dogma #2 arose with advancements in molecular biology, imaging technologies, and biomarker identification, revealing that only a

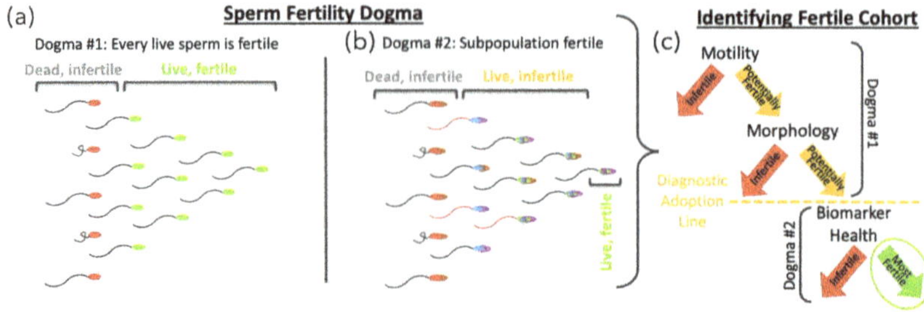

Figure 1 Prevailing sperm fertility dogma and methods of identifying the fertile cohort. (A) Dogma #1, originating centuries ago, posits that every live, morphologically normal spermatozoon is fertile. (B) Dogma #2, developed in recent decades, asserts that only a subpopulation of morphologically normal sperm can achieve fertilization competency. (C) While high-skill, time-consuming, and costly technologies exist to identify fertile cohorts related to Dogma #2, industry daily diagnostic adoption is stuck at Dogma #1.

subpopulation of spermatozoa possesses true fertilization potential (Sutovsky et al., 2020). This perspective acknowledges uncompensable (irreparable) traits – defects such as DNA damage or deficiencies in oocyte activation – which directly affect fertilization and embryonic development and cannot be compensated by increasing sperm numbers (Saacke et al., 2008; Amann & Waberski, 2014). Recent findings further support Dogma #2, demonstrating that only a subpopulation of spermatozoa undergo essential processes like the rapid increase in intracellular calcium or hyperpolarization during capacitation, critical for successful fertilization (Luque et al., 2018; Escoffier et al., 2015). Dogma #2 as a concept is firmly rooted in late twentieth-century advancements, supported by emerging evidence of functional variability among spermatozoa. Despite these advancements, industry diagnostics largely remain entrenched in Dogma #1 due to accessibility and cost constraints for sperm evaluation. Bridging this gap with affordable, high-throughput technologies is essential for modern fertility diagnostics.

4 Sperm analysis: computer-assisted sperm analysis

Understandings in sperm kinematics, morphology, motility, and count has allowed and promoted further adoption of AI in the swine industry. The CASA has helped andrologists have a reliable, repeatable, and rapid means of analyzing semen. It aims to use computational power with the addition of sperm motility tracking and counting via a microscope to calculate real-time sperm parameters (Bartoov et al., 1981). CASA machines use a camera mounted to a microscope to capture time lapses of sperm and then perform various sperm calculations on a computer. It outputs metrics on motility, morphology, and cell count. Through advances in microscopes, cameras, and computational power, there has been an increased growth and adoption in CASA usage (Amann

& Waberski, 2014). Some use a nucleocounter as an automated, sensitive approach for assessing semen concentration. It utilizes a fluorescence-based technology and provides a reliable and consistent way to assess sperm concentration (Sanches et al., 2010). Recent advancements in sperm analysis have led to the development of Androvision eFlow system on the CASA. This is an automated microfluidic system which has demonstrated high concordance with the Nucleocounter SP-100 for sperm concentration measurements in both raw and diluted boar semen. It offers a reliable and efficient method for assessing sperm concentration, motility, and morphology, thereby enhancing the accuracy of semen evaluations in the swine industry (Grossfeld et al., 2022).

Thanks to the adoption of the CASA system at boar studs, individual samples can be analyzed, pooled, and extended to produce optimized semen doses that can be used in AI with known sperm parameters. This allows studs to produce more accurate doses of semen than what was previously possible. The CASA system is a rapid and objective method for evaluating semen quality and providing quantitative reports on motility as opposed to visual technician observation. The utilization of the CASA system has decreased the subjective nature of seminal quality evaluation, resulting in more accurate and precise outcomes.

There are believed to be four distinct motility subpopulations of spermatozoa. In one study, these subpopulations were defined by their velocity and linearity, with SP1 exhibiting high velocity and high linearity, SP2 exhibiting high velocity but low linearity, SP3 exhibiting low velocity but high linearity, and SP4 exhibiting low velocity and low linearity. The proportions of these subpopulations vary significantly based on the season of semen collection, with winter having the most significant effect, having SP2 dominate the proportions conditions that can be exploited in order to increase specific subpopulations of sperm for use in increasing boar fertility (Ibănescu et al., 2018).

5 Sperm analysis: proteomics and lipidomics

Proteomics is the study of protein structure, function, post-translational modifications(PTMs)and interactions(Wrightetal.,2012).PTMsaremodifications in the structure and functionality of a protein that occur after its synthesis and are considered key events for sperm function and fertilization ability (Brohi & Huo, 2017). Studies in several animal species have demonstrated that seminal plasma (SP) and/or sperm proteins influence the response of ejaculates to sperm biotechnologies. These range from the simplest technologies, such as conventional AI with spermatozoa subjected to long-term and liquid storage, to more sophisticated technologies such as freezing and/or sex-sorting, thus helping to identify presumable markers for sperm resilience. Whether this influence is related to PTMs remains unclear (Caballero et al., 2012). A study by

Kwon (2015) was conducted to identify sperm protein markers linked to fertility levels. The research showed that certain proteins are associated with larger or smaller litter sizes. Understanding the proteome profile of fertile sperm and seminal fluids could help lead to new diagnostic strategies related to male reproductive potential (Parrilla et al., 2019).

Sperm membranes are made up of phospholipids which remodel during sperm capacitation in preparation for fertilization success. It has been suggested that lipidomics could be a new frontier for semen evaluation (Gadella et al., 1995; van Gestel et al., 2016). The sperm lipidome of some higher order mammals has been analyzed (Ramal-Sanchez et al., 2020; Rivera-Egea et al., 2018; Wood et al., 2016). The lipidome of boar SP was recently described (Höfner et al., 2020). However, to date, no one has described differences occurring in the boar sperm lipidome after capacitation. This large knowledge gap is, surprising, as cholesterol removal and lipid remodeling of sperm plasma membrane leading to membrane raft formation has been long known as a hallmark of the sperm capacitation process. Cryopreservation is an important emerging technology for boar genetic preservation and dissemination. The low tolerance of boar sperm to low temperatures, and the resulting quality-reducing 'cold-shock' phenomenon, is attributed to the lipid phase transition in the cell membrane (Drobnis et. al., 1993). A better understanding of lipidomics could be a path to overcoming this limiting factor in boar semen cryopreservation.

6 Measurement techniques: flow cytometry

The conventional method for evaluating semen quality involves assessing sperm concentration, viability, motility, and morphology. In some cases, these parameters may not be sufficient to detect altered or non-functional spermatozoa within boar ejaculates. Such spermatozoa can lead to lower reproductive performance. Fluorescent biomarkers allow a deeper understanding of the molecular and cellular status of cells, thereby gaining more insight into the fertility status of boars. These markers can indicate alterations caused by defective spermatogenesis, dysfunctional capacitation, or in sperm handling. Utilizing biomarkers, breeders can better assess the fertilizing ability of boar ejaculate and improve the reproductive performance of swine.

Increasing fertility of semen doses collected for AI has occurred due to advancements in semen handling, dilution, and storage protocols, as well as the use of new technologies to understand the impact on the environment. One such technology that has aided these studies is FC paired with fluorescent fluorochromes to better identify sperm cell health, capacitation status, DNA integrity, and overall cell function in a way that has otherwise been impossible. These are collectively referred to as sperm biomarkers. FC allows rapid and detailed analysis of individual cells in a suspended solution.

This process uses lasers to emit light when certain attributes are present on cells stained with fluorescent fluorochromes, further referred to as biomarkers. The biomarkers are detected by a light sensor and then analyzed by a computer. Using biomarkers, scientists can study the unique characteristics of different cell populations. Recent advancements, such as multicolor-staining panels, enable the simultaneous analysis of multiple parameters, including plasma membrane integrity, acrosomal status, mitochondrial activity, and membrane lipid organization, simplifying the application of fluorochromes and enhancing the efficiency and depth of sperm quality assessments (Quirino et al., 2022). FC has transformed various fields of research, including immunology, molecular biology, bacteriology, virology, cancer biology, infectious disease monitoring, and now andrology, allowing insight into cell biology. Over the past 30 years, there have been significant advancements in this technology, making it a highly valuable and versatile tool for scientific investigation (McKinnon, 2018).

Studies have highlighted the significance of FC in predicting boar reproductive capacity. For instance, correlations have been observed between mitochondrial activity measured via FC (Schulze et al., 2013b) and sperm DNA integrity (Broekhuijse et al., 2012). Despite FC offering advanced semen analysis methods in the boar semen industry for AI and quality control, the adoption of these techniques has been slow. The boar fertility community is in particular need of controlled fertility trials with biomarker algorithms that reflect actual field fertility performance before further adoption can occur. The pairing of FC and the CASA has the potential to revolutionize our understanding of spermatozoa and can enhance the production of high-quality boar semen for AI (Boe-Hansen & Satake, 2019).

6.1 Image-based flow cytometry

Image-based FC (IBFC) is a relatively new technology that has married together the high throughput nature of FC with the imaging capabilities of microscopy, allowing images of multiple biomarkers as well as brightfield imaging to observe morphology simultaneously. IBFC enables the ability to visually validate discoveries and gather unique data that would be otherwise unattainable (McFarlin & Gary, 2017). Sperm phenotype analysis by IBFC will aid the advancement of swine genomics, linking aberrant sperm phenotypes to fertility-influencing gene polymorphisms (Sutovsky, 2015).

Studies utilizing IBFC have already made correlations with field fertility. Specifically, we previously used IBFC to measure the mitochondrial sheath (MS) length, showing a correlation between increased MS length and decreased fertility. Long MS sperm-producing boars may need more cells per dose (Kerns et al., 2020). Work in zinc localization in boar sperm has also been a promising frontier through IBFC. Four distinct types of sperm zinc ion distribution patterns have been noted before and after in vitro capacitation. These discoveries signify a fundamental change in our comprehension of mammalian fertilization, laying the groundwork for enhanced semen analysis, AI, and *in vitro* fertilization (IVF) (Kerns et al., 2018). IBFC is a technology that has the potential to fundamentally shift the way sperm is analyzed, moving andrology away from simply using morphology, motility, and count as the principal indicators of sperm fertility. Pairing IBFC with capacitation can likely be used as a more advanced predictor of boar fertility status (Keller & Kerns, 2023).

7 Factors affecting semen quality: from neonatal development to puberty

7.1 Neonatal development

During neonatal development, the male reproductive system initially consists of a minimally functioning gonad and nonfunctional accessory sex glands. The development of this system is influenced by neonatal sex steroids. Both estrogens and androgens play a role in proper testis formation in neonatal males (Berger et al., 2008). It has been shown through pharmaceutical inhibition of aromatase, causing a reduction in estrogen, that males are born with smaller testis and later onset of spermatogenesis (At-Taras et al., 2006). Furthermore, boars that have suppressed estrogen during development have reductions in sperm quality (McCarthy et al., 2006).

Estrogen plays an important role in male reproduction, critical for sustained fertility. Boars comparatively have very high levels of endogenous estrogen. When boars are postnatally given estrogen blockers, a transient change in boars epidydimal development occurs (Pearl et al., 2007). Factors such as heat stress during mid-gestation have been shown to reduce testicular size and increase in sperm abnormalities in boars (Lugar et al., 2018). Additionally, neonatal factors leading to low birthweight are directly linked to reductions in boar fertility development and longevity (Ashworth et al., 2001).

7.2 Development during the first 2 months

The relationship between birthweight and boar fertility has been widely investigated due to its association with reduced reproductive development in low-birthweight piglets (Rubim et al., 2022; Flowers, 2023). Studies have identified positive correlations between birthweight and key reproductive traits, including total seminiferous tubule length, Sertoli cell count, body weight, and testis weight (Auler et al., 2017; Almeida et al., 2013). Although lighter adult boars may not exhibit detectable semen abnormalities or impaired spermatogenesis, heavier birthweight boars demonstrate a greater capacity for sperm production and can produce more semen doses per ejaculate. This makes them reproductively valuable, contributing to efforts to maximize semen output and enhance genetic utilization and improvement (Rubim Sacramento et al., 2022; Auler et al., 2017).

7.3 Development during puberty

Environmental effects can have profound effects on boar development. Sertoli cell population established before puberty and their functionality after sexual maturity have a cornerstone effect on boar fertility. Factors affecting Sertoli cell mitosis early on in development have a greater effect on sperm production as compared to those affecting spermatogenesis after development has occurred. Other factors such as birthweight can indicate testicular development while prepubertal growth rates positively correlate with testis size and sperm production. Ultrasound scans combined with computer-assisted grayscale analysis have been shown to be an effective method for examining testicular microstructure and examining testicular health (Afsarlar et al., 2019). This approach has been used to detect structural changes in testicular tissue during adolescence and establish diagnostic thresholds for predicting future reproductive success in boars (Schulze et al., 2020). Colostrum intake and pre-weaning nutrition can improve lifetime sperm production (Rahman et al., 2014). During puberty, high temperatures can negatively affect Sertoli cell function and endocrine regulation (Larsson et al., 1983). Identification of genes associated with heat-tolerant phenotypes can help mitigate this challenge (He et al., 2020). Supplementation of vitamins, minerals, and other compounds can help maintain sperm production during stressful periods (Audet et al., 2009). For a comprehensive overview, see Flowers (2022).

8 Factors affecting semen quality: during productive life

A recent retrospective analysis of European AI centers highlighted that substandard semen quality is the leading cause of culling among Pietrain boars. This finding underscores the critical role of semen quality in determining the longevity and productivity of AI boars, as maintaining high reproductive performance is essential for the efficiency of AI programs. The study emphasizes the importance of rigorous semen evaluation protocols to ensure the selection and retention of boars with superior hereditary traits (Henneberg et al., 2023). When building breeding programs, several factors can hinder sperm quality. Genetic selection for growth factors (Robinson & Buhr, 2005), seasonal variation with changes in photoperiod/temperature, and ejaculate frequency are several such factors (Berger et al., 1980). Other considerations, such as selecting appropriate bedding materials and managing bacterial contamination are essential, in addition, it is crucial to consider the potential presence of pesticides and mycotoxins in bedding, as these contaminants can adversely affect boar semen quality (Althouse et al., 2000; Riedel et al., 2023). Cooling systems may also be necessary to avoid heat stress. High collection frequency will negatively influence sperm quality as well (Lopez Rodriguez et al., 2017). Sperm quality can also be supported by feeding strategies aiming to increase sperm resistance to oxidative stress and extend storage duration (de Andrade et al., 2023). Oxidative stress factors arrive in semen from processing, stress and capacitation-like changes potentially reducing their fertility potential (Althouse & Rossow, 2011). Furthermore, dilution rate, frequency of collection, and temperature will affect the quality of the semen doses (López Rodríguez et al., 2012; Schulze et al., 2013a, 2017).

Bacterial infection and viral spread through AI doses are important metrics to monitor (Prieto et al., 1997). If collection is not hygienically performed, it will result in bacterial contamination of the semen doses. Bacterial contamination is important to manage because of its negative effect on sperm quality over time (Kuster & Althouse, 2016). Disease monitoring is likewise important, as many diseases can be transmitted via AI doses.

Reductions in boar fertility arising from warmer seasons are a well-documented phenomenon (Wettemann & Bazer, 1985). Findings from heat stress studies show exposure to heat stress can temporarily decrease the levels of testosterone in the bloodstream. This leads to a clear inhibition of spermatid maturation and significant disruptions in the production of androgens in the testes (Larsson et., al 1983; Wettemann & Desjardins, 1979). Additionally, it is known that heat stress impairs sperm motility parameters detected by CASA, including average path distance, straight-line velocity, straightness, and linearity along with differences in metabolic classes and pathways (Sui et al., 2022). Studies have delved into systemic and local aspects of heat stress. Local scrotal insulation studies, done in porcine, have revealed damage to meiotic germ cells, likely due to alterations in the blood–testis barrier (BTB) integrity (Parrish et al., 2017). In this study an insulator was placed over the boars' testicles, causing there to be a heightened localized temperature. This revealed that there was an increase of abnormal sperm morphologies, nuclear damage, a 15–18% increase in the height of the seminiferous tubule, a 1.8–2.2 fold increase in debris from primary, secondary, and round spermatids in the lumen, a 20–30% reduction in preleptotene/leptotene primary spermatocytes and pachytene primary spermatocytes, an increase in vacuoles in the seminiferous epithelium, and a reduction of germ cell migration across the BTB. Due to this subsequent decrease in boar fertility brought on by heat stress, it is important for boar studs to keep barn climates cool. The prevalence of air conditioning units in U.S. boar studs has become widespread due to the hot climatic conditions experienced during the summer months in pig farming regions.

9 Developments in artificial insemination

The use of AI in pigs began as a means of preventing the transmission of diseases, enhancing work efficiency in sow farms, and accelerating genetic progress by using high-value genetics across a larger population of sows. Early attempts with pig AI were reported in Russia in the early twentieth century, but the evolution of AI in pigs was slower than in cattle. By the beginning of the new millennium, AI rates had increased to around 60% in the United States, 12–60% in South America, and between 40% and 85% in Europe. Prior to the foundation of the International Conference of Boar Semen Preservation in the 1980s, there was minimal research on improving frozen boar sperm quality (Roca et al., 2016). While cryopreservation of bull semen was a success in the 1950s, cryopreservation of boar semen did not meet such success until later.

Several authors suggest that the primary reason frozen boar semen has not found widespread acceptance in commercial swine production is due to its limited efficiency compared to liquid semen (Bolarín et al., 2009; Eriksson et al., 2002; Wongtawan et al., 2006). Due to swine being a polytocous species, the number of piglets per litter is an important economic fertility measure (Knox, 2014). Semen quality is an important measure in improving litter size (Myromslien et al., 2019). As previously cited, the efficiency of liquid semen is superior to that of frozen semen, with less variability between boars, providing less strict AI management and easier handling of semen doses on the farm. As boar cryopreservation stands now, large commercial swine operations are finding it more financially efficient to use fresh semen for production sow servicing. However, gene banking at the nucleus level and distributing genetics across country borders when fresh semen cannot be used favors the use of cryopreserved semen, regardless of expected lower fertility or efficiency (Jung et al., 2015). Research for improving boar semen cryopreservation fertility is needed before it can be widely adopted, specifically methods to make it more economical.

AI has increased breeding efficiency as measured by conception rate and litter size with minimal sperm (Am-in et al., 2010). With China as the exception, in the last decade, AI has become the predominant method of mating in the top ten pork-producing countries, with AI rates peaking at 90–100%, and around 99% in the United States (Waberski et al., 2019b). Additionally, the amount of sperm used per sow or gilt has decreased in recent years. The traditional method of AI involves the transcervical deposition of extended semen, containing 1.5–3.0 billion spermatozoa. This remains the most common method of AI. However, PCAI, which involves the deposition of semen past the cervix, directly into the uterine body, has gained popularity due to its ability to reduce the number of sperm cells needed in a semen dose. In commercial operations, the current threshold values for PCAI are between 1.0 and 2.0 billion sperm in a volume of

30–50 mL (Mezalira et al., 2005; Waberski et al., 2019b). One study, measuring both the reproductive performance and the seminal backflow concluded that PCAI is more efficient than cervical AI, offering the benefits of reduced costs without compromising reproductive efficiency (Hernández-Caravaca et al., 2012).

Quality control of semen doses going out of the boar stud is important (Schulze et al., 2023). The industry standards for high-yield boar studs use 80% total motility, 20–30% abnormal forms as the cutoff point for sperm quality. Bacterial load is also a major point of quality concern. An 8-year study of 28 European AI centers highlighted the benefits of implementing basic hygienic standards, with significant improvements in bacterial contamination control through monitoring nine critical control points. These measures enhanced semen quality, extended shelf life, and improved the efficiency of AI programs (Nitsche-Melkus et al., 2020). As a result, the implementation of stringent hygienic measures has become a crucial component of breeding programs. Regardless of stringent hygienic measures, the use of antibacterial agents or methods to reduce bacterial growth during storage of fresh liquid preserved semen is still necessary. Today, most people utilize antibiotics in their semen extenders. However, antimicrobial resistance to gentamicin and penicillin, the most commonly used antibiotics in boar semen, has already been documented at high rates (Costinar et al., 2021). This has promoted the exploration of alternative antimicrobial approaches for boar semen. Methods to eliminate the use of antimicrobials have included reducing the temperature at which fresh semen is stored to 5°C, referred to as hypothermic storage. This temperature reduces bacterial growth and is successful in reproductive performance (Menezes et al., 2020; Maaßen et al., 2023); however, it requires additives that protect sperm cell membranes at these reduced temperatures (Waberski et al., 2019a). It is important to consider antibiotic elimination, especially given that a significant number of commensal bacteria are naturally transferred into the uterus without adversely affecting fertility under normal physiological conditions (Althouse, 2008; Althouse et al., 2000; Waberski et al., 2019a).

Management techniques to preserve sperm health are continually evolving, but many known methods have yet to be fully integrated into standard practice. Transport-related vibrations during handling can negatively impact sperm motility and viability, emphasizing the need for improved methods to safeguard semen quality (Hafemeister et al., 2022). Temperature management of sperm samples during transport remains a critical control point. Proper temperature management during semen processing is vital for optimizing boar sperm viability and fertility outcomes, keeping temperatures at a consistent 15–17°C (59–63°F) (Schulz et al., 2013a). The temperature during diluting steps has also been discussed, with results indicating that diluting at 32°C, rather than diluting it at 20°C, can negatively affect its performance in *in vitro* capacitation assays. Researchers propose that maintaining sperm at a temperature closer to physiological levels helps preserve its metabolism, reducing changes that could compromise semen quality (Petrunkina et al., 2005). Chemical leakage can also be a concern. Reproductive failures in Spanish sow farms were linked to toxic compounds leaching from plastic semen storage bags, in this instance spermatozoa stored in the contaminated plastic bags showed no differences in routine quality control tests, *in vitro* fecundation, or endocrine profiler panel analysis; however, *in vivo* tests confirmed reproductive failures (Nerin et al., 2014).

AI^2 techniques such as machine and deep learning are transforming reproductive technologies by offering innovative tools for semen analysis and quality control in the boar industry. Studies, such as those by Hürland et al. (2023), have demonstrated the high accuracy of machine learning algorithms in predicting sperm quality, while Keller et al. (2024) showcased the use of deep learning methods to automate the classification of sperm morphology, particularly head abnormalities critical for fertility as well as label-free biomarker utilization to reduce the need for expensive fluorescent biomarker equipment. These AI^2-driven approaches, including convolutional neural networks, enhance the precision and efficiency of semen evaluation, reducing reliance on manual assessments. By enabling a more accurate selection of high-quality sperm, AI^2 holds significant potential to optimize AI programs, improve fertility outcomes, and streamline production processes in swine reproduction.

10 Conclusion and future trends

As the global population continues to grow, the demand for protein-rich foods like pork is increasing. To meet this demand, the swine industry will need to continue to innovate and improve its production methods and production strategies. Research will play a crucial role in this process. Artificial insemination has rapidly become the most important reproductive technique in maximizing genetic usage and managing reproductive performance in the swine industry. Further technologies such as the CASA and FC aid in creating objective analysis of semen quality, resulting in more precise and accurate results. There is a need for further research in key areas such as sperm cryopreservation, infertility diagnostics, and bacterial management to improve productivity in the swine industry and reduce reliance on antibiotics. The identification of subpopulations of sperm and the use of IBFC have the potential to further enhance the quality and functionality of spermatozoa.

The widespread use of sex-sorted sperm in the swine industry has not yet been fully developed as it has in the dairy industry. Development of such techniques is currently focused in the commercial sectors of the United States and Canada. The original sperm sorting method was introduced in 1987

(Johnson et al., 1987), with subsequent improvements made, including the development of a high-speed sorter (Johnson, 2000). Ongoing advances in sorting instrumentation, software, and speed have continued to be made in the commercial sector. However, the use of sexed pig sperm in the industry is limited by the anatomical features of the pig, which require large numbers of sperm for intra-cervical AI. Currently, the production rate of sex-sorted sperm is about 20–25 million sperm per hour (Vishwanath & Moreno, 2018). Under experimental conditions, the use of sexed sperm has been found to compromise both fertility and litter size. This is likely due to the reduced number of sperm used for insemination; however, the effect of the sorting process on the sperm cell and downstream ability to capacitate normally has yet to be fully elucidated. Despite the limitations, there have been some experimental uses of sexed pig semen in nuclear and multiplier herds, with the primary target of the multiplication herd to make an increased number of high genetic value gilts to enter sow barns. Challenges seen in cryopreserving boar semen are even more pronounced with sexed semen (Rath et al., 2003).

An additional development in the future of boar semen AI is nanoparticle separation. Nanoparticle separation is a term referring to the removal of nanoscale particles from an entire population. In the case of sperm nanoparticle separation, sperm with damaged acrosome membranes are removed with peanut agglutinin (PNA) conjugated to magnetic nanoparticles (MNPs). Annexin V binds and detects apoptotic spermatozoa and has also been conjugated to MNPs to remove this apoptotic population. These cells have undergone programmed cell death and expose phosphatidylserine on their surface which Annexin V detects. Lectin-MNP, on the other hand, binds to acrosome-reacted spermatozoa, which have prematurely reacted acrosomes and altered carbohydrate composition on their surface, once bound they can be removed (Durfey et al., 2019).

Diluting boar semen into larger volumes with semen extender and then splitting that volume into doses to breed a larger quantity of sows are the basis for AI in most swine production operations. Increasing healthy sperm count, leading to more sows bred by fewer boars, is the primary goal in AI. Understanding of the true fertile population of spermatozoa in a boar's ejaculate will be paramount in calculating the concentration of sperm per dose in future boar AI. Nano-purification of boar semen is a new opportunity to ensure that only high-quality sperm cells are present in a sample; however, it has not reached wide-scale adoption (Feugang et al., 2019). As a result, the opportunity for nano-purification and its potential remains untapped.

Colloid centrifugation is a technique that is not used commercially but is thought can be used to collect specific subpopulations of sperm from an ejaculate. This is possible because ejaculate samples contain a heterogeneous population of spermatozoa with differing abilities to fertilize. The technique involves layering the ejaculate sample onto a colloid solution with a specific density, typically a discontinuous gradient. The sample is then centrifuged at high speeds, causing the spermatozoa to migrate through the colloid solution and separate into different layers based on their density (Morrell, 2019). The technique of colloid centrifugation has potential uses beyond improving the quality and functionality of spermatozoa. For example, it can be used to reducing the bacterial load in semen through a low-density colloid centrifugation which can potentially provide an alternative to the use of antibiotics in semen extenders. Scaling such in an econonmical manner is likely limiting.

11 References

Afsarlar, C. E., et al. (2019). Novel prognostic grayscale ultrasonographic findings in the testis from a comprehensive analysis of pediatric patients with testicular torsion. *J Pediat Urol*, 15(5), 480.e1–480.e7. https://doi.org/10.1016/j.jpurol.2019.08.002

Almeida, F. R. C. L., et al. (2013). Birth weight and its impacts on testicular development in boars. In H. Rodriguez-Martinez, HM Soede & WL Flowers (Eds.), *Control of pig reproduction IX* (pp. 113–114). Leics, UK: Context Products Ltd.

Althouse, G. C. (2008). Sanitary procedures for the production of extended semen. *Reprod Domest Anim*, 43(Suppl 2), 374–378. https://doi.org/10.1111/j.1439-0531.2008.01187.x

Althouse, G. C., et al. (2000). Field investigations of bacterial contaminants and their effects on extended porcine semen. *Theriogenology*, 53(5), 1167–1176. https://doi.org/10.1016/S0093-691X(00)00261-2

Althouse, G. C. & Rossow, K. (2011). The potential risk of infectious disease dissemination via artificial insemination in swine. *Reprod Domest Anim*, 46(Suppl 2), 64–67. https://doi.org/10.1111/j.1439-0531.2011.01863.x

Am-in, N., et al. (2010). Comparison of artificial insemination with natural mating on smallholder farms in Thailand, and the effects of boar stimulation and distance of semen delivery on sow reproductive performance. *Trop Anim Health Prod*, 42(5), 921–924. https://doi.org/10.1007/s11250-009-9508-3

Amann, R. P. & Waberski, D. (2014). Computer-assisted sperm analysis (CASA): capabilities and potential developments. *Theriogenology*, 81(1), 5–17.e11–13. https://doi.org/10.1016/j.theriogenology.2013.09.004

Ashworth, C. J., et al. (2001). Causes and consequences of fetal growth retardation in pigs. *Reprod Suppl*, 58, 233–246.

At-Taras, E. E., et al. (2006). Reducing estrogen synthesis in developing boars increases testis size and total sperm production. *J Androl*, 27(4), 552–559. https://doi.org/10.2164/jandrol.05195

Audet, I., et al. (2009). Effects of dietary vitamin supplementation and semen collection frequency on reproductive performance and semen quality in boars. *J Anim Sci*, 87(6), 1960–1970. https://doi.org/10.2527/jas.2008–1714

Auler, P. A., et al. (2017). Testicular parameters and spermatogenesis in different birthweight boars. *Reprod Fertil Dev*, 29(9), 1720–1728.

Austin, C. R. (1951). Observations on the penetration of the sperm in the mammalian egg. *Aust J Sci Res B*, 4(4), 581–596. https://doi.org/10.1071/bi9510581

Baker H., et al. (1739). An Account of Mr. Leeuwenhoek's Microscopes; by Mr. Henry Baker F. R. S. 1739. Phil. Trans. 41, 503–519. https://royalsocietypublishing.org/doi/10.1098/rstl.1739.00

Bartoov, B., et al. (1981). Sperm Motility Analyzer (SMA), a practical tool of motility and cell concentration determinations in artificial insemination centers. *Theriogenology*, 15(2), 173–182. https://doi.org/10.1016/s0093-691x(81)80005-2

Berger, T., et al. (1980). Sexual maturation of boars and growth of swine exposed to extended photoperiod during decreasing natural photoperiod. *J Anim Sci*, 51(3), 672–678. https://doi.org/10.2527/jas1980.513672x

Berger, T., et al. (2008). Reducing endogenous estrogens during the neonatal and juvenile periods affects reproductive tract development and sperm production in postpuberal boars. *Anim Reprod Sci*, 109(1–4), 218–235. https://doi.org/10.1016/j.anireprosci.2007.10.003

Boe-Hansen, G. B. & Satake, N. (2019). An update on boar semen assessments by flow cytometry and CASA. *Theriogenology*, 137, 93–103. https://doi.org/10.1016/j.theriogenology.2019.05.043

Bolarín, A., et al. (2009). Use of frozen-thawed semen aggravates the summer-autumn infertility of artificially inseminated weaned sows in the Mediterranean region. *J Anim Sci*, 87(12), 3967–3975. https://doi.org/10.2527/jas.2009–1941

Brohi, R. D. & Huo, L. J. (2017). Posttranslational Modifications in Spermatozoa and Effects on Male Fertility and Sperm Viability. *OMICS*, 21(5), 245–256. https://doi.org/10.1089/omi.2016.0173

Broekhuijse, M. L., et al. (2012). Relationship of flow cytometric sperm integrity assessments with boar fertility performance under optimized field conditions. *J Anim Sci*, 90, 4327–4336.

Caballero, I., et al. (2012). Seminal plasma proteins as modulators of the sperm function and their application in sperm biotechnologies. *Reprod Domest Anim*, 47(Suppl 3), 12–21. https://doi.org/10.1111/j.1439-0531.2012.02028.x

Chang, M. C. (1951). Fertilizing capacity of spermatozoa deposited into the fallopian tubes. *Nature*, 168(4277), 697–698. https://doi.org/10.1038/168697b0

Costinar, L., et al. (2021). Boar Semen Contamination: Identification Of Gram-negative Bacteria And Antimicrobial Resistance Profile. *Anim Open Access J MDPI*, 12(1), 43. https://doi.org/10.3390/ani12010043

De Andrade, A. F. C., et al. (2023). Effects of supplemental antioxidants on in vitro fertility measures for cryopreserved boar spermatozoa. *Theriogenology*, 200, 33–42. https://doi.org/10.1016/j.theriogenology.2023.01.025

Drobnis, et al. (1993). Cold Shock Damage is Due to Lipid Phase Transitions in Cell Membranes: a Demonstration Using Sperm as a Model. *The Journal of experimental zoology*, *265*(4), 432–437. https://doi.org/10.1002/jez.1402650413

Durfey, C. L., et al. (2019). Nanotechnology-based approach for safer enrichment of semen with best spermatozoa. *J Anim Sci Biotechnol*, 10, 14. https://doi.org/10.1186/s40104-018-0307-4

Eriksson, B. M., et al. (2002). Field fertility with exported boar semen frozen in the new flatpack container. *Theriogenology*, 58(6), 1065–1079. https://doi.org/10.1016/s0093-691x(02)00947-0

Escoffier, J., et al. (2015). Flow cytometry analysis reveals that only a subpopulation of mouse sperm undergoes hyperpolarization during capacitation. *Biol Reprod*, 92(5), 121. https://doi.org/10.1095/biolreprod.114.127266

Feugang, J. M., et al. (2019). Treatment of boar sperm with nanoparticles for improved fertility. *Theriogenology*, 137, 75–81. https://doi.org/10.1016/j.theriogenology.2019.05.040
Flowers, W. L. (2022). Factors affecting the production of quality ejaculates from boars. *Anim Reprod Sci*, 246, 106840. https://doi.org/10.1016/j.anireprosci.2021.106840
Flowers, W. L. (2023). Litter-of-origin traits and their association with lifetime productivity in sows and boars. *Mol Reprod Dev*, 90(7), 585–593. https://doi.org/10.1002/mrd.23565
Foxcroft, G. R., et al. (2008). Identifying useable semen. *Theriogenology*, 70(8), 1324–1336. https://doi.org/10.1016/j.theriogenology.2008.07.015
Gadella, B. M., et al. (1995). Glycolipid migration from the apical to the equatorial subdomains of the sperm head plasma membrane precedes the acrosome reaction. Evidence for a primary capacitation event in boar spermatozoa. *J Cell Sci*, 108 (Pt 3), 935–946. https://doi.org/10.1242/jcs.108.3.935
Grossfeld, R., et al. (2022). Comparison of NUCLEOCOUNTER, ANDROVISION with Leja chambers and the newly developed ANDROVISION eFlow for sperm concentration analysis in boars. *Sci Rep*, 12, Article 11943. https://doi.org/10.1038/s41598-022-16280-6
Hafemeister, T., et al. (2022). Boar Semen shipping for artificial insemination: current status and analysis of transport conditions with a major focus on vibration emissions. *Anim Open Access J MDPI*, *12*(10), 1331. https://doi.org/10.3390/ani12101331
He, Y., et al. (2020). Transcriptome analysis identifies genes and co-expression networks underlying heat tolerance in pigs. *BMC Genet*, 21(1), 44. https://doi.org/10.1186/s12863-020-00852-4
Henneberg, S., et al. (2023). Lifetime and removal reasons for Pietrain boars in European AI centers: a retrospective analysis. *J Anim Sci*, 101, skac408. https://doi.org/10.1093/jas/skac408
Hernández-Caravaca, I., et al. (2012). Reproductive performance and backflow study in cervical and post-cervical artificial insemination in sows. *Anim Reprod Sci*, 136(1–2), 14–22. https://doi.org/10.1016/j.anireprosci.2012.10.007
Howards, S. S. (1997). Antoine van Leeuwenhoek and the discovery of sperm. *Fertil Steril*, 67(1), 16–17. https://doi.org/10.1016/s0015-0282(97)81848-1
Höfner, L., et al. (2020). Tolerance of stored boar spermatozoa to autologous seminal plasma: a proteomic and lipidomic approach. *Int J Mol Sci*, 21(18). https://doi.org/10.3390/ijms21186474
Hürland, M., et al. (2023). The use of machine learning methods to predict sperm quality in Holstein bulls. *Theriogenology*, *197*, 16–25. https://doi.org/10.1016/j.theriogenology.2022.11.032
Ibănescu, I., et al. (2018). Cluster analysis reveals seasonal variation of sperm subpopulations in extended boar semen. *J Reprod Dev*, 64(1), 33–39. https://doi.org/10.1262/jrd.2017-083
Johnson, L. A., et al. (1987). Flow sorting of X and Y chromosome-bearing spermatozoa into two populations. *Gamete Res*, 16(1), 1–9. https://doi.org/10.1002/mrd.1120160102
Johnson, L. A. (2000). Sexing mammalian sperm for production of offspring: the state-of-the-art. *Anim Reprod Sci*, 60–61, 93–107. https://doi.org/10.1016/s0378-4320(00)00088-9

Jung, M., et al. (2015). In vitro measures for assessing boar semen fertility. *Reprod Domest Anim Zuchthygiene*, 50(Suppl 2), 20–24. https://doi.org/10.1111/rda.12533

Karamanou, M., et al. (2010). Anton van Leeuwenhoek (1632–1723): father of micromorphology and discoverer of spermatozoa. *Rev Argent Microbiol*, 42(4), 311–314. https://doi.org/10.1590/S0325-75412010000400013

Keller, A., et al. (2024). Deep learning clnerassification method for boar sperm morphology analysis. *Andrology*, 10.1111/andr.13758. Advance online publication. https://doi.org/10.1111/andr.13758

Keller, A. & Kerns, K. (2023). Sperm capacitation as a predictor of boar fertility. *Mol Reprod Dev*. https://doi.org/10.1002/mrd.23690

Kerns, K., et al. (2020). Relationship between the length of sperm tail mitochondrial sheath and fertility traits in boars used for artificial insemination. *Antioxidants (Basel)*, 9(11). https://doi.org/10.3390/antiox9111033

Kerns, K., et al. (2018). Zinc ion flux during mammalian sperm capacitation. *Nat Commun*, 9(1), 2061. https://doi.org/10.1038/s41467-018-04523-y

Knox, R. V. (2014). Impact of swine reproductive technologies on pig and global food production. *Adv Exp Med Biol*, 752, 131–160. https://doi.org/10.1007/978-1-4614-8887-3_7

Kuster, C. E. & Althouse, G. C. (2016). The impact of bacteriospermia on boar sperm storage and reproductive performance. *Theriogenology*, 85(1), 21–26. https://doi.org/10.1016/j.theriogenology.2015.09.049

Kwon, et al. (2015). Increased Male Fertility Using Fertility-related Biomarkers. *Scientific Reports*, 5, 15654. https://doi.org/10.1038/srep15654

Lane, N. (2015). The unseen world: reflections on Leeuwenhoek (1677) 'Concerning little animals'. *Philos Trans R Soc Lond B Biol Sci*, 370(1666). https://doi.org/10.1098/rstb.2014.0344

Larsson, K., et al. (1983). Endocrine effects of heat stress in boars. *Acta Vet Scand*, 24(3), 305–314. https://doi.org/10.1186/BF03546734

Lopez Rodriguez, A., et al. (2017). Boar management and semen handling factors affect the quality of boar extended semen. *Porcine Health Manag*, 3, 15. https://doi.org/10.1186/s40813-017-0062-5

Lugar, D. W., et al. (2018). In utero heat stress causes reduced testicular area at puberty, reduced total sperm production, and increased sperm abnormalities in boars. *Anim Reprod Sci*, 192, 126–135. https://doi.org/10.1016/j.anireprosci.2018.02.022

Luque, G. M., et al. (2018). Only a subpopulation of mouse sperm displays a rapid increase in intracellular calcium during capacitation. *J Cell Physiol*, 234(3), 3745–3756. https://doi.org/10.1002/jcp.26883

López Rodríguez, A., et al. (2012). Effect of dilution temperature on boar semen quality. *Reprod Domest Anim*, 47(5), e63–66. https://doi.org/10.1111/j.1439-0531.2011.01938.

Maaßen, I. K., et al. (2023). Storage of extended Boar Semen at 5°C inhibits growth of multi-drug resistant *Serratia marcescens* and *Klebsiella oxytoca* while maintaining high sperm quality. *Antibiotics (Basel, Switzerland)*, 12(5), 857. https://doi.org/10.3390/antibiotics12050857

Mazzarello, P. (1999). A unifying concept: the history of cell theory. *Nat Cell Biol*, 1(1), E13–15. https://doi.org/10.1038/8964

Menezes, T. A., et al. (2020). Antibiotic-free extended boar semen preserved under low temperature maintains acceptable in-vitro sperm quality and reduces bacterial load. *Theriogenology*, 149, 131–138.
McCarthy, M. J., et al. (2006). Suppression of endogenous estrogen during development affects porcine epididymal sperm maturation. *Mol Reprod Dev*, 73(9), 1122–1128. https://doi.org/10.1002/mrd.20519
McFarlin, B. K. & Gary, M. A. (2017). Flow cytometry what you see matters: enhanced clinical detection using image-based flow cytometry. *Methods*, 112, 1–8. https://doi.org/10.1016/j.ymeth.2016.09.001
McKinnon, K. M. (2018). Flow cytometry: an overview. *Curr Protoc Immunol*, 120, 5.1.1–5.1.11. https://doi.org/10.1002/cpim.40
Mezalira, A., et al. (2005). Influence of sperm cell dose and post-insemination backflow on reproductive performance of intrauterine inseminated sows. *Reprod Domest Anim*, 40(1), 1–5. https://doi.org/10.1111/j.1439-0531.2004.00524.x
Morrell, J. M. (2019). Effect of colloid centrifugation on boar sperm quality during storage and function in in vitro fertilization. *Theriogenology*, 137, 122–126. https://doi.org/10.1016/j.theriogenology.2019.05.046
Myromslien, F. D., et al. (2019). Sperm DNA integrity in Landrace and Duroc boar semen and its relationship to litter size. *Reprod Domest Anim*, 54(2), 160–166. https://doi.org/10.1111/rda.13322
Müller-Wille, S. (2010). Cell theory, specificity, and reproduction, 1837–1870. *Stud Hist Philos Biol Biomed Sci*, 41(3), 225–231. https://doi.org/10.1016/j.shpsc.2010.07.008
Nerin, C., et al. (2014). Compounds from multilayer plastic bags cause reproductive failures in artificial insemination. *Sci Rep*, 4, 4913. https://doi.org/10.1038/srep04913
Nitsche-Melkus, E., et al. (2020). Impact of hygiene on bacterial contamination in extended boar semen: an eight-year retrospective study of 28 European AI centers. *Theriogenology*, 146, 133–139. https://doi.org/10.1016/j.theriogenology.2019.11.031
Parrilla, I., et al. (2019). Boar semen proteomics and sperm preservation. *Theriogenology*, 137, 23–29. https://doi.org/10.1016/j.theriogenology.2019.05.033
Parrish, J. J., et al. (2017). Scrotal insulation and sperm production in the boar. *Mol Reprod Dev*, 84(9), 969–978. https://doi.org/10.1002/mrd.22841
Pearl, C. A., et al. (2007). Reduced endogenous estrogen delays epididymal development but has no effect on efferent duct morphology in boars. *Reproduction (Cambridge, England)*, 134(4), 593–604. https://doi.org/10.1530/REP-06-0239
Petrunkina, A. M., et al. (2005). Detection of cooling-induced membrane changes in the response of boar sperm to capacitating conditions. *Theriogenology*, 63(8), 2278–2299. https://doi.org/10.1016/j.theriogenology.2004.10.008
Prieto, C. et al. (1997). Insemination of susceptible and preimmunized gilts with boar semen containing porcine reproductive and respiratory syndrome virus. *Theriogenology*, *47*(3), 647–654. https://doi.org/10.1016/s0093-691x(97)00023-x
Quirino, M., et al. (2022). A 5-color flow cytometry panel to assess plasma membrane integrity, acrosomal status, membrane lipid organization and mitochondrial activity of boar and stallion spermatozoa following liquid semen storage. *Anim Reprod Sci*, 247, 107076.
Rahman, K. M., et al. (2014). Nursing supports neonatal porcine testicular development. *Domest Anim Endocrinol*, 48, 84–92. https://doi.org/10.1016/j.domaniend.2014.02.006

Ramal-Sanchez, M., et al. (2020). Progesterone induces sperm release from oviductal epithelial cells by modifying sperm proteomics, lipidomics and membrane fluidity. *Mol Cell Endocrinol*, 504, 110723. https://doi.org/10.1016/j.mce.2020.110723

Rath, D., et al. (2003). Birth of female piglets following intrauterine insemination of a sow using flow cytometrically sexed boar semen. *Vet Rec*, 152(13), 400–401. https://doi.org/10.1136/vr.152.13.400

Riedel, A., et al. (2023). Comparison of deep-litter bedding materials and analysis of semen traits in Piétrain boars: a randomized controlled field study. *Anim Reprod Sci*, *259*, 107379. https://doi.org/10.1016/j.anireprosci.2023.107379

Rivera-Egea, R., et al. (2018). Sperm lipidic profiles differ significantly between ejaculates resulting in pregnancy or not following intracytoplasmic sperm injection. *J Assist Reprod Genet*, 35(11), 1973–1985. https://doi.org/10.1007/s10815-018-1284-4

Robinson, J. A. & Buhr, M. M. (2005). Impact of genetic selection on management of boar replacement. *Theriogenology*, 63(2), 668–678. https://doi.org/10.1016/j.theriogenology.2004.09.040

Roca, J., et al. (2016). Will AI in pigs become more efficient? *Theriogenology*, 86(1), 187–193. https://doi.org/10.1016/j.theriogenology.2015.11.026

Rubim Sacramento, R. C., et al. (2022). Birthweight leads to seminal and testicular morphofunctional commitment in sexually mature boars. *Theriogenology*, 189, 237–245. https://doi.org/10.1016/j.theriogenology.2022.06.024

Saacke, R. G., et al. (2008). Relationship of semen quality to sperm transport, fertilization, and embryo quality in ruminants. *Theriogenology*, 70(3), 451–459. https://doi.org/10.1016/j.theriogenology.2008.04.014

Sanches, E. A., et al. (2010). Sperm motility of Rhamdia quelen studied using computer-assisted analysis by open-source software. In E. A. Sanches, R. A. Bombardelli, R. M. Marcos, G. Neumann, C. P. R. de Toledo, & E. Romagosa (Eds.), *Aquaculture research* (Vol. 42, Issue 1, p. 153). Hoboken, NJ: Wiley. https://doi.org/10.1111/j.1365-2109.2010.02598

Schleiden, M. J. (1838). Beiträge zur phytogenesis, Arch. Anat. Physiol. Wiss Med.

Schulze, M., et al. (2013a). Temperature management during semen processing: impact on boar sperm quality under laboratory and field conditions. *Theriogenology*, 80(9), 990–998. https://doi.org/10.1016/j.theriogenology.2013.07.026

Schulze, M., et al. (2013b). Development of an in vitro index to characterize fertilizing capacity of boar ejaculates. *Anim Reprod Sci*, 140(1–2), 70–76. https://doi.org/10.1016/j.anireprosci.2013.05.012

Schulze, M., et al. (2017). Impact of different dilution techniques on boar sperm quality and sperm distribution of the extended ejaculate. *Anim Reprod Sci*, 182, 138–145. https://doi.org/10.1016/j.anireprosci.2017.05.013

Schulze, M., et al. (2020). Relationship between pubertal testicular ultrasonographic evaluation and future reproductive performance potential in Piétrain boars. *Theriogenology*, 158, 58–65. https://doi.org/10.1016/j.theriogenology.2020.09.003

Schulze, M., et al. (2023). Science-based quality control in boar semen production. *Mol Reprod Dev*, 90(7), 612–620. https://doi.org/10.1002/mrd.23566

Schwann T. (1839) Microscopical Researches into the Accordance in the Structure and Growth of Animals and Plants. Sydenham Society; London. Mikroskopische Untersuchungen über die Uebereinstimmung in der Struktur und dem Wachsthum der Thiere und Pflanzen, Reimer, Berlin. Translated in English by H. Smith, 1847

Singleton, W. L. (2000). *Growth and development of AI centers in the USA* (p. 147). Lawrence, KS: Allen Press, Inc.

Smital, J., et al. (2004). Differences among breeds and manifestation of heterosis in AI boar sperm output. *Anim Reprod Sci*, *80*(1–2), 121–130. https://doi.org/10.1016/S0378-4320(03)00142-8

Smital J. (2009). Effects influencing boar semen. *Anim Reprod Sci*, *110*(3–4), 335–346. https://doi.org/10.1016/j.anireprosci.2008.01.024

Sonderman, J. P. & Luebbe, J. J. (2008). Semen production and fertility issues related to differences in genetic lines of boars. *Theriogenology*, 70(8), 1380–1383. https://doi.org/10.1016/j.theriogenology.2008.08.009

Sui, H., et al. (2022). Effects of heat stress on motion characteristics and metabolomic profiles of boar spermatozoa. *Genes (Basel)*, 13(9). https://doi.org/10.3390/genes13091647

Sutovsky, P. (2015). New approaches to boar semen evaluation, processing and improvement. *Reprod Domest Anim*, 50(Suppl 2), 11–19. https://doi.org/10.1111/rda.12554

Sutovsky, P., et al. (2020). Biomarker-based human and animal sperm phenotyping: the good, the bad, and the ugly. *Biol Reprod*, 103(3), 509–524. https://doi.org/10.1093/biolre/ioaa129

Van Gestel, R. A., et al. (2016). Ultrastructure and lipid composition of detergent-resistant membranes derived from mammalian sperm and two types of epithelial cells. *Cell Tissue Res*, 363(1), 129–145. https://doi.org/10.1007/s00441-015-2272-y

Vishwanath, R. & Moreno, J. F. (2018). Review: semen sexing - current state of the art with emphasis on bovine species. *Animal*, 12(s1), s85–s96. https://doi.org/10.1017/S1751731118000496

Waberski, D., et al. (2019a). Sperm function in vitro and fertility after antibiotic-free, hypothermic storage of liquid preserved boar semen. *Sci Rep*, 9(1), 14748. https://doi.org/10.1038/s41598-019-51319-1

Waberski, D., et al. (2019b). Application of preserved boar semen for artificial insemination: past, present and future challenges. *Theriogenology*, 137, 2–7. https://doi.org/10.1016/j.theriogenology.2019.05.030

Wettemann, R. P. & Bazer, F. W. (1985). Influence of environmental temperature on prolificacy of pigs. *J Reprod Fertil Suppl*, 33, 199–208.

Wettemann, R. P. & Desjardins, C. (1979). Testicular function in boars exposed to elevated ambient temperature. *Biol Reprod*, 20(2), 235–241. https://doi.org/10.1095/biolreprod20.2.235

Wolf, J. (2009). Genetic parameters for semen traits in AI boars estimated from data on individual ejaculates. *Reprod Domest Anim*, 44(2), 338–344.

Wollman, A. J., et al. (2015). From animaculum to single molecules: 300 years of the light microscope. *Open Biol*, 5(4), 150019. https://doi.org/10.1098/rsob.150019

Wongtawan, T., et al. (2006). Fertility after deep intra-uterine artificial insemination of concentrated low-volume boar semen doses. *Theriogenology*, 65(4), 773–787. https://doi.org/10.1016/j.theriogenology.2005.07.003

Wood, P. L., et al. (2016). Lipidomics of equine sperm and seminal plasma: identification of amphiphilic (O-acyl)-ω-hydroxy-fatty acids. *Theriogenology*, 86(5), 1212–1221. https://doi.org/10.1016/j.theriogenology.2016.04.012

Wright, P. C., et al. (2012). A review of current proteomics technologies with a survey on their widespread use in reproductive biology investigations. *Theriogenology*, 77(4), 738–765.e752. https://doi.org/10.1016/j.theriogenology.2011.11.012

Chapter 8

Gilt development to improve offspring performance and survivability

Jamil E. G. Faccin, Mike D. Tokach, Robert D. Goodband, Joel M. DeRouchey, Jason C. Woodworth and Jordan T. Gebhardt, Kansas State University, USA

1 Introduction

Modern maternal-line genotypes have impressive farrowing performance, increasing litter size by approximately 0.33 pigs per litter per year since 2006 (Tokach et al., 2019). One of the keys to continually improve litter size is genetic improvement through gilt introductions. However, deficient management

http://dx.doi.org/10.19103/AS.2024.0137.12

strategies in gilt development spanning from the gilt's first day of life until the first farrowing can lead to poor offspring performance and survivability.

There are two main paths to achieving high performance and survivability of gilts' offspring. First, improving gilt longevity in the sow herd aids in stabilizing herd health. An older sow parity structure will benefit newborn pigs by enhanced immunity which has a lasting impact on offspring performance and survivability. The second path to improving offspring survivability is by maximizing colostrum and milk production. Due to lower birth weight and inferior immune response, first-parity litters usually have poorer lifetime performance than litters born to multiparous sows (Piñeiro et al., 2019). Besides poorer growth performance, first-parity progenies have higher nursery and finishing drug costs and mortality (Moore, 2001). Thus, improving parity-one sows' colostrum and milk production is essential to alleviate these disadvantages. One way to accomplish this goal through gilt development is to provide special attention to feeding and management strategies that improve mammary development in prepubertal stages (Farmer, 2018).

Implementing development programs that focus on selecting gilts with the greatest reproductive potential, ensuring the number of gilts needed by the system, and adequate management and nutrition are vital in reaching a high-quality, productive herd. This chapter will review current knowledge on gilt development-related key management and nutritional strategies that improve longevity and their production, as well as enhance offspring performance and survivability.

2 Factors affecting gilt development: birth weight and litter of origin

Gilts must be developed correctly to ensure their longevity in the herd and improve the survivability of their offspring. Gilt management and selection start at birth (Table 1). Birth weight, litter of origin, and lactation and weaning management are points involved in early selection strategies that are indicators of future performance and efficiency (Patterson and Foxcroft, 2019).

Recently, studies have addressed the importance of the litter of origin of a replacement gilt. Considering that birth weight is a phenotypically repeatable trait (Foxcroft et al., 2009), it can be hypothesized that gilts born to sows with a low birth weight potential could pass that trait on to their offspring. Patterson et al. (2018) reported a lower retention rate within 4 d after birth, at 24 and 70 d of age, and at pre-selection when analyzing gilts born to sows classified as having a low birth weight phenotype (BWP; litter average birthweight <1.15 kg) based on at least two successive parities compared to categories with over 1.16 kg birthweight. Natural selection and increased survival rate of the heavy weight gilts played an important role in that result. Conversely, it has also

Table 1 Key aspects of managing and feeding gilts to improve lifetime performance and longevity, and produce high-survival pigs

Area	Action	Reason
Litter of origin	At the multiplier, avoid selecting gilts born from sows farrowing litters with an average birth weight <1.15 kg in two or more farrows.	Gilts born to low birth weight phenotype sows can pass this feature to their progeny.
Birth weight	Avoid selecting gilts weighing less than 1 kg at birth	Gilts born <1 kg of birthweight produce 4.5 fewer piglets over three parities than the other birthweight classes.
Colostrum intake/ cross-fostering	Sows with higher milk production (e.g. multiparity sows) should nurse future replacement gilts.	Gilts consuming at least 250 g of colostrum will be heavier at 42 d of age and gilts heavier at weaning have higher survivability and odd to farrow a litter.
Weaning age	Replacement gilts should be at least 24 d old at weaning.	Gut health, post-weaning survivability, and lifetime productivity are greater for pigs when weaned at 24 days of age or older.
Vitamins and minerals	Add more phosphorous than terminal lines, consider organic trace mineral sources, and add extra choline, pyridoxine, folic acid, and biotin.	Phosphorus recommendation levels for bone mineralization are 8% higher than for growth performance. Organic trace mineral supplementation can reduce osteochondrosis incidence. These vitamins are involved in reproductive functions and are often not included in finishing pig premixes.
Mammary development	From 90 d of age to puberty, do not restrict energy intake too much; avoid overfeeding gilts in late gestation.	Appropriate mammary gland development can be reached without maximizing energy intake, but low intake also reduces mammary DNA. Overfeeding gilts in late gestation will deposit more fat in mammary glands, reducing colostrum and milk production.
Boar exposure	Provide adequate boar exposure with boars at least 10 months of age. Promote as much physical interaction with gilts as possible.	Having more boars is better to avoid fatigue. Mature boars have more *libido* and are large enough to stimulate estrous without being submissive to the gilts in pens.
Breeding targets	Breed on the second estrus and 135–160 kg BW.	Inseminating in the first estrous reduces the farrowing rate and litter size in the first farrowing. Breeding overweight gilts reduces their longevity in the herd.
Flush feed	Only flush feed gilts that might not reach the body weight target at breeding.	Flush feeding only improves reproductive outputs when gilts are below the target weight for breeding.

(*Continued*)

Table 1 (Continued)

Area	Action	Reason
Gestation feeding	Early: Do not feed gilts below maintenance and growth requirements and avoid feeding more than 7.5 Mcal of NE/day.Late: Unless body condition is low, avoid bump-feeding.	Over-conditioned gilts can have lower litter size and have reduced feed intake and milk production in lactation. Bump feeding improves birth weight slightly but increases the stillborn rate and lowers lactation feed intake and colostrum and milk production.
First lactation	Avoid having unsuckled mammary glands.	Mammary glands not suckled during the first lactation produce less colostrum and milk in the subsequent farrowing.

been observed that gilts from high BWP are more likely to grow exceptionally fast, such that they may be bred too heavy, impairing their reproductive performance over four parities (Patterson et al., 2020). Amaral Filha et al. (2010) observed that heavy gilts at first service had decreased farrowing rate in parity two, possibly due to low feed intake in the first lactation, and those gilts bred at >170 kg were at risk of low retention and greater locomotion problems over three parities. Thus, gilts that are heavy at birth and have exceptional growth rates may need to be slowed down to prevent them from becoming too heavy at breeding. Collectively, these data suggest that the selection of replacement gilts should start at birth, avoiding extremely low or high BWP litters of origin.

Birth weight is one of the essential traits that affect lifetime productive performance, both in finishing pigs (Wolter et al., 2002; Douglas et al., 2014) and replacement gilts (Almeida et al., 2015; Magnabosco et al., 2015). In a high-prolific multiplier system, Magnabosco et al. (2016) observed the impact of individual birthweight of gilts on their future reproductive performance up to parity 3. Age at puberty was not influenced by birth weight; however, gilts weighing less than 1.0 kg at birth produced fewer pigs born alive and more stillborn in the first farrowing. These gilts had 4.5 fewer piglets over three parities than the other birthweight classes. The retention rate was also affected by gilt birthweight, with gilts weighing greater than 1.28 kg at birth having greater days in the herd than gilts weighing between 1.0 kg and 1.28 kg at birth, with those weighing less than 1 kg having the fewest days in the herd. The fact that gilts with a birth weight under 1 kg show satisfactory results in terms of onset of estrus and farrowing rate hides the negative impact on retention from selecting light birthweight gilts as future replacements. Considering the long-lasting effect on reproductive performance, gilts less than 1 kg at birth should not be selected to enter a breeding herd.

3 Colostrum and milk intake

Adequate colostrum intake plays a vital role in promoting newborn pig health, growth, and survivability, which can improve subsequent reproductive performance (Vallet et al., 2015; Declerck et al., 2016). Bartol et al. (2013) suggested a positive relationship between colostrum intake and uterine development. Colostrum intake also has an essential role in growth rate, with pigs consuming at least 250 g of colostrum being heavier at 42 d of age than those consuming less than 250 g (Ferrari et al., 2014). Knauer (2016) observed that increasing preweaning growth rate increased the proportion of gilts that ultimately farrow a litter. Due to the impact on pre-weaning survivability and weight gain, strategies to improve milk intake should be considered. In the study by Ferrari et al. (2014), regardless of the parity of the biological mother, pigs fostered on to multiparous sows were heavier at 42 d of age than pigs nursing primiparous sows. Bierhals et al. (2011) observed that pigs born to first parity sows and fostered by parity 5 sows were heavier at weaning than those not cross-fostered. Nucleus and multiplier farms can take advantage of these results to maximize colostrum and milk intake of replacement gilts.

Priority should be placed on individual gilt birth weight, average pig weight in litter of origin, day-one care, and cross-fostering. These early growth parameters initiate a cascade of positive outputs during the lifetime reproduction performance of the replacement gilt (Vallet et al., 2015).

4 Weaning age

Research regarding weaning age has been conducted with a primary focus on economics and growth performance from wean to finish. However, several findings from these studies may be applied to gilt development. From a physiologic standpoint, early weaned pigs had increased intestinal permeability (16 days (d) vs. 20 d; McLamb et al., 2013) and a less prepared immunological response to stress than older weaned pigs (14 d vs. 21 d; Davis et al., 2006). In these studies, the negative effects of early weaning persisted for months, similar to mast cells not degranulated, which is a signal of inactivation, found in early weaned pigs by Pohl et al. (2017; 15 d vs. 28 d), which lasted until adulthood. Together, these studies provide valuable information regarding the relationship between young pig's gut health and gilt's health in the future.

Male and female enteric nervous systems react differently when challenged by weaning stressors like early weaning. Females exhibited higher neural-evoked secretory responses and more pronounced cholinergic activation (Medland et al., 2016) when compared to males. Thus, gilts respond more negatively to being weaned at young ages than male pigs. Therefore, multipliers raising replacement gilts should wean at older ages.

A major influence on survivability is related to weaning age. Studies have shown the significant effect of older weaning ages on the number of pigs that reach market weight (Main et al., 2004, Faccin et al., 2020). Faccin et al. (2020) reported an absolute reduction of 0.63% in nursery losses per day as weaning age increased from 19 d to 28 d. Therefore, a 3 d increase in weaning age can decrease nursery mortality and removals by almost 2.0%. Besides gut health improvement, the swine industry can take advantage of older weaning ages to increase survivability, also of breeding gilts.

Knauer (2016) intended to associate pre and at-weaning factors of 6249 gilts that entered commercial sow farms with lifetime productivity. In this dataset, weaning age ranged from 15 d to 27 d. For each day of increase in weaning age, sow productivity was improved by 0.185 pigs/sow/year through four parities. Thus, increasing the weaning age by 3 d represented an increase of more than 0.5 pigs/sow/year. Also, in this study, weaning age positively affected the retention rate from sow farm entry to parity 2.

The economics of increasing weaning age in individual systems must be considered when producing replacement gilts. Because of the positive benefits on gilt health and reproductive performance, multiplication farms should wean gilts at a minimum of 24 d of age.

5 Target weight

An essential goal for replacement gilt nutrition is to develop prepubertal gilts to their physiological maturity in body weight (BW), tissue composition, structural soundness, and reproductive development. Thus, one main objective of gilt nutrition for longevity and excellent lifetime performance is to maintain growth rates to achieve 115 kg to 140 kg at puberty and 135 kg and 160 kg at breeding (Bortolozzo et al., 2009). Considering 1 kg/d of average daily gain (ADG), gilts need to be a maximum of 140 kg of BW when reaching puberty to be inseminated in the second estrus at less than 160 kg of BW. With *ad libitum* feeding systems, commercial gilts can exceed this weight threshold. Thus, production systems often need to limit growth to increase the proportion of gilts meeting BW targets at puberty and breeding. This is a key concept because inseminating overweight gilts can lead to locomotion and structural problems that result in early removal from the herd (Amaral Filha et al., 2010).

6 Nutrition strategies during the growing phase

Ad libitum feed intake for current high feed intake and fast-growing genotypes is a concern because of potential lameness and leg problems (Farmer, 2018). *Ad libitum* feeding after 10 week of age has been shown to increase the risk of osteochondrosis by 20% for each 100 g increase in ADG according to de Koning et al. (2013). However, limiting feed intake too much can be detrimental to the development of mammary secretory cells (Sorensen et al., 2002). A reduction in the Lys to energy ratio of 17–20% can be used to decrease the growth rate and increase the number of gilts in an optimal body weight range at first estrus (Lents et al., 2020) and reduce the occurrence of lameness and the severeness of joint lesions (Quinn et al., 2015). However, pubertal development can also be delayed by reducing the dietary Lys to energy ratio (Lents et al., 2020) or by a 25% restriction in energy intake (Johnson et al., 2022). These strategies did not have any other negative impacts on reproductive performance. Thus, reduced dietary energy and lysine can be used to alter body weight at puberty with the understanding that puberty may also be delayed. Fiber, like soy hulls, can also be used to restrict the amount of daily metabolizable energy intake (Winkel et al., 2018); however, only high inclusion levels can indeed restrict feed intake (Helm et al., 2021). Extra feeder space and space allowance/gilt can generate a compensatory feed consumption resulting in the same feed intake of gilts fed *ad libitum*. As a tool to increase sow longevity, fiber intake can benefit gastrointestinal development and decrease the incidence of severe esophagogastric ulcers (Dirkzwager et al., 1998).

To minimize the incidence of gastric ulcers, during the developer phase, efforts should also be made to prevent out-of-feed events and use of more coarsely ground grains (800 microns) should also be considered (Mößeler et al., 2012).

Lameness is usually one of the top three reasons for culling gilts and sows (Sasaki and Koketsu, 2008; Wang et al., 2019). From a nutrition standpoint, replacement gilt diets should be formulated differently than finishing gilt diets. Calcium and phosphorus are the most abundant minerals in the body and are related to multiple biological functions like growth performance and development, and maintenance of the skeletal system (Berndt and Kumar, 2009). Increasing levels of calcium and phosphorus can be beneficial because the levels to maximize bone mineralization are 8% greater than that to maximize growth rate (NRC, 2012; Vier et al., 2019).

Vitamins involved in reproduction like choline, pyridoxine, folic acid, and biotin should be added to the diet because finishing premixes usually do not include these vitamins. Recently, in a U.S. pork industry survey, nutritionists reported levels of 446.9 mg/kg, 2.8 mg/kg, 1.9 mg/kg, and 0.3 mg/kg of those vitamins, respectively (Faccin et al., 2023). It is worth mentioning that these are industry values and not requirements estimates. However, data regarding the exact timing to start supplementation are not available. Thus, beginning at approximately 150 d of age is suggested due to proximity to the onset of the reproductive activity. Because of research indicating that supplementation of organic trace minerals (i.e. Zn, Mn, and Cu) reduces osteochondrosis in replacement gilts (Fabà et al., 2019), their addition to the diet should also be considered. Similarly, research demonstrates that dietary addition of 25-hydroxy-D3 can reduce the incidence or severity of osteochondrosis lesions (Sugiyama et al., 2013) and, thus, the use of 25-hydroxy-D3 should be considered for developing gilts.

7 Mammary development and feed management

Over the years, modern gilts have become leaner, more feed efficient, and faster growing than gilts of previous decades. The mammary development of the young gilt is relatively slow, with mammary tissue and DNA accretion being low until approximately 90 d of age. Farmer (2018) summarized that the first stage of mammary development was from 90 d of age until puberty with mammary DNA increasing four to six times during this period (Sorensen et al., 2002). Restricting feed intake by 20–25% during the growing-finishing period in gilts has been shown to negatively affect mammary development at puberty (Farmer et al., 2004; Sorensen et al., 2006). *Ad libitum* feeding from day 90 to puberty increases the parenchymal weight of gilt's mammary gland by 36–46% compared to restricted feed in the same period (Sorensen et al., 2002; Farmer et al., 2004). Since the time of these studies, feed intake potential has increased considerably, such that a 15–20% restriction of feed intake today results in similar intake to *ad libitum* consumption in Sorensen et al. (2002) and Farmer et al. (2004) studies. Recently, Gregory (2021) compared *ad libitum* to 20% feed restriction from day 90 to breeding, and no effect on milk yield was found. Together, these data suggest that mammary development may be impaired if a minimum nutrient intake is not maintained, but *ad-libitum* consumption may not be needed to reach that minimum level of required energy intake.

8 Space allowance and group size

It is well known that stocking density affects growth performance and welfare indicators of growing pigs (Brumm, 2004; Fu et al., 2016). Even with barrows growing faster than gilts, no gender and stocking density interactions appear to exist (Brumm, 2004). In gilt development, there are limited studies regarding space allowance. Young et al. (2008) compared 0.77 m^2/gilt and 22 gilts/pen with 1.13 m^2/gilt and 15 gilts/pen from 75 d to 200 d of age. Growth performance was not affected by space allowance and group size. Still, lower space allowance and larger group size reduced the percentage of gilts in puberty by 200 d of age, which resulted in a greater average age at puberty. Gilts showing late puberty have an increased risk of being bred overweight. Young et al. (2008) observed no significant effects on reproductive performance through parity 3, and the percentage of removals and mortality was not affected by space allowance and group size. However, an increase of cracks on gilt's rear hooves was reported for gilts with greater space allowance and smaller group size, which may indicate some relationship between increased space and more physical activity. Young et al. (2008) did not observe a large impact of space allowance and group size during rearing; however, gilt developer facilities must pay attention to space allowance and the number of gilts per pen because they often impact access to feeders, drinkers, and the quality of boar exposure.

9 Boar exposure

Quality of boar exposure is one of the pillars of gilt development. Failing to detect the first estrus of a gilt automatically adds an extra 20 non-productive days to the system, can impair the target of females inseminated per group, and results in overweight gilts at mating. It is common to find corpora lutea in gilts culled for anestrous. Diehl et al. (2003) found that 76% of gilts culled for anestrous had corpora lutea and/or albicans, the corpus luteum involuted form, at the slaughterhouse. Gilts with a greater growth rate generally reach puberty sooner. Early age at puberty has been linked to greater lifetime productivity (Patterson and Foxcroft, 2019), so boar stimulation should start between 150 and 170 d of age to achieve at least 80% of gilts reaching puberty after 30 d of boar exposure (Amaral Filha et al., 2009). Boars should be at least 11 months of age, express good *libido*, and impose physical presence when in gilt pens. Young and small boars usually become submissive to gilts (Bortolozzo et al., 2006). Having a high number of boars is also important to prevent their exhaustion and locomotor problems. With good boar exposure, gilt development units should breed the majority of the gilts in the second estrus (prior to 220–225 d of age) and delay to the third estrus-only gilts that have not met the target weight. Direct contact with boars in the pens reduces the time to find gilts in estrus by approximately 10 d compared to only providing fenceline contact (Patterson et al., 2002).

10 Flush-feeding

Flush-feeding gilts before breeding is a common practice recommended more than 30 years ago to increase first parity litter size (Cox et al., 1987; Beltranena et al., 1991). Flushing consists of full feed or increasing the amount of feed and consequently energy intake 1–2 weeks before planned insemination. This results in an improved gonadotropin-releasing hormone pulse generator system followed by a higher stimulation of folliculogenesis and ovulation rate by luteinizing hormone (Ashworth and Antipatis, 1999; Prunier and Quesnel, 2000). In the 1990s, studies using Yorkshire × Landrace sows (Beltranena et al., 1991) and Chinese Meishan (Ashworth et al., 1999) confirmed the beneficial effect of increasing pre-breeding feed intake. Improved nutrient uptake and utilization resulted in a greater number of follicles ovulated.

A recent trial addressed insights regarding flush-feeding modern gilts. Mallmann et al. (2020b) inseminated gilts on their third estrus that were provided either 2.1 or 3.6 kg/d from puberty to second estrus or from second to third estrus. On day 30 post-breeding, gilts fed 3.6 kg/d during both cycles were determined to have had a greater ovulation rate. However, the increased feed allowance only increased viable embryos when provided from puberty to second estrus. Gilts that were provided extra feed between the second and third estrus had reduced embryo survival. This study demonstrated that flushing from puberty to the second estrus resulted in two extra viable embryos on d 30 of gestation; however, continuing flushing after the second estrus provided no further benefit. Thus, if gilts are not mated by the second estrus, feed intake should return to basal levels (2.1 kg in this study), representing 30 kg of feed savings per gilt. Another interpretation of these data is that most gilts met breeding weight targets in the second estrus (133–143 kg BW for the 2.1 kg and 3.6 kg treatments, respectively), which means that some gilts were overweight when inseminated in the third estrus, and this may have affected their reproductive performance. The lower percentage of embryo survivability for gilts fed 3.6 kg/d during the second cycle may have been due to lower plasma progesterone during early pregnancy. Progesterone metabolization by the liver in the first few days of pregnancy can significantly impact fertility because progesterone secreted by the ovaries is still low (Langendijk and Peltoniemi, 2013).

Bruun et al. (2021) compared no flush feeding to flushing for 7 d (follicular phase) prior to breeding, flushing for 18 d from day 7–25 prior to breeding (luteal phase), or flushing for both periods (25 d). They observed that the weight of the gilt influenced total litter size. Gilts with low backfat tended to respond more positively to a longer flushing period than heavier gilts with more backfat. Flush feeding during the follicular phase (7 d) provided the greatest benefit in total born without increasing backfat.

In summary, the feed level provided during the first cycle after pubertal estrus is essential to establish the ovulation rate and consequently, potential litter size. In contrast, the feed level in the estrous cycle before insemination affects more embryo survival. Considering bodyweight at breeding, flush-feeding gilts only result in positive reproductive performance until achieving the target weight for breeding. Gilts inseminated above bodyweight recommendations have no improvement in reproductive performance when flush fed.

11 Gestation feeding strategies

Feeding strategies in the first 3–4 weeks of gestation have been discussed and researched for several years because of their capacity to impair embryo survival and, consequently, litter size (Leal et al., 2019). Progesterone metabolism by the liver can cause a reduction in embryo survival after insemination; however, published studies have conflicted on the response to feed intake in early gestation. When gilts had increased average daily feed intake (ADFI) (2.6 kg/d vs. 1.9 kg/d) from breeding to day 15 of gestation, a decrease of 20% embryo survival was found (Jindal et al., 1996). However, other reports have shown no differences when feeding 2.0 kg/d vs. 4.0 kg/d from day 0 to 7 after breeding (Quesnel et al., 2010). In a recent large study on a commercial farm, Mallmann et al. (2020a) fed gilts and parity 1 sows with 1.8 kg/d, 2.5 kg/d, or 3.2 kg/d from day 6 to 30 of gestation and found a linear increase in body weight and backfat thickness at 30 d of gestation, a linear negative impact in litter size, and reduced lactation feed intake. Collectively, these data suggest that besides reducing litter size, providing extra feed intake early in gestation can result in over-conditioned sows at farrowing, which may compromise lactation feed intake and increase lifetime maintenance requirements. Increasing feed intake has only been beneficial when the control was near maintenance feeding levels (e.g. 1.5 kg/d; Athorn et al., 2013).

Between day 22 and 42 of pregnancy, placenta vascularity increases and the transition from embryo to fetus starts (Wu et al., 2004). High energy intake during gestation can positively affect placental efficiency by reducing piglet weight variation (Che et al., 2017). However, increasing the amount of feed from day 22 to 42 of gestation has failed to improve farrowing performance. Mallmann et al. (2019b) compared 1.8–3.5 kg/d of feed in gilts fed from day 22 to 42 of gestation and observed a similar number of total and born alive piglets and no differences in birth weight. High feeding levels (5 kg/d) in mid-gestation gestation have been found to increase secondary muscle fibers in the fetus as compared to feeding 2.5 kg/d (Dwyer et al., 1994). Two subsequent experiments with a larger sample size found no impact on birth weight when feeding 3.6 kg/d compared to sows fed 1.8 kg/d during mid-gestation (Musser et al., 2006).

In the past decade, several studies have examined the effect of increasing feed intake (bump-feeding) in the last 20–30 d of gestation of gilts and sows. A meta-analysis (Gonçalves, 2016) summarized trials that bump-fed gilts in commercial farms. On average, each 1 kg increase in a daily feed intake from day 90 to farrowing results in approximately 7 kg increase in sow body weight. However, it only resulted in a modest increase in piglet birth weight of 13.5 g in gilts. More recently, Mallmann et al. (2019a) conducted a dose-response study (1.8 kg/d, 2.3 kg/d, 2.8 kg/d, and 3.3 kg/d) in gilts from day 90 to farrowing. Similar to other studies, a linear increase in gilt BW and backfat thickness and no influence on litter size was found. However, gilts fed the lowest amount of feed had fewer stillborn pigs than gilts fed other levels. Greater backfat thickness also has been shown by others to be related to fewer piglets born alive (Lavery et al., 2018). Gilts fed 2.3 kg/d tended to farrow 20–30 g heavier piglets than gilts fed 1.8 kg/d, 2.8 kg/d, or 3.3 kg/d with no effect found for birth weight CV. Lactation feed intake may provide the most consistent response across bump feeding trials. Gilts and sows with greater feed intake in late gestation have lower lactation feed intake (Shelton et al., 2009; Mallmann et al., 2020a). These results highlight the concern with high feeding levels at any point in gestation which can result in reduced lactation feed intake increasing gilts' catabolic state and reducing milk production and litter weight gain.

Besides gilt growth rate, farrowing performance, and lactation feed intake, other questions frequently raised around bump feeding concern subsequent reproductive performance and colostrum yield. Mallmann et al. (2019a) observed no effect in subsequent wean-to-estrus interval, farrowing rate, and litter size due to feeding level during the previous gestation period. However, colostrum yield was linearly decreased with increasing feed intake in late gestation. The fact that gilts fed 1.8 kg/d was the only feeding level that lost backfat from day 90 to day 112 seems to suggest that nutrients were prioritized for colostrum yield over fat deposition since colostrum is mainly synthesized in the week before farrowing (Devillers et al., 2007) and shortly after farrowing (Quesnel and Farmer, 2019).

In the last 13 years, the average total born has increased by 4.5 piglets (Tokach et al., 2019). Many efforts have been made to optimize colostrum and milk yield through better mammary gland development. Although mammary tissues undergo a high degree of hyperplasia post-farrowing (Kim et al., 1999b), the protein accretion rate after day 75 of gestation indicates a more significant protein requirement for mammary gland growth (Ji et al., 2006). This does not mean that it can be easily handled with nutrition because, as already mentioned, bump feeding gilts can reduce colostrum yield and decrease lactation feeding intake. A significant determinant of milk production is the number of secretory cells present in the gland (Head and Williams, 1991). Mammary glands of excessively fat gilts in late gestation have more fat deposition as the adipocyte competes for space with secretory cells, which is suggestive evidence for a reduction in colostrum and milk production (Head and Williams, 1991; Howard, 1995; Farmer et al., 2004). During late gestation, mammary glands undergo a major composition shift, going from a high-lipid to a high-protein content (Ji et al., 2006). Although no effect was observed when feeding high levels of crude protein, early research has shown that feeding high energy levels in the last third of gestation decreased secretory tissue, which can be detrimental to colostrum and milk production (Weldon et al., 1991). Interestingly, feeding 150% of the NRC (2012) requirement estimate for valine in gestating gilts improved fat synthesis in the milk, mammary DNA and protein content, and colostrum and milk production (Che et al., 2019, 2020, 2021). The authors revealed several pathways where Val is involved in the expression of proteins related to fatty acid synthesis. Due to the inconsistent results found in the literature, further work around amino acid supplementation in late gestation is needed

Questions about the lysine requirement of highly productive sows during gestation have also recently been addressed. Thomas et al. (2021a) found that increasing dietary standardized ileal digestibility (SID) Lys (11.0 g/d, 13.5 g/d, 16.0 g/d, and 18.5 g/d) during the course of gestation is associated with maternal body protein deposition, supported by increased body weight gain with no change in backfat thickness. Increasing dietary SID Lys increased lean growth of the gilt but did not have an impact in litter size (15.3 pigs born alive across all treatments) or piglet birth weight (1.23-1.26 kg/pig across all treatments). Thomas et al. (2021b) modeled the Lys balance of gilts in their study and observed that those fed 13 g/d were above their Lys requirement for the entire gestation period. Those fed 11 g/d were in excesses for the first 100 d of gestation but fell below their estimated requirements for the last 14 d of gestation.

In summary, in the period immediately after breeding, gilts should not be fed below the base level (minimal energy to meet the requirement for maintenance, growth, and reproduction) but avoid increasing ADFI to more than 10 Mcal of ME/day. In late gestation, bump feeding modestly improves birth weight but results in a higher piglet stillborn rate, lower lactation feed intake, and lower colostrum and milk production. Keeping adequate body condition from breeding through lactation is essential to support sow longevity, minimize the cost of maintenance, and increase lifetime productivity.

12 Avoiding unsuckled mammary glands

Mammary gland development continues during the first lactation (Ford et al., 2003), and maximizing lactation feed intake contributes to this development (Kim et al., 1999a). When not suckled, the gland suffers a process of regression and involution and after 10 d loses around 75% of DNA and wet weight mass (Kim et al., 2001). After weaning, a similar percentage of losses is seen in both suckled and unsuckled glands (Ford et al., 2003). Thus, unsuckled glands have less DNA available for subsequent parities as confirmed by Farmer et al. (2012), who found mammary glands not suckled in the first farrowing have 20% less parenchymal tissue per teat and 22% fewer DNA grams/teat in the second lactation. Glands that are suckled during the first lactation have greater milk production in the subsequent lactation than glands that are not suckled in the first lactation (Kim et al., 2001). Furthermore, piglets can recognize the more productive previously suckled teats than the ones not suckled in the first lactation (Devillers et al., 2016). Additionally, Guo et al. (2019) reported that increasing the suckling intensity through greater litter size and lactation length did not impair farrowing performance and reduced body weight loss in the second lactation. Thus, avoiding having unsuckled mammary glands in the first farrowing has a vital role in reducing the culling of young sows due to poor nursing capacity and is vital to improve colostrum and milk production (Kim et al., 2001).

13 Heat stress

Heat stress is detrimental to the pig's life from both a growth and reproductive performance perspective. Heat stress also causes a long-term impact during gestation, damaging both the sow and its offspring (Lucy and Safranski, 2017). In their study, gestating females under heat stress increased body temperature and respiration rate, reduced activity, and increased back fat at the end of gestation compared to sows in a thermoneutral environment. Heat stress during gestation increases insulin resistance (Lucy and Safranski, 2017), which can lead to a cascade of adverse effects such as lower lactation feed intake (Mosnier et al., 2010), reduced birth weight (Lucy and Safranski, 2017), and higher preweaning mortality (Safranski et al., 2015). Heat-stressing sows during gestation also resulted in offspring with elevated body temperature, increased fat deposition, and impaired gonad development (Lucy and Safranski, 2017). Some older studies found *in-utero* effects like reducing embryo viability, mainly when high environment temperatures occur in the first weeks of gestation (Omtvedt et al., 1971; Wildt et al., 1975). Collectively, these findings highlight the importance of promoting an appropriate environment (temperature and humidity) for gestating gilts and sows in multiplier farms.

14 Other environmental considerations

Although there is a relationship between reproductive activity and photoperiod, varying the intensity of lighting inside the barn, either in the follicular phase or early gestation, does not appear to affect a gilt's reproductive performance (Canaday et al., 2013). To our knowledge, there is limited information regarding air quality for gilts. However, both lighting and air quality must be thought of from a caretaker's point of view. For example, if people involved in managing replacement gilts are exposed to a dark barn with high concentrations of gases, time spent with the gilts will most likely be reduced. Also, the flooring condition must also be considered. Slippery floors or floors needing repair may result in injuries and losses of gilts.

15 Animal-human interaction

The development of the relationship between humans and pigs involves many senses, including hearing, visual, tactile, and chemical (Tallet et al., 2018).

However, the proportion of pigs attended per caretaker increased considerably over the years, reducing the time spent with each animal. Therefore, people involved with gilt management should make every interaction with replacement gilts a positive interaction. In other words, gilts should not be frightened by people when approached because fearful sows during gestation are more likely to savage their piglets in the subsequent lactation (Kraeling and Webel, 2015).

16 Conclusion

Proper replacement gilt development plays a role in offspring survival. Good gilt development promotes an ideal parity structure and having more older sows than gilts improves overall passive immunity creating more viable piglets. Furthermore, gilt development will influence mammary gland development which will play a role in colostrum and milk production over time, again influencing the viability of piglets. Even with gilts only representing 20% of active females in a herd, gilt management will dictate the performance and health status of the whole farm through her productive lifetime. Collectively, these points highlight the vital importance of developing replacement gilts to enhance offspring viability.

17 Conflict of interest statement

The authors declare no conflict of interest.

18 References

Amaral Filha, W. S., et al. 2009. Growth rate and age at boar exposure as factors influencing gilt puberty. *Livest. Sci.* 120:51–57. doi:10.1016/j.livsci.2008.04.009.

Amaral Filha, W. S., et al. 2010. Reproductive performance of gilts according to growth rate and backfat thickness at mating. *Anim. Reprod. Sci.* 121:139–144. doi:10.1016/j.anireprosci.2010.05.013.

Almeida, F. R. C. L., et al. 2015. Effects of birthweight on reproductive system development and onset of puberty in gilts. *Reprod. Fertil. Dev.* 29:254–261. doi:10.1071/RD15036.

Ashworth, C. J. and Antipatis, C. 1999. Effects of pre- and post-mating nutrition on embryo survival in gilts. *Reprod. Domest. Anim.* 34:103–108. doi:10.1111/j.1439-0531.1999.tb01226.x.

Ashworth, C. J., et al. 1999. Effects of pre- and post-mating nutritional status on hepatic function, progesterone concentration, uterine protein secretion and embryo survival in Meishan pigs. *Reprod. Fertil. Dev.* 11:67–73. doi:10.1071/rd99007.

Athorn, R. Z., et al. 2013. Effect of feeding level on luteal function and progesterone concentration in the vena cava during early pregnancy in gilts. *Reprod. Fertil. Dev.* 25:531–538. doi:10.1071/RD11295.

Bartol, F. F., et al. 2013. Lactation biology symposium: Lactocrine signaling and developmental programming. *J. Anim. Sci.* 91:696–705. doi:10.2527/jas.2012-5764.

Beltranena, E., et al. 1991. Effects of pre- and postpubertal feeding on production traits at first and second estrus in gilts. *J. Anim. Sci.* 69:886–893. doi:10.2527/1991.693886x.
Berndt, T. and Kumar, R. 2009. Novel mechanisms in the regulation of phosphorus homeostasis. *Physiology*. 24:17–25. doi:10.1152/physiol.00034.2008.
Bierhals, T., et al. 2011. Performance of litter after crossfostering of piglets between females of parity order 1 and 5. *Acta Sci. Vet.* 39:1–5.
Bortolozzo, F. P., et al. 2009. Growth, body state and breeding performance in gilts and primiparous sows. *Soc. Reprod. Fertil.* 66:281–291.
Bortolozzo, F. P., et al. 2006. Manejo para a indução da puberdade na leitoa. In: F.P. Bortolozzo and I. Wentz, editors, *Suinocultura em ação: A fêmea suína de reposição*. Ed Palotti, Porto Alegre, RS, Brazil. pp. 69–84.
Brumm, M. C. 2004. The effect of space allocation on barrow and gilt performance. *J. Anim. Sci.* 82:2460–2466. doi:10.2527/2004.8282460x.
Bruun, T., et al. 2021. The effects of long- or short-term increased feed allowance prior to first service on litter size in gilts. *Transl. Anim. Sci.* 5:1–12. doi:10.1093/tas/txab005.
Canaday, D. C., et al. 2013. Effect of variability in lighting and temperature environments for mature gilts housed in gestation crates on measures of reproduction and animal well-being. *J. Anim. Sci.* 91:1225–1236. doi:10.2527/jas.2012-5733.
Che, L., et al. 2017. Maternal nutrition modulates fetal development by inducing placental efficiency changes in gilts. *BMC Genom.* 18:213. doi:10.1186/s12864-017-3601-1.
Che, L., et al. 2019. Valine increases milk fat synthesis in mammary gland of gilts through stimulating Akt/mTOR/SREBP1 pathway. *Biol. Reprod.* 101:126–137. doi:10.1093/biolre/ioz065.
Che, L., et al. 2020. Effects of dietary valine supplementation during late gestation on the reproductive performance and mammary gland development of gilts. *J. Anim. Sci. Biotechnol.* 11:1–13. doi:10.1186/s40104-019-0420-z.
Che, L., et al. 2021. Mammary tissue proteomics in a pig model indicates that dietary valine supplementation increases milk fat content via increased de novo synthesis of fatty acid. *Nutr. Food Sci.* 9:6213–6223. doi:10.1002/fsn3.2574.
Cox, N. M., et al. 1987. Enhancement of ovulation rate in gilts by increasing dietary energy and administering insulin during follicular growth. *J. Anim. Sci.* 64:507–516. doi:10.2527/jas1987.642507x.
Davis, M. E., et al. 2006. Effect of weaning age and commingling after the nursery phase of pigs in a wean-to-finish facility on growth, and humoral and behavioral indicators of well-being. *J. Anim Sci.* 84:743–756. doi:10.2527/2006.843743x.
de Koning, D. B., et al. 2013. The influence of dietary restriction before and after 10 weeks of age on osteochondrosis in growing gilts. *J. Anim. Sci.* 91:5167–5176. doi:10.2527/jas.2013-6591.
Declerck, I., et al. 2016. Long-term effects of colostrum intake in piglet mortality and performance. *J. Anim. Sci.* 94:1633–1643. doi:10.2527/jas.2015-9564.
Devillers, N., et al. 2007. Variability of colostrum yield and colostrum intake in pigs. *Animal* 1:1033–1041. doi:10.1017/S175173110700016X.
Devillers, N., et al. 2016. Neonatal piglets are able to differentiate more productive from less productive teats. *Appl. Anim. Behav. Sci.* 174:24–31. doi:10.1016/j.applanim.2015.11.018.
Diehl, G., et al. 2003. Monitoramento ovariano ao abate de leitoas descartadas por anestro ou estro atípico. *Arch. Vet. Sci.* 3:121–125. doi:10.5380/avs.v8i1.4025.

Dirkzwager, A., et al. 1998. Effect of particle size and addition of sunflower hulls to diets on the occurrence of oesophagogastric lesions and performance in growing-finishing pigs. *Livest. Prod. Sci.* 56:53–60. doi:10.1016/S0301-6226(98)00143-2.

Douglas, S. L., et al. 2014. Management strategies to improve the performance of low birth weight pigs to weaning and their long-term consequences. *J. Anim. Sci.* 92:2280–2288. doi:10.2527/jas2013-7388.

Dwyer, C. M., et al. 1994. The influence of maternal nutrition on muscle fiber number development in the porcine fetus and on subsequent postnatal growth. *J. Anim. Sci.* 72:911–917. doi:10.2527/1994.724911x.

Fabà, L., et al. 2019. Effects of supplementing organic microminerals and methionine with or without limiting growth during the rearing phase of replacement gilts on lameness, growth, and body composition. *Transl. Anim. Sci.* 3:717–730. doi:10.1093/tas/txz038.

Faccin, J. E. G., et al. 2020. Impact of increasing weaning age on pig performance and belly nosing prevalence in a commercial multisite production system. *J. Anim. Sci.* 98:skaa031. doi:10.1093/jas/skaa031.

Faccin J. E. G., et al. 2023. Industry survey of added vitamins and trace minerals in U.S. swine diets. *Transl. Anim. Sci.* 7(1):txad035. doi:10.1093/tas/txad035.

Farmer, C., et al. 2004. Impacts of dietary protein level and feed restriction during prepuberty on mammogenesis in gilts. *J. Anim. Sci.* 82:2343–2351. doi:10.2527/2004.8282343x.

Farmer, C. 2018. Nutritional impact on mammary development in pigs: a review. *J. Anim. Sci.* 96:3748–3756. doi:10.1093/jas/sky243.

Farmer, C., et al. 2012. Milk production in sows from a teat in second parity is influenced by whether it was suckled in first parity. *J. Anim. Sci.* 90:3743–3751. doi:10.2527/jas.2012-5127

Ferrari, C. V., et al. 2014. Effect of birth weight and colostrum intake on mortality and performance of piglets after cross-fostering in sows of different parities. *Prev. Vet. Med.* 114:259–266. doi:10.1016/j.prevetmed.2014.02.013.

Ford, Jr. J. A., et al. 2003. Quantification of mammary gland tissue size and composition changes after weaning in sows. *J. Anim. Sci.* 81:2583–2589. doi:10.2527/2003.81102583x.

Foxcroft, G. R., et al. 2009. Prenatal programming of postnatal development in the pig. *Soc. Reprod. Fertil. Suppl.* 66:213–231.

Fu, L., et al. 2016. Stocking density affects welfare indicators of growing pigs of different group sizes after regrouping. *Appl. Anim. Behav. Sci.* 174:42–50. doi:10.1016/j.applanim.2015.10.002.

Gonçalves, M. A. D., et al. 2016. Fact sheet-Impact of increased feed intake during late gestation on reproductive performance of gilts and sows. *J. Swine Health Prod.* 24:264–266.

Gregory, N. 2021. *The effect of moderate energy and protein restriction during gilt development on body composition and subsequent lactation performance*. M.Sc. Thesis, Animal Bioscience Department, University of Guelph, Guelph, ON, Canada, 83 pp.

Guo, J. Y., et al. 2019. Effect of suckling intensity of primiparous sows on production performance during current and subsequent parities. *J. Anim. Sci.* 97:4845–4854. doi:10.1093/jas/skz341.

Head, R. H. and Williams, I. H. 1991. Mammogenesis is influenced by pregnancy nutrition. In: E. S. Batterham, editor, *Manipulating pig production III*. Australas. Pig Sci. Assoc., Werribee, VIC, Australia. p. 33.

Helm, E. T., et al. 2021. Evaluation of increased fiber, decreased amino acids, or decreased electrolyte balance as dietary approaches to slow finishing pig growth rates. *J. Anim. Sci.* 99:skab164. doi:10.1093/jas/skab164.

Howard, K. A. 1995. *Effect of porcine somatotropin, soybean oil and energy intake on mammary development in gestating gilts*. PhD Diss, Department of Animal Science, Michigan State University, East Lansing, MI, USA, 72 pp.

Johnson, R. K., et al. 2022. Effects of restricting energy during the gilt development period on growth and reproduction of lines differing in lean growth rate: responses in reproductive performance and longevity. *J. Anim. Sci.* 100:1–12. doi:10.1093/jas/skab352.

Ji, F., et al. 2006. Characterization of mammary gland development in pregnant gilts. *J. Anim. Sci.* 84:579–587. doi:10.2527/2006.843579x.

Jindal, J., et al. 1996. Effect of nutrition on embryonal mortality in gilts: association with progesterone. *J. Anim. Sci.* 74:620–624. doi:10.2527/1996.743620x.

Kim, S. W., et al. 2001. The regression of unsuckled mammary glands during lactation in sows: the influence of lactation stage, dietary nutrients, and litter size. *J. Anim. Sci.* 79:2659–2668. doi:10.2527/2001.79102659x.

Kim, S. W., et al. 1999a. Effect of nutrient intake on mammary gland growth in lactating sows. *J. Anim. Sci.* 77:3304–3315. doi:10.2527/1999.77123304x.

Kim, S. W., et al. 1999b. Changes in tissue composition associated with mammary gland growth during lactation in sows. *J. Anim. Sci.* 77:2510–2516. doi:10.2527/1999.7792510x.

Knauer, M. 2016. Effects of preweaning factors on sow lifetime productivity. National Pork Board Research Report #11–146.

Kraeling, R. R. and Webel, S. K. 2015. Current strategies for reproductive management of gilts and sows in North America. *J. Anim. Sci. Biotechnol.* 6:1–14. doi:10.1186/2049-1891-6-3.

Langendijk, P. and Peltoniemi, O. 2013. How does nutrition influence luteal function and early embryo survival? *Proc. Int. Conf. Pig. Reprod. IX*. 63:145–158.

Lavery, A., et al. 2018. An association analysis of sow parity, live-weight and back-fat depth as indicators of sow productivity. *Animal*. 13:622–630. doi:10.1017/S1751731118001799.

Leal, D. F., et al. 2019. Effects of post-insemination energy content of feed on embryonic survival in pigs: a systematic review. *Anim. Reprod. Sci.* 205:70–77. doi:10.1016/j.anireprosci.2019.04.005.

Lents, C. A., et al. 2020. Dietary lysine-to-energy ratios for managing growth and pubertal development in replacement gilts. *Appl. Anim. Sci.* 36:701–714. doi:10.15232/aas.2020-02016.

Lucy, M. C. and Safranski, T. J. 2017. Heat stress in pregnant sows: thermal responses and subsequent performance of sows and their offspring. *Mol. Reprod. Dev.* 84:946–956. doi:10.1002/mrd.22844.

Main, R. G., et al. 2004. Increasing weaning age improves pig performance in a multisite production system. *J. Anim. Sci.* 82:1499–1507. doi:10.2527/2004.8251499.

Magnabosco, D., et al. 2016. Low birth weight affects lifetime productive performance and longevity of female swine. *Livest. Sci.* 184:119–125. doi:10.1016/j.livsci.2015.12.008.

Magnabosco, D., et al. 2015. Impact of the birth weight of Landrace × Large White dam line gilts on mortality, culling and growth performance until selection for breeding herd. *Acta Sci. Vet.* 43:1–8.

Mallmann, A. L., et al. 2019a. Impact of feed intake during late gestation on piglet birth weight and reproductive performance: a dose-response study performed in gilts. *J. Anim. Sci.* 97:1262–1272. doi:10.1093/jas/skz017.

Mallmann, A. L., et al. 2019b. Maternal nutrition during early and late gestation in gilts and sows under commercial conditions: impacts on maternal growth and litter traits. *J. Anim. Sci.* 97:4957–4964. doi:10.1093/jas/skz349.

Mallmann, A. L., et al. 2020a. Impact of feed intake in early gestation on maternal growth and litter size according to body reserves at weaning of young parity sows. *J. Anim. Sci.* 98:skaa075. doi:10.1093/jas/skaa075.

Mallmann, A. L., et al. 2020b. Effects of flush feeding strategy before breeding on reproductive performance of modern replacement gilts: impacts on ovulation rate and litter traits. *J. Anim. Sci.* 98:skaa186. doi:10.1093/jas/skaa186.

McLamb, B. L., et al. 2013. Early weaning stress in pigs impairs innate mucosal immune responses to enterotoxigenic E. coli challenge and exacerbates intestinal injury and clinical disease. *PLoS One*. 8:e59838. doi:10.1371/journal.pone.0059838.

Medland, J. E., et al. 2016. Early life adversity in piglets induces long-term upregulation of the enteric cholinergic nervous system and heightened, sex-specific secretomotor neuron responses. *Neurogastroenterol. Motil.* 28:1317–1329. doi:10.1111/nmo.12828.

Moore C. 2001. Segregated production: how far could we go? In: *Proc. A. D. Leman Swine Conf.* pp. 203–206.

Mosnier, E., et al. 2010. The metabolic status during the peri partum period affects the voluntary feed intake and the metabolism of the lactating multiparous sow. *Livest. Sci.* 127:127–136. doi:10.1016/j.livsci.2009.06.023.

Mößeler, A., et al. 2012. Effect of diet grinding and pelleting fed either dry or liquid feed on dry matter and pH in the stomach of pigs and the development of gastric ulcers. *J. Anim. Sci.* 90:343–345. doi:10.2527/jas.53772.

Musser, R. E., et al. 2006. Effects of high feed intake during early gestation on sow performance and offspring growth and carcass characteristics. *Anim. Feed Sci. Technol.* 127.187–199. doi:10.1016/j.anifeedsci.2005.08.017.

NRC. 2012. *Nutrient requirements of swine* 11th ed. Natl. Acad. Press, Washington, DC.

Omtvedt, I. T., et al. 1971. Influence of heat stress during early, mid and late pregnancy of gilts. *J. Anim. Sci.* 32:312–317. doi:10.2527/jas1971.322312x.

Patterson, J., et al. 2002. Impact of boar exposure on puberty attainment and breeding outcomes in gilts. *Theriogenology*. 57:2015–2025. doi:10.1016/S0093-691X(02)00674-X.

Patterson, J. and G. Foxcroft. 2019. Gilt management for fertility and longevity. *Animals* 9:434. doi:10.3390/ani9070434.

Patterson, J., et al. 2018. A low litter birth weight phenotype reduces the retention rate of potential replacement gilts. *J. Anim. Sci.* 96(suppl.2):62. (Abstr.) doi:10.1093/jas/sky073.115.

Patterson, J., et al. 2020. Associations among individual gilt birth weight, litter birth weight phenotype, and the efficiency of replacement gilt production. *J. Anim. Sci.* 98:skaa331. doi:10.1093/jas/skaa331.

Piñeiro, C., et al. 2019. Influence of sows' parity on performance and humoral immune response of the offspring. *Porc. Health Manag.* 5:1–8. doi:10.1186/s40813-018-0111-8.

Pohl, C. S., et al. 2017. Early weaning stress induces chronic functional diarrhea, intestinal barrier defects, and increased mast cell activity in a porcine model of early life adversity. *Neurogastroenterol. Motil.* 29:13118. doi:10.1111/nmo.13118.

Prunier, A. and Quesnel, H. 2000. Nutritional influences on the hormonal control of reproduction in female pigs. *Livest. Prod. Sci.* 63:1–16. doi:10.1016/S0301-6226(99)00113-X.

Quesnel, H. and Farmer, C. 2019. Review: nutritional and endocrine control of colostrogenesis in swine. *Animal.* 13:26–34. doi:10.1017/S1751731118003555.

Quesnel, H., et al. 2010. Post-insemination level of feeding does not influence embryonic survival and growth in highly prolific gilts. *Anim. Reprod. Sci.* 120:120–124. doi:10.1016/j.anireprosci.2010.04.006.

Quinn, A. J., et al. 2015. The effect of feeding a diet formulated for developing gilts between 70kg and ~140kg on lameness indicators and carcass traits. *Livest. Sci.* 174:87–95. doi:10.1016/j.livsci.2014.12.016

Safranski, T. J., et al. 2015. Reproductive performance of gilts having developed in heat stressed dams. *J. Anim. Sci.* 93(Suppl 2):85 (Abstr.).

Sasaki, Y. and Koketsu, Y. 2008. Sows having high lifetime efficiency and high longevity associated with herd productivity in commercial herds. *Livest. Sci.* 118:140–146. doi:10.1016/j.livsci.2007.12.029.

Shelton N. W., et al. 2009. *Effects of increasing feeding level during late gestation on sow and litter performance.* Kansas State University, Agricultural Experiment Station and Cooperative Extension Service, Manhattan, KS.

Sorensen, M. T., et al. 2006. Mammary development in prepubertal gilts fed restrictively or ad libitum in two sub-periods between weaning and puberty. *Livest. Sci.* 99:249–255. doi:10.1016/j.livprodsci.2005.08.001.

Sorensen, M. T., et al. 2002. Mammary gland development in gilts. *Livest. Prod. Sci.* 75:143–148. doi:10.1016/S0301-6226(01)00310-4.

Sugiyama, T., et al. 2013. Effects of 25-hydroxy-cholecalciferol on the development of osteochondrosis in Swine. *Anim. Sci. J.* 84:341–349. doi:10.1111/asj.12000.

Tallet, C., et al. 2018. Pig-human interactions: creating a positive perception of humans to ensure pig welfare. In: M. Spinka, editor, *Advances in pig welfare*. Woodhead Publishing, Duxford, UK, p. 381–398.

Thomas, L., et al. 2021a. Effects of increasing standardized ileal digestible lysine during gestation on reproductive performance of gilts and sows. *Animal.* 15:100221. doi:10.1016/j.animal.2021.100221.

Thomas, L. L., et al. 2021b. Modeling standardized ileal digestible lysine requirements during gestation on gilts and sows. *Livest. Sci.* 248:104500. doi:10.1016/j.livsci.2021.104500.

Tokach, M. D., et al. 2019. Review: nutrient requirements of the modern high-producing lactating sow, with an emphasis on amino acid requirements. *Animal.* 13:2967–2977. doi:10.1017/S1751731119001253.

Vallet, J. L., et al. 2015. Relationships between day one piglet serum immunoglobulin immunocrit and subsequent growth, puberty attainment, litter size, and lactation performance. *J. Anim. Sci.* 93:2722–2729. doi:10.2527/jas.2014-8535.

Vier, C. M., et al. 2019. Calcium to phosphorus ratio requirement of 26-to 127-kg pigs fed diets with or without phytase. *J. Anim. Sci.* 97:4041–4052. doi:10.1093/jas/skz257.

Wang, C., et al. 2019. An analysis of culling patterns during the breeding cycle and lifetime production from the aspect of culling reasons for gilts and sows in Southwest China. *Animals*. 9:skab189. doi:10.3390/ani9040160.

Weldon, W. C., et al. 1991. Effects of increased dietary energy and protein during late gestation on mammary development in gilts. *J. Anim. Sci.* 69:194–200. doi:10.2527/1991.691194x.

Wildt, D. E., et al. 1975. Physiological temperature response and embryonic mortality in stressed swine. *Am. J. Physiol.* 229:1471–1475. doi:10.1152/ajplegacy.1975.229.6.1471.

Winkel, S. M., et al. 2018. Effects of energy restriction during gilt development on milk nutrient profile, milk oligosaccharides, and progeny biomarkers. *J. Anim. Sci.* 96:3077–3088. doi:10.1093/jas/sky212.

Wolter, B. F., et al. 2002. The effect of birth weight and feeding of supplemental milk replacer to piglets during lactation on preweaning and postweaning growth performance and carcass characteristics. *J. Anim. Sci.* 80:301–308. doi:10.2527/2002.802301x.

Wu, G., et al. 2004. Maternal nutrition and fetal development. *J. Nutr.* 134:2169–2172. doi:10.1093/jn/134.9.2169.

Young, M. G., et al. 2008. Effect of space allowance during rearing and selection criteria, on performance of gilts over three parities, in a commercial swine production system. *J. Anim. Sci.* 86:3181–3193. doi:10.2527/jas.2007-0600.

Chapter 9

Understanding ovarian function in pigs

María Estefanía González-Alvarez[1], *Zöe E. Kiefer*[1], *Jamie M. Studer*[1], *Jason W. Ross and Aileen F. Keating, Iowa State University, USA*

1 Introduction

This chapter summarizes the basic physiology of porcine female reproduction, including:

- ovarian structure and function;
- ovarian follicle development (folliculogenesis);
- the role of the corpus luteum (CL) in the establishment and maintenance of pregnancy;

1 Co-first authors.

http://dx.doi.org/10.19103/AS.2024.0137.13

- biosynthesis of steroid hormones (steroidogenesis); and
- the role of the hypothalamic–pituitary–gonadal (HPG) axis in regulating reproduction.

It also describes stages of the porcine estrous cycle as well as puberty and gilt development. The chapter goes on to discuss a range of environmental factors which can impair reproductive performance, including the effects of pesticide residues and mycotoxins. Finally, the chapter reviews research on the way heat stress can affect porcine female reproductive function.

2 The ovary: structure and function

The female reproductive system is comprised of the oviducts, uterus, cervix, vagina, ovaries, and external genitalia (Senger, 2012; Martini, 1998). The ovary is localized in the pelvic cavity near the lateral walls (Martini, 1998) and is composed of somatic and germ cells (Bahr and Milich, 2013). The two primary functions of the ovary are:

1. production and release of oocytes through the processes of oogenesis and folliculogenesis; and
2. production and secretion of hormones (17β-estradiol (E_2) and progesterone (P_4)) that are essential for the proper functioning of the female reproductive system (Senger, 2012; Hoyer, 2010; Keating and Hoyer, 2009) and female general health (Hoyer and Devine, 2001).

The tunica albuginea is an outer connective tissue that surrounds the ovary. Under the tunica albuginea is the ovarian cortex, where the majority of primordial follicle oocytes and the corpora lutea (CL) reside (Senger, 2012). The ovarian medulla is the central part of the ovary containing the major vasculature, nerves, and lymphatics (Senger, 2012). Oogenesis begins when the fetus is developing *in utero* (Martini, 1998; Bahr and Milich, 2013) and includes both mitosis and the onset of meiosis (Bahr and Milich, 2013).

In mammals, during embryonic development, primordial germ cells are formed and travel from the yolk sac to the indifferent gonad (Monk and McLaren, 1981; Borum, 1961). Once they are in the ovary, germ cells or oogonia undergo mitosis multiplying in number after losing their motile function and differentiating into oocytes during gestation when meiosis begins (Hirshfield, 1991; McGee and Hsueh, 2000; Wylie et al., 1986; Donovan et al., 1986). Ovarian cords are formed due to the interaction of oogonia with somatic cells (Hirshfield, 1991; Borum, 1961; Byskov, 1986; Guigon and Magre, 2006). Germ cell nests are then formed once the oocytes cluster together, becoming the precursors of primordial follicles that are formed once the germ cell nests break down (Hirshfield, 1991; Borum, 1961; Barnett et al., 2006), the timing of which is species-specific (Bahr and Milich, 2013; Barnett et al., 2006). The entire ovarian reserve of oocytes is produced during fetal development (Hirshfield, 1997; Faddy et al., 1987). Meiosis I resumes in the oocyte at the time of ovulation (Bahr and Milich, 2013; Hirshfield, 1991) and once completed, meiosis II begins; however, this is arrested in the second meiotic metaphase until fertilization occurs (Bahr and Milich, 2013; Hirshfield, 1991).

3 Follicle development: folliculogenesis

The process of folliculogenesis describes how immature oocyte-containing follicles develop and mature to be ovulated or die by atresia (Senger, 2012) and is illustrated in Figure 1. Primordial follicles are the most immature follicle stage present in the ovary (Senger, 2012). Primordial follicles are comprised of the oocyte surrounded by a single layer of flattened or squamous granulosa cells that are surrounded by a basal lamina (Bahr and Milich, 2013; Senger, 2012; Barnett et al., 2006). Local ovarian factors activate primordial follicles to develop into primary follicles, and an increase in the oocyte size is observed

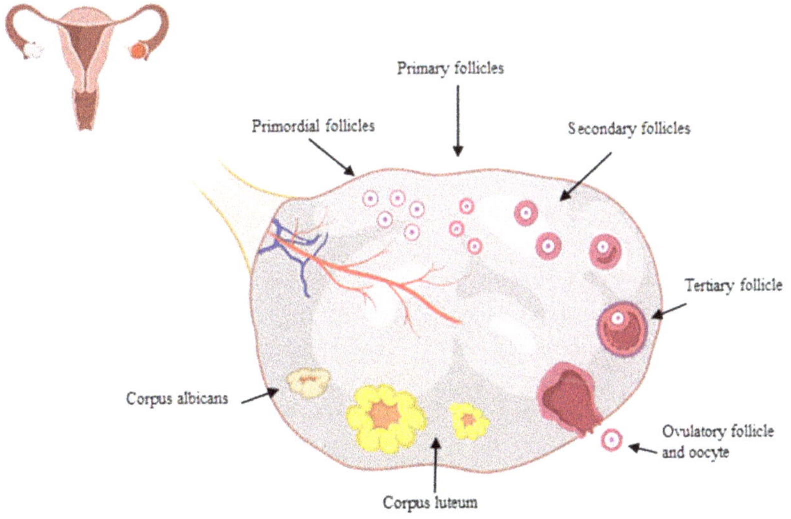

Figure 1 Ovarian folliculogenesis. The ovary contains a pool of primordial follicles surrounded by a layer of flattened granulosa cells. Primordial follicles are activated by ovarian factors to grow into primary follicles with bigger oocyte and cuboidal granulosa cells. Primary follicles continue growing until they have two or more layers of granulosa cells and a new layer of theca cells; they are then referred to as secondary follicles. Tertiary follicles are subsequently formed when the antrum is filled with follicular fluid. Ovulation happens after an LH surge. The cumulus–oocyte complex is released. The remaining granulosa and theca cells form the corpus luteum, which then becomes a corpus albicans and finally will disappear. Most follicles will undergo atresia. Created with BioRender.com

(Martini, 1998; Barnett et al., 2006). The flattened granulosa cells surrounding the oocyte become cuboidal granulosa cells (Bahr and Milich, 2013; Senger, 2012; Barnett et al., 2006). The zona pellucida, a non-cellular layer, also appears at this stage of follicular development, surrounding the oocyte, and it is preserved until ovulation (Bahr and Milich, 2013; Barnett et al., 2006). Granulosa cells are essential for the nutrition of the oocyte, synthesizing nutritive material that is transported to the oocyte by diffusion through the zona pellucida (Gondos, 1970).

Secondary follicles develop from primary follicles and are referred to as pre-antral follicles, containing multiple layers of granulosa and theca cells (Bahr and Milich, 2013). The newly recruited theca cell layer contains blood vessels and nerves, and provides support for the follicle (Bahr and Milich, 2013). Secondary follicles also have a network of gap junctions, consisting of connexins proteins which are necessary for follicular development, and connect adjacent cells allowing nutrients, small metabolites, and second messengers to pass from cell to cell (Barnett et al., 2006). Only a portion of follicles proceed to the next developmental stage when they are referred to as tertiary or antral follicles, which are dependent upon follicle-stimulating hormone (FSH) (Bahr and Milich, 2013; Barnett et al., 2006). Follicular fluid is formed from filtered blood circulating in the thecal capillaries (Rodgers and Irving-Rodgers, 2010) and accumulates to separate the inner and outer layers of the follicle to form an antral cavity (Martini, 1998; Bahr and Milich, 2013; Barnett et al., 2006). Follicular fluid is composed of electrolytes, water, serum proteins, and steroid hormones (Barnett et al., 2006).

Most of these follicles will undergo atresia; the remaining will grow to the preovulatory stage (Barnett et al., 2006). Described more fully in the next section, the preovulatory follicles produce E_2, which rises in concentration, resulting in positive feedback on the hypothalamus and pituitary to precipitate the luteinizing hormone (LH) surge, critical for ovulation of the oocyte (Barnett et al., 2006; Bahr and Milich, 2013). The LH surge will cause a decrease in E_2 production and an increase in P_4 secretion (Bahr and Milich, 2013). The remaining granulosa and theca cells luteinize to form the ovarian CL.

4 The corpus luteum

The CL is a temporary endocrine structure formed from the ruptured follicle following ovulation. The presence of functional CL is essential to the establishment and maintenance of pregnancy until parturition (Ziecik et al., 2018). The CL is a highly efficient gland that produces large amounts of P_4, responsible for providing an intrauterine environment that supports implantation, placentation, and fetal–placental growth and development (Meidan, 2016). The life cycle of the CL has three phases during the estrous cycle, formation, maintenance, and regression (Ziecik et al., 2018). In the event of pregnancy, a fourth phase, rescue and sustained function during pregnancy, ensues (Meidan, 2016). Mammalian CL can be divided into three categories based on the length of post-ovulation CL function and the effect pregnancy has on the lifespan of the CL (Stouffer and Hennebold, 2015). Many polyestrous species (including pigs) have short-lived CL, in which the CL develops and functions for a finite length of time during the ovarian cycle, but the lifespan of the CL is substantially increased if pregnancy occurs (Stouffer and Hennebold, 2015; Rothchild, 1981).

The development of an intricate network of blood vessels via the process of angiogenesis is crucial for the development and function of the CL (Robinson and Woad, 2017). As luteal formation progresses, LH stimulates several angiogenic processes (Garrido et al., 1993), where factors such as vascular endothelial growth factor (VEGF) facilitate the formation of the capillary network of the corpus hemorrhagicum, the early luteal structure (Senger et al., 1997; Schams and Berisha, 2004). Additionally, LH initiates differentiation of theca and granulosa cells from the ovarian follicle into the small and large luteal cells forming the CL, inducing the steroidogenic shift from E_2 to P_4 production, and eventual intense remodeling of follicular tissue into the fully developed CL (Robinson and Woad, 2017). Luteal steroidogenesis relies on the formation of this dense capillary network that enables luteal cells to obtain nutrients needed to synthesize and release substantial quantities of P_4 required for the establishment and maintenance of pregnancy (Fraser and Wulff, 2003).

5 Ovarian steroidogenesis and steroid hormone changes during late gestation

Steroidogenesis is defined as the biosynthesis of steroid hormones from cholesterol through a series of enzymatic steps as summarized in Fig. 2. Cholesterol can be obtained from the bloodstream or synthesized *de novo* in the ovary. The enzymatic steps required for ovarian steroidogenesis are split between granulosa and theca cells that form the follicle (Miller and Auchus, 2011). Steroid hormone patterns vary throughout the cycle, with E_2 production occurring predominantly during the follicular phase and P_4 production primarily occurring during the luteal phase. Production of E_2 by preovulatory follicles during the follicular phase requires both granulosa and theca cells since neither cell has the complete set of enzymes necessary for E_2 synthesis. For this reason, this process is known as the two-cell two-step theory of steroidogenesis. Both cell types are stimulated by gonadotropins; FSH acts on granulosa cells while LH mainly acts on theca cells, however, granulosa cells also express LH

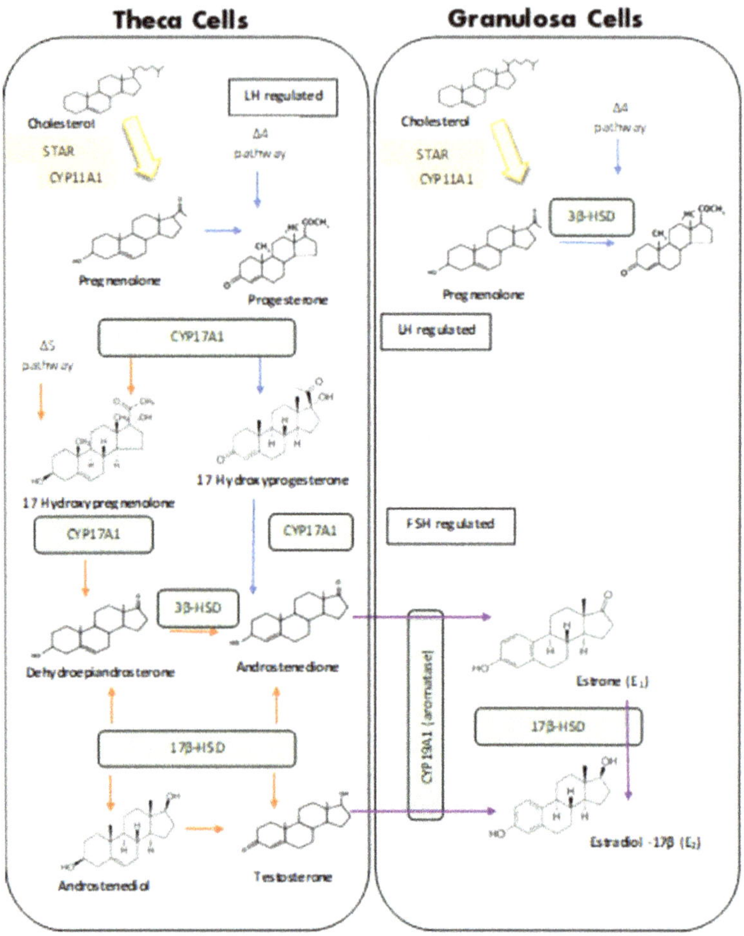

Figure 2 Two-cell two-step theory of ovarian steroid hormone production. Ovarian steroid hormones are produced from cholesterol in granulosa and theca cells. LH and FSH bind to the receptors initiating a signal cascade. Cholesterol is transported to the inner mitochondrial matrix by StAR, where it is converted to pregnenolone. Pregnenolone can be converted to progesterone via the Δ4 pathway and further transformed into androgens and estrogens or via the Δ5 pathway, where it is changed to DHEA and finally transformed into estrogens. The conversion from androgens to estrogens happens in granulosa cells by CYP19A1.

receptors in antral follicles at later stages of development (Erickson et al., 1979; Méduri et al., 2002).

The first step which is rate limiting in the steroidogenic pathway, is the transport of cholesterol into the inner mitochondrial matrix by steroid acute regulatory protein (StAR) (Christenson and Devoto, 2003). This step can occur in both cell types. Theca cells are near the blood vessels and can uptake low-density lipoprotein (LDL) cholesterol from the bloodstream in response to stimulation from LH (Jones, 2009). Conversely, granulosa cells are far from blood vessels and must obtain cholesterol through *de novo* synthesis.

Once cholesterol is transported into the inner mitochondrial matrix by StAR, it is converted into pregnenolone by cholesterol side chain cleavage enzyme (CYP11A1). In the theca cell, pregnenolone diffuses out of the mitochondria to the smooth endoplasmic reticulum and is converted into dehydroepiandrosterone by cytochrome P450 17A1 (CYP17A1) and then into androstenedione by 3β-hydroxysteroid dehydrogenase (3βHSD). Theca cells lack the enzyme cytochrome P450 19A1 (CYP19A1; also known as aromatase), needed to convert androstenedione into E_2, so androstenedione diffuses through the basement membrane into the granulosa cell (Jones, 2009). In the granulosa cell, androstenedione is converted into testosterone by 17β-hydroxysteroid dehydrogenase (17βHSD) and then into E_2 by CYP19A1 (Jones, 2009).

After ovulation, theca and granulosa cells from the ruptured follicle differentiate into theca-lutein (small luteal cells), and granulosa-lutein cells (large luteal cells), respectively, to form the CL (Pitzel et al., 1990). During the luteal phase, P_4 is produced by both small and large luteal cells and synthesis begins with transport of cholesterol into the inner mitochondrial matrix by StAR (Christenson and Devoto, 2003). Cholesterol for luteal steroidogenesis can be obtained from multiple sources: *de novo* synthesis from acetate, circulating lipoproteins, and intracellular stores such as cholesteryl esters stored in lipid droplets and cholesterol within the plasma membrane (Gwynne and Strauss, 1982). The major source of cholesterol for luteal cells is high-density lipoproteins (HDL) and LDL in circulation (Grummer and Carroll, 1988; Wiltbank et al., 1990). Only two enzymatic steps are required for P_4 biosynthesis; the conversion of cholesterol into pregnenolone by CYP11A1 followed by conversion into P_4 catalyzed by 3βHSD in the smooth endoplasmic reticulum (Christenson and Devoto, 2003). Progesterone then diffuses out of the luteal cell and into the bloodstream for transport to target tissues.

5.1 Steroid hormone changes during late gestation

Steroid hormones play a crucial role in female health, and there are many changes in hormones during pregnancy and leading up to parturition. In sows, P_4 produced by the CL is necessary for pregnancy establishment and maintenance for the entire duration of pregnancy (Senger, 1997). It has been observed that P4 within circulation begins a slight decline approximately 5 days pre-partum with a rapid decline just before farrowing (Molokwu and Wagner, 1973; Baldwin and Stabenfeldt, 1975). Levels do not decrease further until after farrowing, but plateau to a low concentration during lactation (Molokwu and Wagner, 1973; Baldwin and Stabenfeldt, 1975). Estrogen levels are inversely related to those of P_4 as a steady increase in E_2 has been observed approximately 6–11 days prior to farrowing, which is followed by a dramatic drop in concentration during farrowing, then remains low throughout lactation (Molokwu and Wagner, 1973; Baldwin and Stabenfeldt, 1975).

6 The hypothalamic-pituitary-gonadal axis

The HPG axis is a complex system of positive and negative feedback between reproductive organs and the brain and is a key component of sexual maturation in mammalian species (Fig. 3). Regulation of reproduction begins in the brain at the level of the hypothalamus where GnRH, localized in the arcuate nucleus and pre-optic area, is synthesized and released in a pulsatile fashion into the hypothalamic–hypophysial–portal vein to the anterior pituitary (Clarke and Cummins, 1982; Klein, 2003).

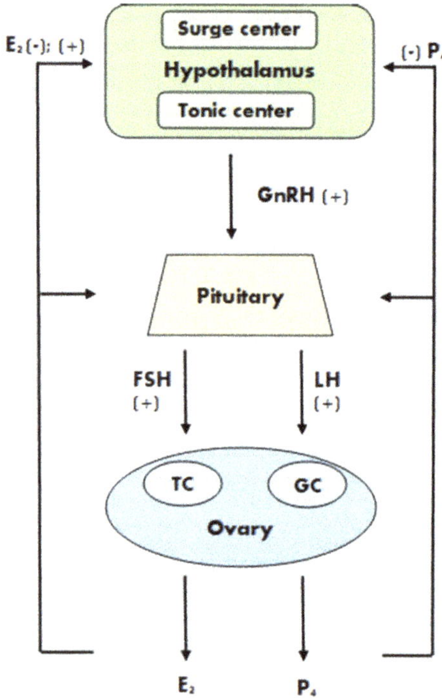

Figure 3 Hypothalamic–pituitary–ovarian axis. GnRH is produced in the hypothalamus and induces the production and secretion of the gonadotropins, FSH, and LH. In the granulosa (GC) and theca cells (TC), FSH and LH induced the production of E_2 or P_4 depending on the stage of the menstrual cycle. E_2 has positive feedback to the surge center at high concentrations and negative feedback to the tonic center at low concentrations. P_4 negatively feeds back to the hypothalamus.

The binding of GnRH to receptors on gonadotrophs in the anterior pituitary results in the synthesis and secretion of the gonadotropins LH and FSH. Basal GnRH pulsatile release from neurons in the arcuate nucleus is negatively regulated by tonic levels of E_2 and P_4 which maintains levels of LH in serum (Richards, 2018). Steroid hormones act on kisspeptin neurons in the arcuate nucleus to regulate basal LH release (Herbison, 2016). On the other hand, increased E_2 from preovulatory follicles causes increased pulsatile GnRH release from neurons in the pre-optic area leading to increased pulsatile release and the ovulatory surge of LH from the anterior pituitary (Richards, 2018). Estrogen-mediated action on kisspeptin neurons in the pre-optic area controls the surge mode of LH release (Herbison, 2016). Progesterone also acts on kisspeptin neurons in the arcuate and pre-optic areas to negatively regulate both the basal and surge modes of GnRH and LH release (Herbison, 2016). The release of FSH from the anterior pituitary is also controlled by E_2 and P_4 regulation of GnRH release (Richards, 2018).

The gonadotropins travel to the ovary through systemic circulation where they interact with receptors on granulosa and theca cells. The LH receptor is expressed on theca cells while the FSH receptor is expressed on granulosa cells of the developing follicle. On the days leading up to ovulation, the LH receptor is also expressed on granulosa cells to prepare the follicle for the LH surge, ovulation, and eventually luteinization for the formation of the CL (Erickson et al., 1979; Méduri et al., 2002).

7 The estrous cycle in swine

The estrous cycle refers to the reoccurring physiological changes induced by hormonal changes that occur cyclically in females and enable fertility. The synthesis and secretion of the ovarian steroid hormones 17β-estradiol (E_2) and progesterone (P_4) regulate cyclicity and facilitate the female to produce offspring.

Pigs achieve sexual maturity at 5 to 8 months of age and are sexually receptive during all times of the year due to spontaneous ovulation and polyestrous cycles (Anderson, 2000). The estrous cycle in pigs can be classified into two phases:

- the follicular phase; and
- the luteal phase.

In pigs, the estrous cycle length ranges from 18 to 24 days, with an average of 21 days (Soede et al., 2011).

7.1 Follicular phase

The follicular phase consists of two stages, proestrus and estrus, and lasts 4–6 days (Soede et al., 2011). During proestrus (days 17–21), P_4 is low due to

CL regression while E_2 increases steadily as growing follicles reach maturity and become antral follicles. Increasing concentrations of E_2 and the reduced negative feedback effect of E_2 on the basal medial nuclei of the hypothalamus allows for the chronic release of GnRH into the hypophyseal portal system causing the pulsatile release of FSH. Estrogen levels peak (~45 pg/ml) during estrus (days 0–2), the time when the gilt or sow is sexually receptive to the boar. In pigs, this is characterized by the 'standing response' that occurs when a boar is present. The sow stands immobilized, arches her back (also referred to as 'locking up'), and allows the boar to mount her (Soede et al., 2011). Changes to the vulva are also observed during estrus, including redness, swelling, and production of mucus or discharge. Estrus typically lasts between 48 and 72 h in pigs with ovulation occurring approximately 36 h after onset of estrus, although estrus length can range from 24 to 96 h (Soede et al., 2011) and timing of ovulation relative to the onset of estrus can be quite variable.

Peak E_2 levels result in positive feedback regulation on the pre-optic nuclei of the hypothalamus resulting in the LH surge which causes the luteinization of the theca interna and granulosa cells resulting in the degradation of the junctional complexes holding the granulosa cells together as well as the oocyte's resumption of meiosis and the expulsion of a polar body. Continued proteasomal degradation of the theca interna outward near the apex of the follicle results in follicular rupture and expulsion of the oocyte. The ovulation of 16–18 oocytes is representative of a follicular pool of as many as 85 tertiary follicles present on the surface of the ovary during the mid-luteal phase of the estrous cycle (Guthrie et al., 1995).

7.2 Luteal phase

After ovulation and following estrus, the gilt/sow is in the luteal phase of the estrous cycle which consists of metestrus and diestrus. The luteal phase is characterized by extensive reorganization of tissue resulting in the formation of fully functional CL by day 4 or 5 of the estrous cycle (Guthrie et al., 1972). Metestrus (days 2–5) is characterized by declining levels of E_2 while P_4 levels rise in response to the formation of CL from the granulosa and theca cells of ruptured follicles, a process known as luteinization. The final stage of the estrous cycle, diestrus (days 5–16), is characterized by high concentrations of P_4 produced from fully functional CL and low concentrations of E_2. The CL reaches peak P_4 production by days 8–12 of diestrus, with levels in circulation ranging from 20 ng/ml to 40 ng/ml (Stabenfeldt et al., 1969; Henricks et al., 1972; Anderson, 2009; Soede et al., 2011). In the absence of pregnancy, the uterus secretes prostaglandin $F_{2\alpha}$ ($PGF_{2\alpha}$) into the bloodstream resulting in regression of the CL (a process known as luteolysis) and reduction of circulating P_4 by day 16 of the estrous cycle, thus allowing for recruitment and maturation of the next group of developing follicles (Moeljono et al., 1977; Ziecik et al., 2017).

Alternatively, if fertilization does occur, the conceptus synthesizes and secretes multiple factors from days 12 to 17 of gestation to establish pregnancy by supporting and maintaining the CL past the length of a normal estrous cycle (Geisert et al., 2021). Traditionally, E_2 produced by the elongating conceptus has been considered the signal for the maternal recognition of pregnancy in pigs, causing $PGF_{2\alpha}$ to be released into the uterine lumen rather than into circulation, thereby preventing luteolysis (Perry et al., 1973, Bazer et al., 1986, Geisert and Bazer, 2015). However, it has since been discovered that conceptus production of interleukin 1 beta 2 (IL1B2), prostaglandins (PG), and interferons (IFN) are critical factors in conceptus development and survival (Geisert et al., 2021). Conceptus expression of IL1B2 activates a cascade of events including E_2 secretion by the conceptus, endometrial and conceptus PGE2 production, and conceptus expression of IFND and IFNG, needed to prevent the maternal immune system from rejecting the developing conceptus (Geisert et al., 2021). In the pig, the persistence of the CL throughout gestation is essential for the maintenance of pregnancy, as the placenta does not produce sufficient P_4 to maintain pregnancy alone (Ziecik et al., 2011).

8 Puberty and gilt development

In a sow's reproductive lifetime, several critical events precede inclusion into the sow herd beginning with puberty onset. Puberty is defined as the biological changes associated with sexual maturation and development enabling a female of being capable of reproduction. On average, it takes 5–8 months for a gilt to reach puberty (Graves et al., 2019), and in commercial settings, gilts are typically 6–8 months of age when entering the breeding pool with the objective of producing her first litter within a year of age. Vulva width has been correlated to a gilt's ability to achieve behavioral estrus by 200 days of age and is an indicator of a sow's lifetime productivity (Graves et al., 2019; Romoser et al., 2020; 2019).

Achievement of puberty is controlled by steroid hormones and neurotransmitters, which induce the GnRH (Delemarre-van de Waal, 2002). Kisspeptin, a neuropeptide, initiates GnRH release by acting directly on the hypothalamic neurons (Seminara and Crowley, 2008). Disruption in kisspeptin signaling delays puberty onset due to its regulation of GnRH (de Roux et al., 2003). The secretion of GnRH occurs in a pulsatile fashion and involves multiple feedback mechanisms of inducing and suppressive signals (Navarro et al., 2009; Livadas and Chrousos, 2016). Once experiencing their first estrous cycle, the gilt is considered sexually mature, and being a polyestrous animal, the reproductive cycle repeats approximately every 21 days.

Following pregnancy and parturition, sows in most commercial U.S. herds will nurse their piglets until weaned at 18–24 days of age. At this time, piglets are transferred from the sow farm and raised at a secondary location to minimize the risk of disease spread. Most weaned pigs from commercial farms are reared until reaching market weight and then harvested. Gilts produced from genetic nucleus and multiplier farms are reared in separate production facilities (i.e. a gilt development unit) until being transferred back to the sow farm to enter the breeding herd. A sow's profitability is dependent on her reproductive success and longevity (Stalder et al., 2003). The earlier a gilt can obtain puberty, the more productive she tends to be through two parities (Romoser et al., 2020).

9 Environmental factors impairing swine reproduction: pesticide residues

The impacts of pesticides have been investigated in both production and wild pigs to understand impacts of agriculturally used chemicals on swine reproduction. Within the ovary, direct ovarian impacts of a toxicant largely depend upon the cell type targeted and these are described separately below.

9.1 Follicular impacts of pesticide exposure

There are follicular impacts of pesticide exposures on the pig ovary and these exposures can be via ingestion, inhalation, or absorption (Laws et al., 2021). Exposure to testosterone propionate, flutamide, 4-tert-octylphenol, the E_2 antagonist ICI 182, or methoxychlor from postnatal days 1–10 determined that testosterone propionate reduced the percentage of oocytes that were present within germ cell nests and increased the abundance of the pro-atretic active caspase 3 – potentially as a mode of action. Exposure to flutamide reduced ovarian cellular proliferation and altered the ratio of primordial to primary follicles while conversely, exposure to 4-tert-octylphenol increased the percentage of primordial and developing follicles while E_2 antagonism decreased primordial follicle activation and reduced ovarian cellular proliferation (Knapczyk-Stwora et al., 2018). Exposure of pigs to the pesticide methoxychlor reduced the developing follicle number and impacted the porcine ovarian transcriptome potentially interfering with folliculogenesis (Knapczyk-Stwora et al., 2022). Thus, there are direct impacts on ovarian folliculogenesis that could impair cell viability and ovarian hormone production.

9.2 Granulosa cell and endocrine impacts of pesticide exposure

The effect of pesticide exposure can also impair granulosa cell viability and steroidogenic function. The fungicide vinclozolin causes granulosa cell apoptosis through pro-atretic caspase-3 and caspase-9 (Knet et al., 2015), while atrazine impairs steroidogenesis and angiogenesis (Basini et al., 2012). DNA damage and oxidative stress were induced due to malathion and diazinon exposure (Wang et al., 2018; 2019) while cell death was also observed with malathion exposure (Wang et al., 2018). Additionally, in granulosa cells from pigs, methoxychlor exposure decreased P_4 synthesis indicating a steroidogenic impact (Jorge Chedrese and Feyles, 2001).

Other evidence of steroidogenic effects of pesticides are from exposure to the organochlorine hexachlorobenzene which inhibited E_2 secretion and reduced CYP isoform 11, CYP 17, 17β-hydroxysteroid dehydrogenase (HSD), and CYP 19 protein levels in porcine follicles (Kamarianos et al., 2003; Gregoraszczuk et al., 2011). Also, exposure to dichlorodiphenyltrichloroethane and dichlorodiphenyldichloroethylene disrupted porcine steroidogenesis and CYP 19 activity in the steroidogenic granulosa and theca cells (Wójtowicz et al., 2007).

9.3 Oocyte effects of pesticide exposures

Direct oocyte effects could impair female germline viability and maturation. Exposure to atrazine inhibited the maturation of porcine oocytes (Casas et al., 2010) and decreased porcine oocyte quality by inducing DNA damage (Yuan et al., 2017). Also, exposure to paraquat reduced cumulus cell expansion and the rate of first polar body extrusion in cumulus–oocyte complexes (Zhou et al., 2022). Porcine oocyte developmental competence was negatively affected by exposure to the organophosphorus pesticide, chlorpyrifos, resulting in reduced blastocyst cleavage and MII oocyte maturation and misaligned chromosomes, with accompanying alterations to messenger RNA abundance of genes encoding cell cycle, the response to oxidative stress and apoptosis (Jiang et al., 2021).

Malathion, another organophosphate pesticide, decreased first polar body extrusion, disrupted spindle assembly and chromosome alignment, in porcine oocytes, and also increased the level of reactive oxygen species and altered levels of genes involved with energy metabolism, apoptosis, and lipid metabolism (Chen et al., 2020). Malathion exposure also altered antioxidant enzyme levels leading to oxidative stress in the cumulus–oocyte complex in pigs (Flores et al., 2017). Impaired viability and maturation of porcine oocytes (Casas et al., 2010) and reduced cell viability were consequences of exposure to the organophosphate, diazinon (Wang et al., 2019). Acetamiprid and imidacloprid, two neonicotinoid insecticides, impaired nuclear maturation and formation of normal chromosomes in porcine oocytes (Ishikawa et al., 2015). *In vitro* exposure to pyrethroid chemicals impaired the meiotic maturation of porcine oocytes (Petr et al., 2013). Thus, exposure to pesticides could affect reproduction in swine possibly leading to infertility.

10 Environmental factors impairing swine reproduction: mycotoxins

10.1 Zearalenone

One of the most extensively studied group of chemicals relevant to swine reproduction is mycotoxins. Zearalenone (ZEA) is a mycotoxin produced by *Fusarium* fungal species, which has nonsteroidal estrogenic properties (Hueza et al., 2014), and is an endocrine disruptor. Negative impacts of ZEA exposure through contaminated feed on female reproduction and specifically in swine, have been recognized for decades (Chang et al., 1979). Chemical biotransformation is the process by which chemicals are enzymatically transformed into metabolites to facilitate excretion from the body (Parkinson et al., 2013; Rourke and Sinal, 2014). Biotransformation of ZEA occurs via hepatic 3α- and 3β-hydroxysteroid dehydrogenases to produce the estrogenic metabolites α-zearalenol (α-ZOL) and β-zearalenol (β-ZOL) (Olsen et al., 1981) which are then conjugated with glucuronic acid by uridine diphosphate glucuronyl transferases for excretion (Olsen et al., 1981). Prepubertal gilts are considered the most susceptible to ZEN exposure due to their bioactivation of ZEA to α-ZOL (Liu et al., 1985).

Acute and chronic ZEA exposure causes reproductive abnormalities in animals manifested as pseudopregnancy (Etienne and Jemmali, 1982; Young and King, 1986), aberrant estrus display (Blaney et al., 1984; Edwards et al., 1987b,a), lengthened estrous cycle (Edwards et al., 1987a,b; Flowers et al., 1987), altered puberty onset (Edwards et al., 1987a, b), reduced meiotic competence (Alm et al., 2006), embryonic loss (Long et al., 1983), and developmental effects in offspring (Long et al., 1992). ZEA is detectable in follicular fluid (Sambuu et al., 2011) supporting the possibility for direct oocyte exposure. Histologically, lipid-filled, atretic ovarian follicles occur in ovaries from pre-pubertal gilts exposed to ZEA (Zwierzchowski et al., 2005). Further, increased oxidative stress evidenced by increased production of reactive oxygen species and reduced abundance of antioxidative stress enzymes in granulosa cells is documented (Qin et al., 2015). Gestational ZEA exposure is a concern since it negatively affects follicular integrity in piglets indicating potential transgenerational concerns for ovarian function due to in utero ZEA exposure (Schoevers et al., 2012). In addition to the concerns for ovarian function, other effects caused by hyper-estrogenism have been observed in piglets after *in utero* exposure (Ványi et al., 1994).

10.2 Reproductive toxicity of other mycotoxins

Reproductive toxicity of mycotoxins is not limited to ZEA, and other mycotoxins can also cause negative reproductive effects in swine, however, the mechanistic details are more limited. Deoxynivalenol (DON) and T-2 toxin are *Fusarium*-produced mycotoxins (EFSA Panel on Contaminants in the Food Chain (CONTAM), 2011, 2017) and ingestion of DON contaminated feed can lead to acute gastrointestinal symptoms and induce acute vomiting in pigs (EFSA Panel on Contaminants in the Food Chain (CONTAM), 2017). Exposure to DON impairs porcine oocyte maturation, inhibits cumulus cell expansion, and induces cumulus cell death, reducing metaphase II and causing malformations of the meiotic spindle in porcine ovaries (Alm et al., 2002; Schoevers et al., 2010). Post-exposure to DON, ovaries of 5-month-old pigs had decreased healthy follicle number and increased numbers of damaged oocytes (Gerez et al., 2017). T-2 is also known to cause negative reproductive effects in swine and exposure during the last third of gestation caused ovarian degeneration and atrophy (Kanora and Maes, 2009) and T-2 exposure in pregnant sows can result in abortion (Weaver et al., 1978). T-2 exposure also induced failure of polar body extrusion, altered meiotic spindle formation and actin distribution, and caused oxidative stress, apoptosis and autophagy in porcine oocytes (Zhang et al., 2016). In addition to *Fusarium*-produced mycotoxins, ergot alkaloids produced by Claviceps species can impair swine reproduction causing oxytocin-resistant agalactia, small-sized litters, premature farrowing, mummification, impaired estrus display, metritis, and mastitis (Kanora and Maes, 2009). Exposure of porcine oocytes to aflatoxin B1, produced by Aspergillus species, negatively altered oocyte maturation, and decreased polar body extrusion rate, with most oocytes being arrested at germinal vesicle breakdown or meiosis I (Liu et al., 2015). Aflatoxin exposure also induced oxidative stress, and cell death in porcine oocytes (Liu et al., 2015), and supplementation with melatonin had the capacity to provide protection of porcine oocytes from the aflatoxin-induced reduced developmental competence and oxidative stress induction (Cheng et al., 2019).

11 Environmental factors impairing swine reproduction: other endocrine-disrupting chemicals

In addition to pesticides and mycotoxins, there are several additional reproductive toxicants that have been determined to alter porcine ovarian function and exposure could arise through medicinal exposure, food containers, or food contamination. Bisphenol A (BPA), used in the production of plastics and 4-chloro-3-methyl phenol (CMP), a preservative added to skin care products and pharmaceuticals, impair oocyte meiotic maturation and also decrease the number of oocytes undergoing germinal vesicle breakdown and the rate of oocytes at metaphase II stage (Mlynarcíková et al., 2009; Wang et al., 2016). Impaired cumulus expansion of FSH-stimulated cumulus-oocyte-complexes was also a consequence of BPA, CMP, and benzyl butyl phthalate (BBP) exposures (Mlynarcíková et al., 2009). In addition, BPA exposure reduced polar body extrusion (Wang et al., 2016), disrupted spindle morphologies and misaligned chromosomes, decreased p-MAPK, H3K4me2 and 5mC levels, and increased early apoptosis and autophagy markers (Wang et al., 2016) in porcine oocytes.

Acting as endocrine disruptors, exposure to BPA and di(2-ethylhexyl) phthalate (DEHP) altered P_4 production in FSH-stimulated cumulus-oocyte complexes (Mlynarcíková et al., 2009). Similarly, BPA and bisphenol A-dimethacrylate (DMA) altered P_4 and E_2 production in porcine granulosa cells *in vitro* (Grasselli et al., 2010; Mlynarcíková et al., 2005). Pentachlorobenzene which has been used as a fire retardant and in the synthesis of pesticides (Liu, 2014), alters *in vitro* secretion of testosterone, P_4 and E_2 and CYP 11, CYP 17, and CYP 19 protein abundance in porcine follicles (Gregoraszczuk et al., 2011).

12 Environmental factors impairing swine reproduction: heat stress

12.1 Physiological effects of heat stress

Heat stress (HS) occurs when environmental conditions (ambient temperature, relative humidity, and solar radiation) coupled with endogenous heat production generate a thermal condition culminating in increased core body temperature (Bernabucci et al., 2010). HS negatively impacts pig production by altering production traits such as lactation, growth, and body composition (Baumgard and Rhoads, 2013). The studies described below have included both constant and cyclical exposure to HS as well as a variety of HS durations.

To attempt to maintain euthermia, HS animals reduce feed intake to minimize the thermic effect of digestion. Despite reduced nutrient consumption, circulating insulin is increased during HS and this is especially apparent when compared to thermal-neutral animals on a similar plane of nutrition (Victoria Sanz Fernandez et al., 2015). Blood flow to the splanchnic bed of the gastrointestinal tract is reduced to maintain blood pressure (Mayorga et al., 2018; Rowell et al., 1968) with loss of intestinal barrier integrity (Glover et al., 2016) ensuing (Al-Attas et al., 2009; Amar et al., 2008; Hawkesworth et al., 2013) causing metabolic endotoxemia (Cani et al., 2007; Luche et al., 2013) and inflammation.

12.2 Heat stress and seasonal infertility

HS compromises swine reproduction by causing infertility and poor pregnancy outcomes, phenotypically noted as anovulation, reduced conception rates, and low pregnancy maintenance (Love, 1978; Love et al., 1993; Xue et al., 1994). Studies examining the ovarian impacts of HS in pigs have discovered that HS alters ovarian steroidogenic enzyme abundance (Nteeba et al., 2015), induces an ovarian heat shock protein response (Seibert et al., 2019), increases granulosa cell apoptotic signaling (Luo et al., 2016) and impairs oocyte quality and gonadotrophin production (Tseng et al., 2006). In addition, HS reduces follicular dominance following recruitment (Ozawa et al., 2005), increases ovarian autophagy (Hale et al., 2017) and reduces follicular gonadotropin receptor expression and steroid hormone production (Ozawa et al., 2005). Further, HS reduces CL weight and size in gilts (Bidne et al., 2019) and oxidative stress occurs during HS leading to oocyte death (Roth and Hansen, 2005). Consequently, there are multiple avenues by which HS can compromise ovarian function and deleteriously influence swine reproduction.

13 Conclusion

In conclusion, proper porcine reproduction requires regulation by the HPO axis and systemic alterations can impair reproduction. Examples of perturbations influencing reproduction in swine include toxicant exposures, endogenous endocrine disruption and altered ambient temperature. Each of these can independently impair fecundity in swine.

14 References

Al-Attas, O. S., et al. 2009. Changes in endotoxin levels in T2DM subjects on anti-diabetic therapies. *Cardiovascular Diabetology*, 8, 20.

Alm, H., et al.. 2002. The influence of the mycotoxins deoxynivalenol and zearalenol on in vitro maturation of pig oocytes and in vitro culture of pig zygotes. *Toxicol in Vitro*, 16, 643–648.

Alm, H., et al. 2006. Influence of Fusarium-toxin contaminated feed on initial quality and meiotic competence of gilt oocytes. *Reproductive Toxicology*, 22, 44–50.

Amar, J., et al. 2008. Energy intake is associated with endotoxemia in apparently healthy men. *The American Journal of Clinical Nutrition*, 87, 1219–1223.

Anderson, L. L. 2000. Pigs. In: Hafez, B. & Hafez, E.S.E. (eds.) *Reproduction in Farm Animals*. Wiley. https://doi.org/10.1002/9781119265306.ch13

Anderson, L. L. 2009. *Reproductive Biology of Pigs. Animal Industry Report*. Ames, IA: Iowa State University.

Bahr, J. & Milich, K. 2013. Ovarian Physiology. In: Hoyer, P. B. (ed.) *Ovarian Toxicology*. 2nd ed. Boca Raton, FL: CRC Press.

Baldwin, D. M. & Stabenfeldt, G. H. 1975. Endocrine changes in the pig during late pregnancy, parturition and lactation. *Biology of Reproduction*, 12, 508–415.

Barnett, K. R., et al.. 2006. Ovarian follicle development and transgenic mouse models. *Human Reproduction Update*, 12, 537–555.

Basini, G., et al. 2012. Atrazine disrupts steroidogenesis, VEGF and NO production in swine granulosa cells. *Ecotoxicology and Environmental Safety*, 85, 59–63.

Baumgard, L. H. & Rhoads, R. P. Jr. 2013. Effects of heat stress on postabsorptive metabolism and energetics. *Annual Review of Animal Biosciences*, 1, 311–337.

Baumgard, L.H., et al., 2015. Effects of heat stress on the immune system, metabolism and nutrient partitioning: implications on reproductive success. Rev. Brazil Reprod. Anim. 39: 173–183.

Bazer, F. W., et al. 1986. Role of conceptus secretory products in establishment of pregnancy. *Journal of Reproduction and Fertility*, 76, 841–850.

Bernabucci, U., et al. 2010. Metabolic and hormonal acclimation to heat stress in domesticated ruminants. *Animal*, 4, 1167–1183.

Bidne, K. L., et al. 2019. Heat stress during the luteal phase decreases luteal size but does not affect circulating progesterone in gilts1. *Journal of Animal Science*, 97, 4314–4322.

Blaney, B. J., et al. 1984. Zearalenone intoxication of pigs. *Australian Veterinary Journal*, 61, 24–27.

Borum, K. 1961. Oogenesis in the mouse: a study of the meiotic prophase. *Experimental Cell Research*, 24, 495–507.

Byskov, A. G. 1986. Differentiation of mammalian embryonic gonad. *Physiological Reviews*, 66, 71–117.

Cani, P. D., et al. 2007. Metabolic endotoxemia initiates obesity and insulin resistance. *Diabetes*, 56, 1761–172.

Casas, E., et al. 2010. Differential effects of herbicides atrazine and fenoxaprop-ethyl, and insecticides diazinon and malathion, on viability and maturation of porcine oocytes in vitro. *Toxicol in Vitro*, 24, 224–230.

EFSA Panel on Contaminants in the Food Chain (CONTAM). 2011. Scientific opinion on the risks for animal and public health related to the presence of T-2 and HT-2 toxin in food and feed. *EFSA Journal*, 9, 2481.

EFSA Panel on Contaminants in the Food Chain (CONTAM). 2017. Risks to human and animal health related to the presence of deoxynivalenol and its acetylated and modified forms in food and feed. *EFSA Journal*, 15, e04718.

Chang, K., et al. 1979. Effects of the mycotoxin zearalenone on swine reproduction. *American Journal of Veterinary Research*, 40, 1260–1267.

Chen, L., et al. 2020. Melatonin protects against defects induced by malathion during porcine oocyte maturation. *Journal of Cellular Physiology*, 235, 2836–2846.

Cheng, L., et al. 2019. Melatonin protects in vitro matured porcine oocytes from toxicity of aflatoxin B1. *Journal of Pineal Research*, 66, e12543.

Christenson, L. K. & Devoto, L. 2003. Cholesterol transport and steroidogenesis by the corpus luteum. *Reproductive Biology and Endocrinology*, 1, 90.

Clarke, I. J. & Cummins, J. T. 1982. The temporal relationship between gonadotropin releasing hormone (GnRH) and luteinizing hormone (LH) secretion in ovariectomized ewes. *Endocrinology*, 111, 1737–1739.

De Roux, N., et al. 2003. Hypogonadotropic hypogonadism due to loss of function of the KiSS1-derived peptide receptor GPR54. *Proceedings of the National Academy of Sciences of the United States of America*, 100, 10972–10976.

Delemarre-Van de Waal, H. A. 2002. Regulation of puberty. *Best Practice & Research: Clinical Endocrinology & Metabolism*, 16, 1–12.

Donovan, P. J., et al. 1986. Migratory and postmigratory mouse primordial germ cells behave differently in culture. *Cell*, 44, 831–838.

Edwards, S., et al. 1987a. The effects of zearalenone on reproduction in swine. II. The effect on puberty attainment and postweaning rebreeding performance. *Theriogenology*, 28, 51–58.

Edwards, S., et al. 1987b. The effects of zearalenone on reproduction in swine. I. The relationship between ingested zearalenone dose and anestrus in non-pregnant, sexually mature gilts. *Theriogenology*, 28, 43–49.

Erickson, G. F., et al. 1979. Gonadotropin binding and aromatase activity in granulosa cells of young proestrous and old constant estrous rats1. *Biology of Reproduction*, 20, 182–190.

Etienne, M. & Jemmali, M. 1982. Effects of zearalenone (F2) on estrous activity and reproduction in gilts2. *Journal of Animal Science*, 55, 1–10.

Faddy, M. J., et al. 1987. The kinetics of pre-antral follicle development in ovaries of CBA/Ca mice during the first 14 weeks of life. *Cell Proliferation*, 20, 551–560.

Flores, D., et al. 2017. Oxidative stress as a damage mechanism in porcine cumulus-oocyte complexes exposed to malathion during in vitro maturation. *Environmental Toxicology*, 32, 1669–1678.

Flowers, B., et al. 1987. A comparison of effects of zearalenone and estradiol benzoate on reproductive function during the estrous cycle in gilts2. *Journal of Animal Science*, 65, 1576–1584.

Fraser, H. M. & Wulff, C. 2003. Angiogenesis in the corpus luteum. *Reproductive Biology and Endocrinology*, 1, 88.

Garrido, C., et al. 1993. Transcriptional regulation of vascular endothelial growth factor gene expression in ovarian bovine granulosa cells. *Growth Factors*, 8, 109–117.

Geisert, R. D. & Bazer, F. W. 2015. *Regulation of Implantation and Establishment of Pregnancy in Mammals: Tribute to 45 Year Anniversary of Roger V. Short's "Maternal Recognition of Pregnancy".* Cham: Springer International Publishing.

Geisert, R. D., et al. 2021. Gene editing to investigate the role of conceptus factors in the establishment of pregnancy in the pig. *Reproduction*, 161, R79–R88.

Gerez, J. R., et al. 2017. Deoxynivalenol induces toxic effects in the ovaries of pigs: an ex vivo approach. *Theriogenology*, 90, 94–100.

Glover, L. E., et al. 2016. Oxygen metabolism and barrier regulation in the intestinal mucosa. *Journal of Clinical Investigation*, 126, 3680–3688.

Gondos, B. 1970. Granulosa cell-germ cell relationship in the developing rabbit ovary. *Development*, 23, 419–426.

González-Alvarez, M.E., et al., 2023. Scrambled eggs-Negative impacts of heat stress and chemical exposures on ovarian function in swine. Mol Reprod Dev. doi: 10.1002/mrd.23669. Epub ahead of print. PMID: 36652419.

González-Alvarez, M.E., et al., 2024. Effects of per- and polyfluoroalkylated substances on female reproduction. Toxics. 12(7):455. doi: 10.3390/toxics12070455.

Grasselli, F., et al. 2010. Bisphenol A disrupts granulosa cell function. *Domestic Animal Endocrinology*, 39, 34–39.

Graves, K. L., et al. 2019. Identification of measures predictive of age of puberty onset in gilts. *Translational Animal Science*, 4, 285–292.

Gregoraszczuk, E. Ł., et al. 2011. Differential accumulation of HCBz and PeCBz in porcine ovarian follicles and their opposing actions on steroid secretion and CYP11, CYP17, 17β-HSD and CYP19 protein expression: a tissue culture approach. *Reproductive Toxicology*, 31, 494–499.

Grummer, R. R. & Carroll, D. J. 1988. A review of lipoprotein cholesterol metabolism: importance to ovarian function. *Journal of Animal Science*, 66, 3160–3173.

Guigon, C. J. & Magre, S. 2006. Contribution of germ cells to the differentiation and maturation of the ovary: insights from models of germ cell depletion. *Biology of Reproduction*, 74, 450–458.

Guthrie, H. D., et al. 1972. Plasma estrogen, progesterone, and luteinizing hormone prior to estrus and during pregnancy in pigs. *Endocrinology*, 91, 675–679.

Guthrie, H. D., et al. 1995. Follicular atresia in pigs: measurement and physiology. *Journal of Animal Science*, 73, 2834–2844.

Gwynne, J. T. & Strauss, J. F. 3rd. 1982. The role of lipoproteins in steroidogenesis and cholesterol metabolism in steroidogenic glands. *Endocrine Reviews*, 3, 299–329.

Hale, B. J., et al. 2017. Heat stress induces autophagy in pig ovaries during follicular development. *Biology of Reproduction*, 97, 426–437.

Hawkesworth, S., et al. 2013. Evidence for metabolic endotoxemia in obese and diabetic Gambian women. *Nutr Diabetes*, 3, e83.

Henricks, D. M., et al. 1972. Plasma estrogen, progesterone and luteinizing hormone levels during the estrous cycle in pigs1. *Biology of Reproduction*, 6, 210–218.

Herbison, A. E. 2016. Control of puberty onset and fertility by gonadotropin-releasing hormone neurons. *Nature Reviews Endocrinology*, 12, 452–466.

Hirshfield, A. N. 1991. Development of follicles in the mammalian ovary. *International Review of Cytology*, 124, 43–101.

Hirshfield, A. N. 1997. Overview of ovarian follicular development: considerations for the toxicologist. *Environmental and Molecular Mutagenesis*, 29, 10–15.

Hoyer, P. B. 2010. 11.16 – Female Reproductive Toxicology. In: Mcqueen, C. A. (ed.) *Comprehensive Toxicology*. 2nd ed. Oxford: Elsevier.

Hoyer, P. B. & Devine, P. J. 2001. Endocrine toxicology: the female reproductive system.

Hoyer, P.B., Keating, A.F. 2014. Xenobiotic effects in the ovary: temporary versus permanent infertility. Drug Metabolism and Toxicology (Expert Opinion). 10(4):511–23.

Hueza, I. M., et al. 2014. Zearalenone, an estrogenic mycotoxin, is an immunotoxic compound. *Toxins*, 6, 1080–1095.

Ishikawa, S., et al. 2015. The effects analysis of two neonicotinoid insecticides on in vitro maturation of porcine oocytes using hanging drop monoculture method. *The Journal of Veterinary Medical Science*, 77, 725–728.

Jiang, Y., et al. 2021. Exposure to chlorpyrifos leads to spindle disorganization and mitochondrial dysfunction of porcine oocytes during in vitro maturation. *Theriogenology*, 173, 249–260.

Jones, E. 2009. The Female Reproductive System. In: Boron, W. F. & Boulpaep, E. L. (eds.) *Medical Physiology: A Cellular and Molecular Approach*. 2nd International ed. Philadelphia, PA: Saunders/Elsevier.

Jorge Chedrese, P. & Feyles, F. 2001. The diverse mechanism of action of dichlorodiphenyldichloroethylene (DDE) and methoxychlor in ovarian cells in vitro. *Reproductive Toxicology*, 15, 693–698.

Kamarianos, A., et al. 2003. The presence of environmental pollutants in the follicular fluid of farm animals (cattle, sheep, goats, and pigs). *Reproductive Toxicology*, 17, 185–190.

Kanora, A. & Maes, D. 2009. The role of mycotoxins in pig reproduction: a review. *Veterinarni Medicina*, 54, 565–576.

Keating, A. F. & Hoyer, P. B. 2009. Mechanisms of Reproductive Toxicity. In: *Drug Metabolism Handbook*. Hoboken, NJ: John Wiley & Sons, Inc, pp. 697–736.

Keating, A.F., Ross, J.W., Baumgard, L.H. 2024. Systemic and Ovarian Impacts of Heat Stress in the Porcine Model. Reproduction. Nov 11;168(6):e240217. doi: 10.1530/REP-24-0217.

Klein, C. 2003. The Hypothalamic-Pituitary-Gonadal Axis. In: Kufe, D. W., et al., (eds.) *Holland-Frei Cancer Medicine.* 6th ed. Hamilton, ON: BC Decker.

Knapczyk-Stwora, K., et al. 2018. The impact of sex steroid agonists and antagonists on folliculogenesis in the neonatal porcine ovary via cell proliferation and apoptosis. *Theriogenology*, 113, 19–26.

Knapczyk-Stwora, K., et al. 2022. Effects of neonatal methoxychlor exposure on the ovarian transcriptome in piglets. *Animal Reproduction Science*, 238, 106956.

Knet, M., et al. 2015. Analysis of porcine granulosa cell death signaling pathways induced by vinclozolin. *Theriogenology*, 84, 927–939.

Laws, M. J., et al. 2021. Chapter Five – Endocrine Disrupting Chemicals and Reproductive Disorders in Women, Men, and Animal Models. In: Vandenberg, L. N. & Turgeon, J. L. (eds.) *Advances in Pharmacology.* Cambridge, MA: Academic Press.

Liu, J. 2014. Pentachlorobenzene. In: Wexler, P. (ed.) *Encyclopedia of Toxicology.* 3rd ed. Oxford: Academic Press.

Liu, J., et al. 2015. Aflatoxin B1 is toxic to porcine oocyte maturation. *Mutagenesis*, 30, 527–535.

Liu, M. T., et al. 1985. Indirect enzyme-linked immunosorbent assay for the mycotoxin zearalenone. *Applied and Environmental Microbiology*, 50, 332–336.

Livadas, S. & Chrousos, G. P. 2016. Control of the onset of puberty. *Current Opinion in Pediatrics*, 28, 551–558.

Long, G. G., et al. 1983. Effect of Fusarium roseum (Gibberella zea) on pregnancy and the estrous cycle in gilts fed molded corn on days 7–17 post-estrus. *Veterinary Research Communications*, 6, 199–204.

Long, G. G., et al. 1992. Effect of zearalenone on days 7 to 10 post-mating on blastocyst development and endometrial morphology in sows. *Veterinary Pathology*, 29, 60–67.

Love, R. J. 1978. Definition of a seasonal infertility problem in pigs. *Veterinary Record*, 103, 443–446.

Love, R. J., et al. 1993. Seasonal effects on fertility in gilts and sows. *Journal of Reproduction and Fertility*, 48, 191–206.

Luche, E., et al. 2013. Metabolic endotoxemia directly increases the proliferation of adipocyte precursors at the onset of metabolic diseases through a CD14-dependent mechanism. *Molecular Metabolism*, 2, 281–291.

Luo, M., et al. 2016. Heat stress impairs mice granulosa cell function by diminishing steroids production and inducing apoptosis. *Molecular and Cellular Biochemistry*, 412, 81–90.

Martini, F. 1998. The Reproductive System. In: Rizzo, D. C. (ed.) *Fundamentals of Anatomy and Phisiology.* 4th ed. Prentice Hall, Inc. Simon and Schuster.

Mayorga, E. J., et al. 2018. Heat stress adaptations in pigs. *Animal Frontiers*, 9, 54–61.

Mayorga E.J., et al., 2020. Biology of heat stress; the nexus between intestinal hyperpermeability and reproduction in swine. Theriogenology. 154:73–83.

McGee, E. A. & Hsueh, A. J. 2000. Initial and cyclic recruitment of ovarian follicles. *Endocrine Reviews*, 21, 200–14.

Méduri, G., et al. 2002. Follicle-stimulating hormone receptors in oocytes? *The Journal of Clinical Endocrinology and Metabolism*, 87, 2266–2276.

Meidan, R. 2016. *The Life Cycle of the Corpus Luteum*. Switzerland: Springer.

Miller, W. L. & Auchus, R. J. 2011. The molecular biology, biochemistry, and physiology of human steroidogenesis and its disorders. *Endocrine Reviews*, 32, 81–151.

Mlynarcíková, A., et al. 2005. Alterations in steroid hormone production by porcine ovarian granulosa cells caused by bisphenol A and bisphenol A dimethacrylate. *Molecular and Cellular Endocrinology*, 244, 57–62.

Mlynarcíková, A., et al. 2009. Effects of selected endocrine disruptors on meiotic maturation, cumulus expansion, synthesis of hyaluronan and progesterone by porcine oocyte-cumulus complexes. *Toxicol in Vitro*, 23, 371–377.

Moeljono, M. P. E., et al. 1977. A study of prostaglandin F2alpha as the luteolysin in swine: II characterization and comparison of prostaglandin F, estrogens and progestin concentrations in utero-ovarian vein plasma of nonpregnant and pregnant gilts. *Prostaglandins*, 14(3), 543–555.

Molokwu, E. C. & Wagner, W. X. 1973. Endocrine physiology of the puerperal sow. *Journal of Animal Science*, 36, 1158–1163.

Monk, M. & Mclaren, A. 1981. X-chromosome activity in foetal germ cells of the mouse. *Journal of Embryology and Experimental Morphology*, 63, 75–84.

Navarro, V. M., et al. 2009. Regulation of gonadotropin-releasing hormone secretion by kisspeptin/dynorphin/neurokinin B neurons in the arcuate nucleus of the mouse. *The Journal of Neuroscience*, 29, 11859–11866.

Nteeba, J., et al. 2015. Heat Stress alters ovarian insulin-mediated phosphatidylinositol-3 kinase and steroidogenic signaling in gilt ovaries1. *Biology of Reproduction*, 92, 1–8.

Olsen, M., et al. 1981. Reduction of zearalenone to zearalenol in female rat liver by 3 alpha-hydroxysteroid dehydrogenase. *Acta Pharmacol Toxicol (Copenh)*, 48, 157–161.

Ozawa, M., et al. 2005. Alterations in follicular dynamics and steroidogenic abilities induced by heat stress during follicular recruitment in goats. *Reproduction*, 129, 621–630.

Parkinson, A., et al. 2013. Biotransformation of Xenobiotics. In: Klaasen, C. (ed.) *Casarett and Doull's Toxicology The Basic Science of Poisons*. 8th ed. New York, NY: McGraw-Hill.

Perry, J. S., et al. 1973. Steroid hormone production by pig blastocysts. *Nature*, 245, 45–47.

Petr, J., et al. 2013. Pyrethroids cypermethrin, deltamethrin and fenvalerate have different effects on in vitro maturation of pig oocytes at different stages of growth. *Animal*, 7, 134–142.

Pitzel, L., et al. 1990. Effects of oxytocin on in vitro steroid release of midstage small and large porcine luteal cells. *Endocrinology*, 126, 2343–2349.

Qin, X., et al. 2015. Oxidative stress induced by zearalenone in porcine granulosa cells and its rescue by curcumin in vitro. *PLoS One*, 10, e0127551.

Richards, J. S. 2018. The Ovarian Cycle. *Vitamins and Hormones*, 107, 1–25.

Robinson, R. S. & Woad, K. J. 2017. Luteal Angiogenesis. In: Meidan, R. (ed.) *The Life Cycle of the Corpus Luteum*. Cham: Springer International Publishing.

Rodgers, R. J. & Irving-Rodgers, H. F. 2010. Formation of the Ovarian Follicular Antrum and Follicular Fluid1. *Biology of Reproduction*, 82, 1021–1029.

Romoser, M. R., et al. 2019. Methods for reproductive tract scoring as a tool for improving sow productivity12. *Translational Animal Science*, 4, 275–284.

Romoser, M. R., et al. 2020. Evaluating the efficacy of visual assessment of gilt vulva size prior to puberty on subsequent reproductive performance *. *Applied Animal Science*, 36, 694–700.

Ross J.W., et al., 2015. Physiological consequences of heat stress in pigs. Animal Production Science. 55: 1381–1390.

Ross, J.W., et al., 2017. Physiological mechanisms through which heat stress compromises reproduction in pigs. Molecular Reproduction and Development. 84(9): 934–945.

Roth, Z. & Hansen, P. J. 2005. Disruption of nuclear maturation and rearrangement of cytoskeletal elements in bovine oocytes exposed to heat shock during maturation. *Reproduction*, 129, 235–244.

Rothchild, I. 1981. The regulation of the mammalian corpus luteum. *Recent Progress in Hormone Research*, 37, 183–298.

Rourke, J. L. & Sinal, C. J. 2014. Biotransformation/Metabolism. In: Wexler, P. (ed.) *Encyclopedia of Toxicology*. 3rd ed. Oxford: Academic Press.

Rowell, L. B., et al. 1968. Splanchnic blood flow and metabolism in heat-stressed man. *Journal of Applied Physiology*, 24, 475–484.

Sambuu, R., et al. 2011. Detection of zearalenone and its metabolites in naturally contaminated porcine follicular fluid by using liquid chromatography-tandem mass spectrometry. *Journal of Reproduction and Development*, 57, 303–306.

Schams, D. & Berisha, B. 2004. Regulation of corpus luteum function in cattle – an overview. *Reproduction in Domestic Animals*, 39, 241–251.

Schoevers, E. J., et al. 2010. Porcine oocytes are most vulnerable to the mycotoxin deoxynivalenol during formation of the meiotic spindle. *Theriogenology*, 74, 968–978.

Schoevers, E. J., et al. 2012. Transgenerational toxicity of Zearalenone in pigs. *Reproductive Toxicology*, 34, 110–119.

Seibert, J. T., et al. 2019. Differentiating between the effects of heat stress and lipopolysaccharide on the porcine ovarian heat shock protein response1. *Journal of Animal Science*, 97, 4965–4973.

Seminara, S. B. & Crowley Jr, W. F. 2008. Kisspeptin and GPR54: discovery of a novel pathway in reproduction. *Journal of Neuroendocrinology*, 20, 727–731.

Senger, D. R., et al. 1997. Angiogenesis promoted by vascular endothelial growth factor: regulation through alpha1beta1 and alpha2beta1 integrins. *Proceedings of the National Academy of Sciences of the United States of America*, 94, 13612–13617.

Senger, P. L. 1997. *Pathways to Pregnancy and Parturition*. Pullman, WA: Current Conceptions.

Senger, P. L. 2012. *Pathways to Pregnancy & Parturition*. Pullman, WA: Current Conceptions.

Soede, N. M., et al. 2011. Reproductive cycles in pigs. *Animal Reproduction Science*, 124, 251–258.

Stabenfeldt, G. H., et al. 1969. Peripheral plasma progesterone levels in pigs during the oestrous cycle. *Journal of Reproduction & Infertility*, 20, 443–449.

Stalder, K., et al. 2003. Financial impact of average parity of culled females in a breed-to-wean swine operation using replacement gilt net present value analysis. *Journal of Swine Health and Production*, 11, 69–74.

Stouffer, R. L. & Hennebold, J. D. 2015. Chapter 23: Structure, Function, and Regulation of the Corpus Luteum. In: Plant, T. M. & Zeleznik, A. J. (eds.) *Knobil and Neill's Physiology of Reproduction*. 4th ed. San Diego, CA: Academic Press.

Tseng, J. K., et al. 2006. In vitro thermal stress induces apoptosis and reduces development of porcine parthenotes. *Theriogenology*, 66, 1073–1082.

Ványi, A., et al. 1994. Perinatal oestrogen syndrome in swine. *Acta Veterinaria Hungarica*, 42, 433–446.

Victoria Sanz Fernandez, M., et al. 2015. Effects of heat stress on carbohydrate and lipid metabolism in growing pigs. *Physiological Reports*, 3, e12315.

Wang, T., et al. 2016. The toxic effects and possible mechanisms of Bisphenol A on oocyte maturation of porcine in vitro. *Oncotarget*, 7, 32554–32565.

Wang, W., et al. 2019. Cytotoxicity and DNA damage caused from diazinon exposure by inhibiting the PI3K-AKT pathway in porcine ovarian granulosa cells. *Journal of Agricultural and Food Chemistry*, 67, 19–31.

Wang, W., et al. 2018. Toxic effects and possible mechanisms following malathion exposure in porcine granulosa cells. *Environmental Toxicology and Pharmacology*, 64, 172–180.

Weaver, G. A., et al. 1978. Mycotoxin-induced abortions in swine. *Canadian Veterinary Journal*, 19, 72–74.

Wiltbank, M. C., et al. 1990. Regulation of the corpus luteum by protein kinase C. II. Inhibition of lipoprotein-stimulated steroidogenesis by prostaglandin F2 alpha. *Biology of Reproduction*, 42, 239–245.

Wójtowicz, A. K., et al. 2007. DDT- and DDE-induced disruption of ovarian steroidogenesis in prepubertal porcine ovarian follicles: a possible interaction with the main steroidogenic enzymes and estrogen receptor beta. *Journal of Physiology and Pharmacology*, 58, 873–885.

Wylie, C. C., et al. 1986. Primordial germ cell migration. *Developmental Biology (NY 1985)*, 2, 433–448.

Xue, J. L., et al. 1994. Multiple manifestations of season on reproductive performance of commercial swine. *Journal of the American Veterinary Medical Association*, 204, 1486–1489.

Young, L. G. & King, G. J. 1986. Low concentrations of zearalenone in diets of mature gilts. *Journal of Animal Science*, 63, 1191–1196.

Yuan, B., et al. 2017. Toxic effects of atrazine on porcine oocytes and possible mechanisms of action. *PLoS One*, 12, e0179861.

Zhang, Y., et al. 2016. Exposure to HT-2 toxin causes oxidative stress induced apoptosis/autophagy in porcine oocytes. *Scientific Reports*, 6, 33904.

Zhou, N., et al. 2022. Paraquat exposure impairs porcine oocyte meiotic maturation. *Theriogenology*, 179, 60–68.

Ziecik, A. J., et al. 2011. Mechanisms for the establishment of pregnancy in the pig. *Reproduction in Domestic Animals*, 46, 31–41.

Ziecik, A. J., et al. 2017. Corpus Luteum Regression and Early Pregnancy Maintenance in Pigs. In: Meidan, R. (ed.) *The Life Cycle of the Corpus Luteum.* Cham: Springer International Publishing.

Ziecik, A. J., et al. 2018. Regulation of the porcine corpus luteum during pregnancy. *Reproduction*, 156, R57–R67.

Zwierzchowski, W., et al. 2005. Level of zearalenone in blood serum and lesions in ovarian follicles of sexually immature gilts in the course of zearalenone micotoxicosis. *Polish Journal of Veterinary Sciences*, 8, 209–218.

Chapter 10

Tools and strategies for optimizing the reproductive performance of sows

Robert V. Knox, University of Illinois, USA

1 Introduction

Reproductive performance in swine breeding herds is very high based on the average number of pigs born per litter and the average farrowing rate for mated sows. Litter size has continued to increase (Fig. 1) in the modern sow over years (Pigchamp, 2022). This trend is a likely result of genetic selection for higher litter size at birth that originates from increases in the correlated traits of ovulation rate and number of embryos (Rutherford et al., 2013; Soede and Kemp, 2019).

Within the population of sows in commercial production, there is an approximate normal distribution for litter size. For economic reasons, farms must focus on minimizing the occurrence of low litter sizes and make improvements in average herd litter size. However, much of the interest in recent years has been on the sows giving birth to very large litters, a concept termed 'hyperprolificacy'. The term is not new and has been used to describe increased litter size relative

http://dx.doi.org/10.19103/AS.2024.0137.14

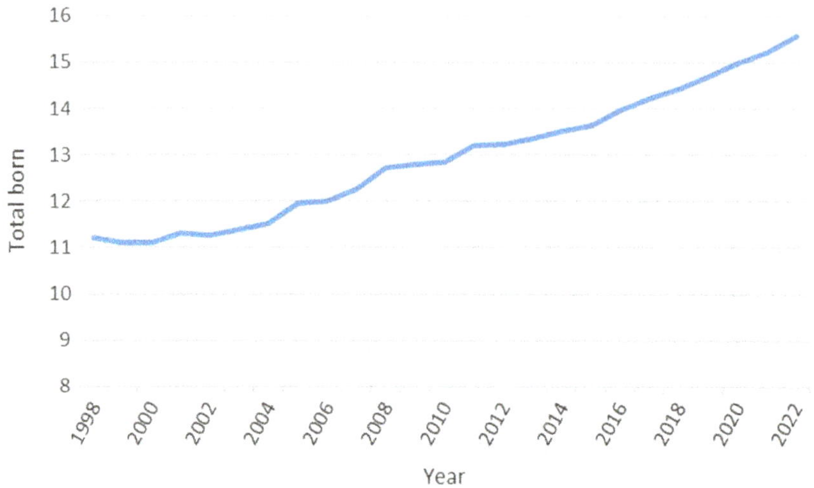

Figure 1 Data for total born pigs for the 1998–2022 PigCHAMP benchmark summaries.

to the population at that time. Today, the hyperprolific sow may include females that can farrow a litter of 14–16 pigs, 17–20 pigs, and more than their nursing capability or can include the capacity to produce 27–30 or >30 pigs each year (Kemp and Soede, 2012; Baxter et al., 2020; Oliviero, 2022). The ability to reach hyperprolificacy status will vary within sows on a farm and among different breeding farms and can depend upon an array of factors that can interact such as genetics, herd health, parity, environment, and breeding management.

The majority of established breeding herds are composed of sows that have had one or more litters (Houska, 2009; Tani et al., 2018). The parity distribution of a herd (Fig. 2) is important to its overall herd productivity, and as sows age, their increase in reproductive maturity and body size also associates with a greater capacity for litter size. Reports indicate that peak productivity occurs in parities 3–4 (Koketsu and Dial, 1997). However, new information also suggests that parity 1 litter size can impact litter sizes in the following parities (Koketsu et al., 2017) and strongly implicates gilt development and management as a factor in lifetime reproductive performance. Because sow replacement costs cannot be recouped until after the third litter (Stalder et al., 2003; Rodriguez-Zas et al., 2006; Sasaki et al., 2011) and maximal litter sizes are observed in parities 3–6 (Bergman et al., 2018), the goal is to ensure that females produce four or more litters before voluntary replacement. Involuntary removal of females from the breeding herd occurs for a variety of reasons but are most often removed due to reproductive failure (33%), followed by poor litter performance (21%) and then locomotion problems (13.3%). The frequency of these removals is related to parity with reproductive failures highest in the earlier parities (Lucia Jr et al., 2000). Optimal herd performance relies on the retention of younger females and culling of older less productive sows. Early

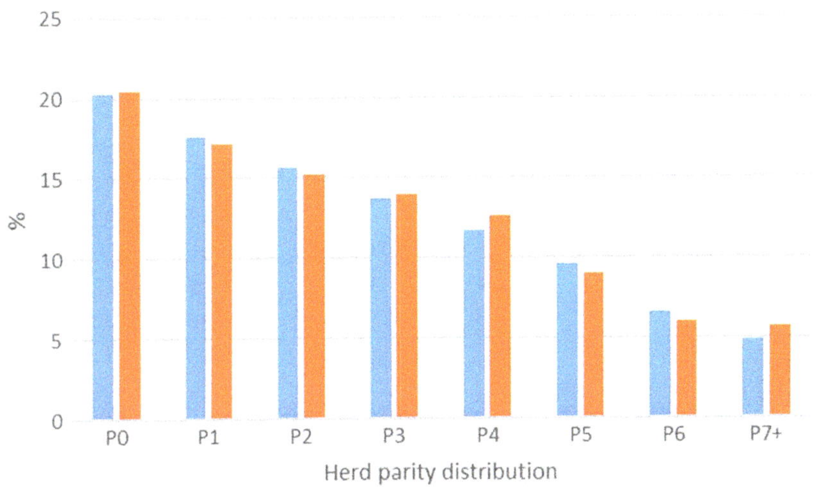

Figure 2 Illustration of parity distributions from industry summaries in Spain (blue) and from North America (orange) using data from multiple years and herds. Source: Adapted from Tani et al. (2018); Ketchum and Rix (2013).

culling for the mentioned reasons is a major problem for farms, and thought to be related to improper gilt development and management from selection until weaning their first litter. Addressing these problems is proposed as the key to retention and lifetime pig production.

2 Sow herd performance metrics

Farm records are essential for assessment of breeding herd performance and can be generated to help identify problems for individual breeding groups, production trends for periods of the year, and for benchmarking herd production averages. Selected data can provide insight into sources of variation in reproductive measures and help producers understand the extent of variation, patterns, potential sources, and opportunities for improvement. In the breeding herd, most farms rely on physical production records that remain and follow the animal to different locations throughout its productive life. An example of some important reproductive measures are shown in Table 1 and includes average measures for helping farms compare their herd performance to other farms (Knox, 2022b). This type of analysis classifies farm performance by individual measure on each line (e.g. farrowing rate %) but could also be generated to output all measures associated with the highest farrowing class. This type of analysis can help identify multiple factors for areas in opportunity.

Breeding herd performance can be assessed by the percent of sows bred by 7 days after weaning and is expected to be >90%. Measures 5–10% below this target would warrant investigation into the source of the delay. To help in

Table 1 Measures of swine reproductive performance relative to industry averages based on breeding herd record summaries from 2019 and including data from ~1500 farms with an average herd size >2000 sows and representing fertility data for >2.5 million sows

Measure	Intervention level		
	Below Average	Average	Above Average
Weaned sows bred by 7 days (%)	<84%	86–90%	91–95%
Wean to service interval (days)	≥7–10	>6.0–6.9	4.5–6.0
Day 30 conception rate (%)	<85%	88–92%	>92%
Farrowing rate (%)	<82%	83–87%	88–95%
Total born/litter	<13.5	13.9–14.9	15.0–16.0
Live pigs born/litter	≤12.6	12.9–13.9	14.0–15.0
Number of stillborn pigs per litter	≥1.2–1.5	0.91–1.1	0.5–0.90
Number of mummified fetuses per litter	≥0.5–0.6	0.3–0.4	0.1–0.2
Pigs weaned per sow farrowed	≤9.8–10.8	11.0–11.5	≥11.6–12.2
Litters/mated female/year	≤2.20–2.25	2.29–2.34	≥2.37–2.45
Pigs weaned/mated female/year	≤22.0–24.4	25.2–26.8	≥27.4–32.0
Weaning age (days)	17.5–19.5	20.1–20.9	21.3–23.1
Gestation length (days)	<114.5 or >116.5	115.9	116.0

Sources: MetaFarms, Production Index Analysis 2019; Swine Management Services, LLC; PigCHAMP, Benchmarking Summaries 2019 USA.

this assessment, wean to service interval can be used, as the first day of estrus and service should coincide. Most sows are expected to be in estrus on days 4–6 after weaning, and averages would reflect herd performance. However, deviations from the target could be related to parity, season, and management procedures. First parity sows are more likely to show delayed return to estrus, and this is exacerbated in the summer months and would associated with increases in wean-to-service interval. Early weaning at <18 days may also associate with delays and the percent of sows bred by 7 days. The combination of these factors can be additive and create larger problems for younger parity sows, weaned early, especially in summer and early fall.

Most commercial pigs worldwide are produced using artificial insemination (AI) (Riesenbeck, 2011). The use of this technology depends upon the detection and breeding of females on each day in estrus (Knox, 2016). The mature female pig typically displays estrus for 2–3 days and one important measure includes the percent of sows receiving multiple matings. When troubleshooting poor conception rates or low litter sizes, this can be an important indicator, as single services are often associated with reduced fertility (Lamberson and Safranski, 2000; Koketsu et al., 2017). The reason for this is that in weaned sows, the interval from estrus to ovulation can vary by more than 24 h and reproductive fertility is related to at least one insemination occurring within the 24 h before ovulation (Soede et al., 1995). In addition, most breeding groups are heterogeneous with mixed parity weaned sows (70%), replacement gilts (20%), and repeat breeder sows (10%). The repeat breeders would have been previously inseminated in an earlier breeding group but failed to conceive and returned to estrus. But this is important for evaluation, as these females have conception failure 10–15% higher compared to the other females in the breeding group (Koketsu, 2003; Vargas et al., 2009a; Vargas et al., 2009b; Tummaruk et al., 2010).

Conception success for inseminated females can be realized as early as 21 days following service. Sows that fail to establish pregnancy may return to estrus at 18–25 days after service and would be classified as regular returns (Koketsu et al., 1997). However, due to limited labor and other priorities, many farms do not routinely test for regular returns to estrus and wait for the ultrasound assessment at 4 weeks (Knox et al., 2013). Part of the issue is that accurate estrous detection requires daily boar exposure and application of the back-pressure test for each female. For mated females that show regular return to estrus (~21 or 42 days after service), it is presumed that these sows had no fertilized eggs or lost all embryos in the first week. For reasons that are not clear, only ~50% of sows that fail to conceive display regular return to estrus while the remainder are detected as not pregnant (open) using ultrasound at 4–5 weeks after breeding (Koketsu et al., 1997; Rodibaugh, 2002; Knox, 2022b). Unfortunately, if not detected at ~21 days, these sows will accumulate a greater number of non-productive days before breeding or culling. A smaller proportion of mated sows are observed to return to estrus at irregular intervals (<18 days or 26–38 days after service). Sows returning to estrus <18 days after service may have been incorrectly identified in estrus or had a problem with cystic ovaries. Those sows returning at 26–38 days are presumed to have had viable embryos, at least up to day 14 after insemination but then lost most or all of the embryos and pregnancy thereafter. The extended cycle is likely a result of conceptus estrogen which acts to extend the lifespan of the corpora lutea (Geisert et al., 2015).

Today, the majority of large production farms use transabdominal B-mode ultrasound to diagnose pregnancy at 4–5 weeks following service. The technology is quick and accurate and allows visual detection of embryonic fluids inside the uterus as early as 25–30 days following breeding. While some farms use both estrus detection and ultrasound, many farms are limited in labor and rely predominantly on ultrasound. Accurate detection of pregnancy at 4–5 weeks is highly correlated to the farrowing rate as very few sows confirmed pregnant lose their pregnancies before farrowing. But farms do report real-time ultrasound fallouts and include sows confirmed pregnant by ultrasound but then do not farrow or are diagnosed open in later gestation. The source of the problem can be related to false positives associated with timing for pregnancy assessment, ovarian cysts, uterine infection, or fluid accumulation within the uterus unrelated to pregnancy. Improper timing of assessment can lead to errors in diagnosis if performed too early with too little fluid (<25 days) or too late (days 38–52) when fetal growth increases and fluid temporarily declines (Knight et al., 1977; Miller et al., 2003). There is also the possibility that some animals had too few embryos or were undergoing degeneration, making diagnosis difficult and pregnancy failures more likely. Cases of pregnancy loss in mid-to-late gestation that occur without evidence of abortion are difficult to confirm. Mid-to-late pregnancy loss would likely involve declining ovarian progesterone production from uterine release of prostaglandin. In the case of disease, toxins, or stress where placental attachment and function is disrupted and/or fetal death occurs, uterine-induced luteolysis might result to terminate progesterone production and allow dilation of the cervix, uterine contractions, and passage of the uterine contents (Taverne, 1992).

Farrowing rate and litter size are the key measures used for assessing performance of the breeding herd. Based on all sows and gilts detected in estrus and inseminated, farrowing rates average between 83% and 87% but can reach 90% for some farms. Measures for farrowing rate are expected to change slightly between breeding groups, but significant declines from previous performance would indicate problems. The earliest indicator of impending farrowing rate declines would be reduced conception rates by day 30. This early measure would allow for investigation into semen quality issues, estrus detection and breeding failures, environmental stressors, and problems in herd health. Declines in farrowing rate have been associated with heat stress and photoperiod in summer and in fall (De Rensis et al., 2017) and with stress with group housing (Peltoniemi et al., 1999). Litter size assessments for average total born and born alive are key measures for assessing the genetic potential of the herd and farm reproductive management. Litter size is highly associated with profitability, and although lowly heritable, it has been a priority in genetic selection. With emphasis on the selection for larger litter sizes for gilts, correlated traits such as ovulation rate and uterine size also increase (Gama and Johnson, 1993). While sow body size (Mcglone et al., 2004) and litter size increase with parity, body size itself is unrelated to litter size (Nielsen et al., 2018). The increase in litter size with parity then is likely due to the increase in ovulation rate (Kemp and Soede, 2012; Soede and Kemp, 2019) and embryos (King and Williams, 1984) that challenge uterine capacity. In gilts, the ovulation rate can vary from 16 to 22 ova with responses affected by feed intake (Mallmann et al., 2020) and genetics (Da Silva et al., 2017) and can increase to 25 ova in mature sows of higher parity (Town et al., 2005; Belstra et al., 2020). In research settings where insemination is controlled to occur within an optimal window before ovulation and semen fertility is classified as very good, the fertilization rate is expected to be >90% and not a limitation to litter size (Soede et al., 1995). But 20% of embryos can be lost before day 10 and further loss to day 30 (Perry and Rowlands, 1962; Lambert et al., 1991), where 30–40% of the total litter size may be lost. As the placental membranes for each of the conceptuses expand to create a large surface area for attachment to optimize gas and nutrient exchange, space becomes limiting. The membranes from adjacent conceptuses touch and limit further expansion. Too many embryos or limited space just before or after day 30 can lead to embryo degeneration (Soede and Kemp, 2019). While most uterine space limits do not seem to appear until days 25–30, as evidenced by underdeveloped or degenerating embryos, effects could manifest themselves on placental, embryo, or fetal sizes (Van Der Waaij et al., 2010) and result in an increase in fetal size variation or low birthweight pigs (Quesnel et al., 2008; Patterson et al., 2020; Riddersholm et al., 2021). The data are clear, that as litter size has increased over years, average piglet birthweight has declined, and the incidence of pigs <1 kg has also increased (Bauer et al., 1998; Beaulieu et al., 2010; Moreira et al., 2020). With larger litter sizes comes the risk for a higher incidence of stillborn pigs (Leenhouwers et al., 1999; Vanderhaeghe et al., 2010). This has been attributed to a longer duration of farrowing and perhaps hypoxia from extended exposure to contractions or early breakage of the umbilical cord (Vanderhaeghe et al., 2013) and this occurs most often in the last third of the litter.

3 Gilt selection and development impacts first parity sow fertility

An extensive review of important factors relating to gilt development and their impact on lifetime performance and longevity is provided by Patterson and Foxcroft (2019). But for discussion purposes, a few critical elements relating to gilts will be addressed here as a starting point for sow reproductive management. In the previous discussion of herd parity structure, culling of sows occurred for a variety of reasons with replacement rates ranging between 35% and 55% (Lucia Jr et al., 2000; Engblom et al., 2007; Ketchum and Rix, 2013; Tani et al., 2018; Knox, 2022b). Much of the attention has been focused on the reasons for removal and problems related to reduced lifetime productivity, welfare, and economics. Some of the most pressing issues involves culling gilts before puberty, after farrowing their first litter, or as sows before producing their third litter. Culling of these females has been linked to genetics, low litter size, fast or slow growth, or problems with reproductive management. Of both served and non-served gilts, leading reasons for removal can fluctuate between reproductive failures and locomotor issues (Lucia Jr et al., 2000; Ulguim et al., 2014; Boripun et al., 2021) while culling occurs at an average of 234 days of age in non-served gilts and >300 days for served gilts. Culling of gilts for anestrus or pregnancy failure leads the reasons for reproductive removal (Hughes et al., 2010; Wang et al., 2019; Boripun et al., 2021) and can account for 10–20% of all culled females. Importantly, the data suggest that culling in the first three parities due to reproductive failure, poor litter performance, and locomotor problems may reflect back to problems in gilt development. Gilt selection, age, weight and maturity at mating can all influence subsequent longevity and lifetime pig production (Patterson and Foxcroft, 2019). Gilts mated at heavier weights as a result of very fast growth rate or delayed age at puberty are more likely to have body condition, feed intake, and structural problems as sows. The fertility of gilts and their lifetime herd performance can be linked to body measures at first service that are also influenced by their growth rate and age at start of boar exposure (Kummer et al., 2006; Roongsitthichai et al., 2014). The data show variable proportions (6 to 30%) of gilts may fail to display pubertal estrus following the start of boar exposure (Eliasson et al., 1991, Calderón Díaz et al., 2015, Patterson et al., 2016). While the reasons for the wide range of failure in puberty are not clear, associated factors can include genetics, health, and parity of the dam at birth (Craig et al., 2017), as well as growth rate, and quality of boar stimulation (Pearce and Paterson, 1992; Patterson et al., 2002b). Early puberty can relate to improved retention (Knox et al., 2021) and herd productivity (Koketsu and Iida, 2020). Depending upon farm location around the world, gilts may develop for variable periods of time under conditions of heat stress and humidity (Tummaruk, 2012), which can be associated with lower feed intake, slower growth, and delayed puberty (Ross et al., 2017). Restricted feed (Quesnel et al., 2000) and slower growth rates have been associated with delayed puberty and anestrus in some cases (Patterson et al., 2002a; Kummer et al., 2009), while other studies show effects of age at boar exposure (Amaral Filha et al., 2009), season, genetics, and group size (Cronin et al., 1983). While concerns over slow growth rates in gilts will remain, the major focus is now shifting to the consequences of very fast growth rates and their association with heavy weights at breeding and at farrowing. While genetic emphasis has increased gilt growth rate, feed efficiency, and body leanness, age at puberty has not been a priority for selection. As a result, farms must deal with very fast-growing gilts but still adhere to conventional age timelines for puberty induction, breeding, and farrowing. It is not unexpected to see greater variation in the population for age and weight at these reproductive events and animals mated at the extremes are less likely to reach optimal productivity and remain in the herd. This presents a dilemma for how modern farms should handle the new genotypes and whether reproductive management approaches should be evaluated and revised.

4 Managing factors during lactation to impact sow fertility

The majority of females within a breeding group are comprised of sows that have recently farrowed and nursed a litter of pigs. Most breeding groups are established with females receiving inseminations within a limited number of days of each other which allows for farrowing to occur with greater synchrony and aids in weaning a more uniform batch of pigs. It can also help in the coordination of labor for animal movement and facility cleaning. Unfortunately, day of farrowing based on day of service can vary by as much as 7 days (Vanderhaeghe et al., 2011), making synchronized farrowing management often elusive. In many locations around the world, lactation length averages 21 days (Table 1) while elsewhere it may be closer to 28 days (Schwarz et al., 2021), where animal welfare regulations are in place. The 3–4-week length of lactation appears to be near optimal when considering post-wean sow reproductive performance and weaned pig health and growth (Pierozan et al., 2020). From a sow fertility perspective, several key factors have been associated with reproductive function in sows following weaning. One of the most important measures associated with extended wean to service intervals and poor fertility has been compromised body condition of sows at weaning (Vargas et al., 2009a; Schenkel et al., 2010). This can result from excessive loss of body fat and weight during lactation and can associate with daily feed intake, parity, season, and number of pigs nursing (Koketsu et al., 1996; Tantasuparuk et al., 2001; Anil et al., 2006). How sows are fed and conditioned during gestation can impact their body weight at farrowing and influence lactation feed intake and body weight loss before weaning (Kim et al., 2016; Mallmann et al., 2019a). Sows or gilts that are over-fed and over-conditioned during gestation may have problems with stillbirths, feed intake, and weight loss in lactation. Reports suggest that reduced feed intake or consecutive days off feed in the first weeks of lactation can have detrimental consequences on wean to estrous interval (Koketsu et al., 1996; Anil et al., 2006). This may be particularly problematic for young parity sows, under high temperatures, and high metabolic demands when nursing a large litter (Eissen et al., 2003). Modern sows show variation in feed and water intake (Kruse et al., 2011) and individual sows show that weight and backfat can be reduced, increase or stay the same during lactation but in some studies without a link to post-wean fertility (Guo et al., 2019; Arend, 2022).

In various locations around the world where animal welfare regulations have mandated lactation lengths for 4 weeks, approaches to comply and optimize performance have evaluated induction and breeding during lactation. The approaches have attempted to remove some of the pressure for high milk production in young sows to minimize their weight loss. Partial removal of the heaviest pigs in the litter from day 18 onward, when combined with boar exposure, can induce lactational estrus and allow good subsequent fertility for farrowing rate and litter size (Tarocco et al., 2000; Terry et al., 2013). The fertility advantages of split weaning are promising, but there has been some evidence that disruption of nursing stimuli with intermittent suckling or split weaning can initiate follicle development with or without estrus. This may be associated with delayed estrus after weaning and cystic follicles (Knox and Probst-Miller, 2004; Gerritsen et al., 2014) and would warrant identifying and tracking fertility in these sows.

5 Management of reproduction in weaned sows

The ovarian physiology of the sow is an important component of weaned sow fertility and will be mentioned only as it pertains to reproductive management. The reproductive physiology of the female pig is controlled primarily by hormone release from the hypothalamic–pituitary axis (HPX) which regulates ovarian function (Schwarz et al., 2008; Knox, 2015). Post-weaning fertility in sows is determined by the release of gonadotropins from the pituitary and the development of ovarian follicles (Lucy et al., 2001; Knox, 2022a). However, there is evidence to suggest that the ovaries respond to an array of other hormones and metabolites to regulate reproductive function in addition to HPX control (Knox, 2022a). The event of weaning removes the neuro-endocrine inhibition to HPX-stimulated gonadotropin release and initiates the start of the follicle phase (Quesnel, 2009). However, HPX and ovarian function after weaning is influenced by the physiology of the sow during lactation. The metabolic changes necessary for sows to produce milk for a large litter, maintain body function and growth, and repair the reproductive system following the previous gestation place a large stress on the female. Proper body condition entering farrowing helps optimize feed intake to allow sows to meet and balance the nutrient requirements for these functions. However, many factors such as illness, injury, stress, and hot weather can lead to off-feed events, and days of low feed intake, that can alter metabolic balance and limit meeting the requirements for lactation and fertility (Britt et al., 1983; Koketsu and Dial, 1997; Eissen et al., 2003; Anil et al., 2006; Auvigne et al., 2010). To lessen the possibilities of post-wean fertility problems, farms should optimize sow body condition by mid-gestation, manage the farrowing rooms to minimize heat stress, and provide feed in a manner to maximize intake during lactation.

Following weaning, sows initiate the process of follicle development with the majority displaying estrus on days 4–6 (Kemp and Soede, 1996; Steverink et al., 1999; Knox et al., 2018). However, different factors can intervene to accelerate or delay follicle development. One of the most important items that stimulates follicle development after weaning is boar exposure (Dyck, 1988; Knox et al., 2004; Gerritsen et al., 2005; Kemp et al., 2005). The data suggest that the method of application can help advance the onset and detection of estrus in sows. However, while the mechanism is not entirely clear, the stimulatory effects likely involve combinations of auditory, olfactory, visual, and physical cues (Hughes et al., 1990; Gerritsen et al., 2005). Yet other elements can affect the response as well, such as interval rather than continuous boar exposure (Hemsworth and Hansen, 1990; Knox et al., 2004). Interestingly, some factors, such as increased feed intake after weaning have no major effect on follicle development or estrus (Gianluppi et al., 2020). In contrast, inhibitors to follicle development may arise from reduced gonadotropin release or from other factors that affect the health and development of follicles on the ovary. Low energy intake throughout lactation delays follicle development (Bracken et al., 2003) and post-wean estrus (Reese et al., 1982). Specifically, reduced feed intake during lactation can impact early antral follicles and those at selection (Chen et al., 2012) with effects noted on steroidogenesis (Costermans et al., 2020). Low feed effects on follicles appear to be mediated at least partly through reduced follicle-stimulating hormone (FSH) and luteinizing hormone (LH) during lactation and after weaning (Kauffold et al., 2008).

Questions pertaining to what factors regulate follicle development during lactation and after weaning are complex and involve numerous follicles in various stages of development and at various intervals from ovulation. Sows maintain corpora lutea producing high levels of progesterone up until the time of luteolysis just before farrowing. Prior to parturition, several hormones are elevated to control the process for delivery of live pigs (Sherwood et al., 1976; Ellendorff et al., 1979; Foisnet et al., 2010). Soon after the last pig, the reproductive system begins the process of placental expulsion, uterine involution, and repair (Meile et al., 2020). Meanwhile, the ovaries of sows contain regressing corpora lutea that decrease in size by day (Kunavongkrit et al., 1982). Sows also appear to show numerous follicles that can vary in largest size within and among sows in the weeks of lactation (Britt et al., 1985; Lopes et al., 2020). The origin of the diversity in ovarian follicles in the first week is uncertain but might reflect differences in the volatile endocrine changes before and after farrowing. At weaning, when largest follicle size is classified as small, some have linked this to delayed estrus and lower litter size (Lopes et al., 2020), while variable follicle sizes throughout lactation can associate with poor estrus (Arend, 2022). Another study indicated that the presence of small–medium follicles in lactation and at weaning associated with optimal estrus (Arend et al., 2023), while another suggested size at weaning could not explain differences in interval to estrus (Lucy et al., 2001). There is not enough data linking endocrine patterns with different follicle patterns during lactation in the sow, but both appear somewhat similar to that of the cyclic gilt and sow (Ryan et al., 1994; Knox et al., 2003; Knox, 2005; Noguchi et al., 2010). In essence, gonadotropins are elevated in circulation, and populations of small and medium follicles are present during lactation, but LH pulses are suppressed and follicle selection is prevented. The endocrine physiology that regulates the process of follicle development in weaned sows has been reviewed (Knox, 2022a) and begins with selection of a cohort of healthy follicles that can competitively respond to FSH and LH to advance. The gonadotropins drive follicle development to the large ovulatory size within ~3–6 days, depending upon rate of development. During this period, the selected follicles produce and release estrogen into blood, which acts at the HPX to regulate estrus expression and ovulation (Soede et al., 1994).

A considerable number of farms around the world house sows in stalls following weaning and utilize these for breeding and gestation. Depending upon farm location and regulatory issues, other farms may utilize a combination of stalls or pens for breeding and group gestation. There are numerous types of stall-pen designs that can provide flexibility, but the objective is the same for the detection of estrus and insemination within 4–6 days of weaning. Based on the rate of follicle development to the large ovulatory size, the first day of estrus is typically day 4 or 5, but it is not unusual to have some sows display estrus on day 3 or day 6. But, sows that express estrus early on day 1 or 2 after weaning or late on day 7 or later do present problems for farms. Early estrus expression on days 1–2 is not normal and suggests errors in estrus detection or loss of nursing inhibition during lactation. These issues should be addressed to confirm the source of the problem, as these sows have lower fertility. The delayed returns are more common and typically involve problems in younger parity sows in response to metabolic or environmental stress.

Regardless of housing, stimulation and detection of estrus are equally important and farms may use different approaches based on the barn design, labor availability, and practicality. Industry surveys have been shared which indicate practices used (De Jong et al., 2012; Knox et al., 2013), and reviews are available on breeding herd management (Knox, 2016). There is variation in when farms start boar exposure after weaning, and this could explain some estrous variation. Female pigs appear to show the best fertility responses to boar stimuli when daily interval exposure is used in comparison to no exposure or continuous contact. Farms also vary in the duration of boar exposure per female and the number of times exposure occurs each day. Most sows are checked in the AM, but if checked only once a day, time of day is likely not that important. For sows, once daily exposure is likely all that is required, as only a limited number of females will be detected with the second check 6–8 h following the AM check. There is actually no major fertility advantage with AI timing in sows with greater frequencies of detection (Knox et al., 2002) and suggests labor may be utilized more effectively elsewhere. From a stimulatory and detection standpoint, factors that optimize the sight, sound, and smell of the boar for sows would have the greatest potential for fertility. This would involve using boars with high libido that are vocal and produce androgens and pheromones. Allowing adequate time of boar exposure is probably the most limiting aspect of the procedure as herds are larger but have limited labor. In stall housing, a boar is typically moved in front of 3–4 sows for 1–3 min. In stalls, control of the boar and stimulation and application of the back-pressure test often requires two persons unless robotic boar movers are employed. If pen exposure is used, then fenceline or physical exposure can be utilized. In pen exposure, it is difficult to expose sows for timed intervals as animals have the freedom to move about, and so times of exposure for the pen are based on the number of sows in the pen but not for individual sows. Some producers use multiple boars at the same time to increase the stimulus level or use a daily rotation of boars to keep libido high. Some maintain high boar libido by allowing detection boars to naturally mate cull females periodically. Although not common, some farms use no boars for detection of estrus and breeding. This approach is based on data indicating that some proportion of sows will display standing estrus to the back-pressure test alone and that this response may be very sensitive and tied with progression of days in estrus. However, while data shows standing alone to only the back-pressure test is possible, the absence of a boar is likely to be less effective for detecting all sows in estrus (Langendijk et al., 2000). Sows display standing estrus for variable lengths of time (Weitze et al., 1994) and has been linked to the wean to estrous interval (Steverink et al., 1999; Knox et al., 2002). Accurate detection of estrus is a critical element for breeding farms, as onset is used for timing inseminations. Further, delays in estrus and failed detection of estrus are two of the primary problems involved with reduced breeding herd performance and culling.

6 Use of artificial insemination technologies for breeding pigs

Most commercial breeding farms breed today use AI technologies to breed sows based on detection of estrus following weaning (Maes et al., 2011; Knox, 2016). The objective is often to breed groups of sows within only a few days of each other to minimize the days of farrowing and increase the age of pigs weaned on the same day. In addition, a pool of mature replacement gilts is maintained to replace sows to be culled and to utilize all available farrowing crates. For synchronized breeding, optimal semen fertility must coincide with peak sow fertility. Boar sperm collected and extended in liquid form typically has an optimal-use window of 3 days from collection but with certain extenders to reach until day 5. Some farms receive multiple deliveries of semen each week to ensure breeding can occur on all days of the week if desired or to cover breeding if there are multiple groups of sows weaned each week.

Procedures for collection and processing of semen from boars in AI centers to optimize fertility have been described (Schulze et al., 2019). The goals are to ensure that the AI dose contains the proper number of high motility sperm with normal morphology and no bacterial contamination (Waberski et al., 2019). Most AI doses are expected to contain ~80% motile sperm with normal morphology. Sperm can be processed for use in liquid extenders that increase the volume for creating AI doses and to prolong the fertile life of sperm while in storage at 17°C. There are different classes of semen extenders available with variable components that provide energy for sperm, membrane protection, pH balance, and other additives such as antibiotics to extend fertile life. The extenders can be classified as short (3 days), mid (5 days), and long (>5 days) term for their targeted ability to extend shelf-life. Examination of sperm under a microscope during storage and the fertility results when used for AI have defined their classification for use (Waberski et al., 1994; Kuster and Althouse, 1999). Many of the more complex extenders with longer shelf-life are used when shipping times and distances are greater. And while many AI centers extend semen in the long-term extenders, most breeding farms target the use of semen within 3 days from collection.

New priorities for the swine breeding industry have emerged over the years to include low sperm insemination, single, fixed-time AI, post-cervical AI (PCAI), and use of antibiotic-free extenders (Waberski et al., 2019). The objective to minimize the use of antibiotics used in human medicine is based on limiting the incidence of antibiotic resistance. This has been a priority for some time in livestock production systems and eliminating its use in semen extenders can help as well. New approaches that advance this objective include improved semen-processing techniques to minimize bacteria load, new packaging that inhibits bacterial growth, and a combination of extender and storage at lower temperatures (5°C) to prohibit bacterial growth. Another important trend has been the use of lower numbers of sperm in the AI dose, with the incentive based on the ability to more efficiently distribute the sperm from boars with higher genetic value. Over years, viable sperm numbers per AI dose have declined from 4 billion to as low as 1.5 billion in certain countries (Roca et al., 2016). In general, with conventional cervical AI, where a catheter is locked into the cervix, the dose typically contains ~3.0 billion viable sperm in a volume of ~80 mL. There are reports that sperm numbers can be lowered to 1.5 billion in cases where boar and semen fertility are monitored and high (Broekhuijse et al., 2012). Over years, improvements have occurred in boar management, semen processing, sperm evaluation, and quality control. In addition, in the sow breeding herd, improvements in health, management, and estrus detection and AI have collectively resulted in high farrowing rates and increased litter size. Because many farms have reached targeted levels for herd fertility, attention can be re-focused on approaches that enable the use of lower numbers from higher value boars. The use of PCAI allows for direct deposition of semen into the uterine body using a modified catheter (Peltoniemi et al., 2009; Kraeling and Webel, 2015). With this technique, numbers of sperm have been reduced from 1.0 to 2.5 billion in a 40–60 mL volume of extender (Rozeboom et al., 2004; Araújo et al., 2009; Hernández-Caravaca et al., 2012; Sbardella et al.,

2014; Bortolozzo et al., 2015). With PCAI, catheter insertion takes longer but insemination time is shorter. With PCAI, there is also some risk for injury to the sow cervix, with blood typically used as an indicator. Insemination with PCAI can also differ from conventional AI, in that, once the female is detected in estrus, the boar should be removed for at least 30 min before insemination to avoid cervical contractions which could make catheter insertion through the cervix more difficult (Ulguim et al., 2018). With conventional AI, catheter insertion may require only seconds, while insemination can take 5–7 min using gravity flow, vacuum from uterine contractions, and mild hand pressure. With PCAI, catheter insertion into the uterus can require 1–2 min, but insemination will only take a few seconds using hand pressure. Lastly, conventional AI is prone to leakage during AI with flow obstacles and disruption, while this problem does not frequently occur with PCAI.

Detection of estrus is used as the primary indicator for when ovulation may occur. However, the interval from onset of estrus to ovulation is quite variable (Weitze et al., 1994). Because of this, the standard procedure for sows is to check for estrus and then inseminate on each day they are in standing estrus to ensure one insemination occurs within 24 h before ovulation, a key measure for fertility (Soede et al., 1995). In weaned sows, while estrus can last 2–3 days, most females receive two inseminations. In commercial production systems, the majority of insemination doses are created from a pool of ejaculates from 3 to 6 boars to minimize effects of boar fertility. While sows do show considerable variation in duration of estrus, and most sows ovulate 60–70% of the way through their duration of estrus, this observation is not helpful in AI timing, since the duration of estrus is not known in real time. However, on average, weaned sows ovulate ~45 h after onset of estrus, and changes only slightly with frequency of estrous detection (Knox et al., 2002). Much of the data for natural service and even AI suggests that sperm maintain fertilizing potential for ~24 h following insemination. In competitive fertilization studies, semen used from two different sires and inseminated using the same numbers of sperm with similar fertility, and at the same time, compete for binding sites within the oviduct and for the opportunity to fertilize an egg, often result in 50% of the pigs sired by each boar. But as sperm numbers are reduced, sperm fertility declines, or the interval from insemination becomes too far or too close to ovulation for one of the sires, the proportion of the litter sired is reduced (Dziuk, 1996; Flowers, 1998). Certain events can reduce the fertile lifespan of sperm following insemination and can include poor semen quality, semen age, temperature stress during shipping, storage, or handling, and bacteria load in semen. To establish a functional sperm reservoir following insemination, sperm are transported to the oviduct over 1–3 h. During this period, only a small proportion of sperm advance to the oviduct, while most are eliminated by phagocytosis in the uterus or backflow through the cervix. Of the billions of sperm that are inseminated, only hundreds of thousands can be accounted for in the oviduct and anterior horn of the uterus. In the late follicular phase, when circulating estrogen increases and reaches a threshold, estrus is expressed, the LH surge occurs, and ovulation occurs (Soede et al., 1994). Ovulation of all the follicles is estimated to occur over 1–3 h (Flowers and Esbenshade, 1993). Immediately following ovulation, some of the sperm in the reservoir are stimulated for hyper-motility and release, to swim, contact and fertilize eggs, with numbers of sperm limited to prevent the occurrence of polyspermy (Miller, 2018). Following fertilization, early embryo development proceeds in the oviduct until the embryos pass into the uterus a few days after fertilization.

7 Use of hormones to control female pig fertility

Hormones are used to induce or synchronize estrus, aid in sperm transport, and synchronize ovulation for breeding in pigs. Induction of a follicle phase and estrus is an important management tool used to initiate puberty in acyclic gilts and synchronize estrus in sows after weaning. The procedures and hormones used include PMSG alone or in combination with hCG (Kraeling and Webel, 2015; De Rensis and Kirkwood, 2016; Knox, 2019). For success with induction in gilts, the female must be prepubertal but approaching mature weight and age. If the gilts are too immature, they may grow and ovulate follicles but could fail to express estrus and cycle a second time. With hormone-induced puberty, most of the gilts (90%) respond and grow follicles and ovulate within 6 days, but 10–20% of these often fail to display estrus (Lewchalermwong et al., 2020). Alternative approaches for puberty induction have included the use of PG600 in gilts failing to display estrus following 2–3 weeks of boar exposure (Patterson et al., 2016; Knox et al., 2021). This has worked well, in younger gilts, as most anestrual females are prepubertal and display estrus within 5 days. Use of gonadotropins at weaning can also advance and improve estrus, especially when problems arise in younger parity sows within summer or fall seasons. Generally, since most sows will display estrus within 6 days after weaning throughout the year, the use of gonadotropins on all sows is not likely to be cost effective, except when problems for the entire group are expected in summer or fall. Generally, there may be only 10–15% of sows that may be compromised as a result of metabolic state, parity, season, or stress (Vargas et al., 2006).

Over many decades, the synthetic progestogen, altrenogest, has been approved and widely used for control of reproduction in pigs. It has been effective for synchronizing estrus in gilts (Davis et al., 1985; Kraeling and Webel, 2015) and has been used in lactating (Lopes et al., 2017; Am-In and Kirkwood, 2019) and in weaned sows (Fernandez et al., 2005; Werlang et al., 2011) to synchronize estrus. The hormone is typically administered as a top dress on the daily feed for 14–18 days for gilts fed individually, while some have successfully transitioned to oral dosing when females are housed in stalls or in group pens. The more recent efforts with altrenogest has been investigating its use in primiparous sows during lactation or post-weaning as a method to control follicle development to improve sow fertility (Van Leeuwen et al., 2011a; Werlang et al., 2011; Lopes et al., 2017). From a practical standpoint, some concerns persist that some synchronization failures, or even abnormalities such as cysts, originate due to partial dosing or a missed altrenogest dosing day, regardless of whether females are in stalls or in pens. In the future, perhaps new formulations or even alternative delivery methods could provide extended biological activity and reduce the need for individual daily dosing. The effect of altrenogest mimics progesterone at the HPX to suppress LH and prevent final maturation of follicles until treatment ends. During altrenogest treatment, a pool of small–medium-sized follicles grow on the ovary (Hazeleger et al., 2005, Van Leeuwen et al., 2011b), similar to those populations observed during lactation in sows and in the luteal phase of cyclic gilts. In recent years, there has been interest in use of altrenogest to synchronize sows post-farrowing in cases of disease where entire litters were lost at various stages after farrowing. The intent was to re-establish breeding synchrony in these groups that lost litters within the weeks after farrowing. The information in early wean studies suggests estrus synchrony could be established, and fertility could be good as long as sows had progressed beyond the second week after farrowing (Dos Santos et al., 2004).

Although not widely used, the use of hormones in semen has been reported on the effects of fertility following AI (Knox, 2016). Seminal plasma has numerous components, some of which are hormones, that can regulate uterine immune function and sperm transport (Claus, 1990). Several studies have evaluated seminal plasma in relation to fertility (Flowers et al., 2013), and others have used it to improve the fertility of compromised sperm (Kirkwood et al., 2008). Specifically, prostaglandin (Pena et al., 2000; Cheng et al., 2001; De Rensis et al., 2012) or oxytocin (Peña et al., 1998; Duziński et al., 2014; Okazaki et al., 2014) have been added to semen or injected into sows to improve fertility with AI. The fertility reports are generally positive and assumes the effect is based on improved sperm transport, in cases of seasonal infertility and with sperm of lower quality. Perhaps one of the technologies of greatest interest involves the use of a single, fixed-time artificial insemination (SFTAI) as a method to accelerate genetic improvement with use of semen from higher indexing sires. This technology also includes the use of less sperm, labor, and costs per litter produced. The concept is based on the ability to use lower numbers of sperm in a single insemination if applied within the optimal period before ovulation (Knox, 2015). Different combinations of hormones can be used to synchronize the start of a follicular phase and ovulation. If synchronization of a follicular phase is used, this may involve the use of PMSG or PG600 or can follow the use of altrenogest in gilts or weaned sows. The critical element for the procedure is the induction of ovulation and the timing of the insemination (Brussow et al., 2009; Knox et al., 2017). The technique works well provided that females are at the proper stage of follicular development (Knox et al., 2011). The rationale for the use of gonadotropins to initiate a follicle phase reduces the risk of failed follicle development and lower fertility with SFTAI. Induction of ovulation can occur with GnRH, its analogues, hCG, or pLH. Ovulation induction works reasonably well with good follicle development and AI occurring within 24 h before ovulation. However, because follicle development failures are unpredictable and likely not synchronous in all females on commercial farms, the technology has had limited adoption due to fertility levels lower than breeding multiple times based on estrus. Nevertheless, in the future, advances in diagnostics could help identify fertile animals for success.

8 Managing sows during gestation for fertility

A female that expressed estrus and was inseminated is subsequently managed as a gestating sow. Gestating sows may remain in stalls or be moved into group pens. Management in gestation involves feeding to allow for body maintenance, growth, and development of the fetuses in utero. Most sows are managed using limited feed intake, but since gilts are still growing, and some sows may need to regain body condition following lactation, individual feeding in gestation can be used to improve weight gain and body condition. There have been concerns about the effects of high feeding levels occurring in the first week of gestation that cause reduced progesterone levels and lower embryo survival (Jindal et al., 1996). The high feeding level reduces systemic progesterone as a result of hepatic metabolism. However, more recent studies with gilts of greater prolificacy indicate no reduction in systemic progesterone or no detrimental effects on embryo survival, suggesting this period could still be used for body condition adjustment (Quesnel et al., 2010; Athorn et al., 2013). Many farms also increase feed intake in the last 3 weeks before farrowing, with the goal to improve litter parameters. But effects can vary, and impact on birth traits limited (Gonçalves et al., 2016). Nevertheless, increased feed intake in late gestation had no effect on sow body condition but did reduce the incidence of light birthweight pigs for gilts (Mallmann et al., 2019b). Gestation management also involves procedures for confirmation of pregnancy status as soon as possible to minimize the number of non-productive days. Regular returns to estrus at 18–24 days after service suggests fertilization and conception failures. Some non-pregnant sows can be detected for regular returns to estrus by moving boars near the sows while others may only be detected with use of ultrasound at 4 weeks following service (Kraeling and Webel, 2015). Sows detected as open with ultrasound can be checked for return to estrus in the next weeks or culled.

But non-pregnant sows should be monitored, as repeat breeder females also have a higher likelihood of subsequent failure (Tummaruk et al., 2010).

Following fertilization, embryos develop in the oviduct before passing into each of the uterine horns on days 3–4 after the first day of estrus and AI. The embryos float and space themselves and continue cell division but without an increase in embryo size until blastocyst hatching occurs after the first week. Following hatching, the embryos grow in size and remain spherical until they begin rapid expansion after day 12. Around this time, elongating embryos release endocrine signals to facilitate maternal recognition of pregnancy, which prevents prostaglandin release from the uterine endometrium and allows continued function of the corpora lutea. This allows for high progesterone and permits pregnancy to proceed. During days 13–18 of gestation, the embryos elongate to contact the surfaces of the uterus, and again releases hormone signals for the uterus (Geisert et al., 2015). Degenerating embryos can be visibly identified during the third week (Belstra et al., 2020), while healthy embryos continue to develop. Early conceptuses display increased fluid volume in the chorioallantois at day 30 (Knight et al., 1977, Miller et al., 2003) and is an optimal time for pregnancy diagnosis. Soon after this time, uterine space becomes limiting and the conceptuses contact each other and compete for attachment sites (Pope, 1994). The transition from the embryo to fetal stage coincides with skeleton calcification and limb formation after day 40 (Ullrey et al., 1965; Wrathall et al., 1974). After this time, the fetus continues to grow until birth (Ashworth et al., 2001) and fluid in the membranes increases and peaks around day 70 before declining (Knight et al., 1977). There are limited instances of individual fetal loss during gestation, and when they occur may be identified as mummified fetuses at farrowing. Farms are also aware of entire loss of litters in cases of abortion, where passage of embryos, fetuses, and membranes are observed in gestation. Most documented cases of abortion involve bacterial or viral diseases that can affect multiple sows, and all or some of the pigs in the litter (Givens and Marley, 2008; Salogni et al., 2016). In some cases, where abortion does not occur, but infection did, individual fetuses may appear at farrowing as mummies, dead pigs, stillbirths, or weak and small birthweight pigs.

During gestation, progesterone remains elevated in blood from. Progesterone stimulates the uterus to provide nourishment for the conceptuses, while suppressing uterine contractions, and keeping the cervix tightly sealed. Before the end of pregnancy at 115–116 days from first service, notable changes in hormones occur. In the week and days before farrowing, estrogen, cortisol, and relaxin are observed to increase in blood and act to prepare the birth canal (Ellendorff et al., 1979). The signal to initiate farrowing is thought to originate from the fetal HPX which induces fetal cortisol release and acts to alter placental function and uterine prostaglandin release (Taverne and Van

Der Weijden, 2008). In the day before farrowing, progesterone declines with regression of the corpora lutea (Baldwin and Stabenfeldt, 1975, Foisnet et al., 2011), and once cleared, oxytocin increases and uterine contractions originate and are coordinated from local fetal-placental units and systemic control from the sow.

9 Housing and environmental effects on pig fertility

Around the world, dramatic changes are occurring in the housing of pigs in breeding, gestation, and farrowing. For breeding pigs, this involves the ability to house gilts and sows in individual crates continuously or for only a limited period of time. This may also include requirements for housing in pens for certain periods of time and with specified space requirements (Galli et al., 2021). In many parts of the world, there may be no welfare restrictions related to housing the breeding herd in crates or pens but this could change in the future (Garcia and Mcglone, 2022). For farms that utilize stalls, the size of the sow must be evaluated. Sows that are too large for the crate will not be comfortable, may have issues accessing feed and water, and show difficulty in standing or lying. In some countries around the world, the use of crates is limited to only a few days after weaning, around breeding, and for farrowing (Maes et al., 2016). While management of females in crates is very similar around the world, there are numerous differences in the management of females in pen and group housing systems. The practical options for managing sows in groups can be related to farm size, facility costs, labor, fertility, and welfare (Levis and Connor, 2013). Sows that are weaned into groups can be stimulated and detected for estrus and inseminated, with a high potential for farrowing and producing a large litter. But there are greater challenges in these systems which require a greater level of management (Kemp and Soede, 2012). The opportunities for group housing involve discovering optimal approaches that are practical and efficient in production with different sizes of static or dynamic groups, and with using conventional or electronic feeding systems (Knox et al., 2014). Because there are numerous combinations for stall and pen housing for the breeding herd, each farm will need to determine what procedures work best for reproductive management for fertility and welfare (Bracken et al., 2003). Regardless of the housing system, each should allow for optimal feed intake, minimized stress and fighting, comfort, and effective reproductive management. With the variety of different farm sizes and systems, optimizing each is difficult on a practical basis and continued efforts are needed to discover improved ways to meet sow requirements while meeting labor, economic and practical considerations for the farms. Some of the major concerns in group housing systems involves feed intake regulation, minimizing stress, and injuries and fighting. Other considerations involve the ability to effectively perform reproductive management procedures in large group pens. This includes procedures to stimulate and detect estrus, diagnose pregnancy, quickly identify animals, and individually manage and treat them. Certain considerations, such as body size disparity associated with parity, and day of grouping, may require adjustments in housing and management.

When considering the effects of housing, the control of light, temperature, humidity and air quality are a priority. Depending upon location of farm around the world, seasonal changes in each these can affect herd performance. The major concern during the year are the effects of heat stress on estrus expression and conception rates. Number of days and hours each day above the critical temperature associate more with the problems of heat stress (Knox et al., 2019). Farm cooling systems such as fans, evaporators, misters, curtains, and natural ventilation can all help. Evaporative cooling is effective but becomes unable to lower heat under very high humidity. Concerns about indoor temperatures should be monitored when outside temperatures are elevated, especially in the afternoon. In periods of high temperature, management procedures should increase frequency of checks for operation of fans, supplemental cooling systems, and adequate supply of water. Animal procedures should be scheduled to avoid the hottest times of the day. Cold temperatures are often not a major problem to fertility in winter with supplemental heating, but there is evidence of large changes in temperatures within the breeding barns in winter. In certain locations within the barn, cold outdoor weather means that certain females may be exposed to very cold temperatures, drafts or flooring during day and night. It is important to walk through the barns to evaluate these potential stressors and minimize their effects for the well-being and fertility of individual sows. There has been much interest in selecting optimal lighting for barns and the effects on reproductive fertility. However, the data on lighting duration and intensity on estrus, conception, and other reproductive measures are elusive. While the European wild boar is a common ancestor of domestic pigs, and is a seasonal breeder and photoperiod responsive, only remnants of this trait remain in modern breeds. Nevertheless, some commercial herds and individual females show significant reductions in fertility in summer and early fall, while a majority of pigs in the herd are fertile year round. For lighting recommendations, 10–12 h should be provided each day at an intensity level at or above 100 lux. Limited data suggests positive effects for photoperiod adjustment in confinement that can clearly prevent seasonal infertility. However, there is data to suggest that very low levels of light in breeding and gestation (<50 lux) may be associated with reduced fertility and would warrant increasing illumination level (Arend et al., 2019). Continuous lighting might also not be advantageous as sows and gilts cannot synchronize their daily rhythm as a group to lights on and off, and as a result establish independent and random circadian hormone patterns.

10 Conclusion and future trends

Modern swine breeding herd reproductive performance has the potential to be very high, based on estrus and farrowing rates as well as litter size. Nevertheless, there are inefficiencies and cases of infertility that increase costs, reduce pig production, and disrupt animal flow. These problems occur in the face of limited labor and increasing herd sizes. It is difficult to predict the future for genetic progress and what traits may be of priority. Whether litter size will continue to increase in the future is uncertain but emphasis on survivability is expected. It is likely that semen numbers used per litter produced will continue to decline and the use of PCAI will increase. With the changes in limited labor availability for managing the breeding herd, it is possible that electronic management technologies will increase. Reproductive diagnostics could help advance certain technologies such as fixed time AI.

11 Where to look for further information

- Kraeling, R. R., and S. K. Webel. 2015. Current strategies for reproductive management of gilts and sows in North America. J. Anim. Sci. Biotechnol. 6:3: 1-14. 10.1186/2049-1891-6-3.Einarsson, S., Y. Brandt, N. Lundeheim, and A. Madej. 2008. Stress and its influence on reproduction in pigs: a review. Acta Veterinaria Scandinavica 50: 1-8. https://doi.org/10.1186/1751-0147-50-48.Knox, R. V. 2016. Artificial insemination in pigs today. Theriogenology 85: 83-93. http://dx.doi.org/10.1016/j.theriogenology.2015.07.009.
- https://www.merckvetmanual.com/management-and-nutrition/management-of-reproduction-pigs/overview-of-management-of-reproduction-pigsKnox, R. V. 2022. Overview of Management of Reproduction: Pigs Merck Manual: Veterinary Manual. Merck & Co., Inc., Rahway, NJ USA.Flowers, W. L. 2023. Litter-of-origin traits and their association with lifetime productivity in sows and boars. Mol Reprod Dev 90: 585-593. https://doi.org/10.1002/mrd.23565.
- DOI: 10.5772/20630Maes, D., A. Lopez Rodriguez, T. Rijsselaere, P. Vyt, and A. Van Soom. 2011. Artificial Insemination in Pigs. In: M. Manafi (ed.) Artificial Insemination in Farm Animals. InTechOpen Limited, London, UK.
- Kemp, B., and N. M. Soede. 2012. Reproductive issues in welfare-friendly housing systems in pig husbandry: A review. Reprod. Domest. Anim. 47: 51-57. 10.1111/j.1439-0531.2012.02108.x.

12 References

Amaral Filha, W. S., et al. 2009. Growth rate and age at boar exposure as factors influencing gilt puberty. *Livest. Sci.*, 120, 51–57.

Am-In, N. and Kirkwood, R. N. 2019. Feeding altrenogest during late lactation improves fertility of primiparous sows nursing smaller litters. *Can. J. Vet. Res.*, 83, 228–230.

Anil, S. S., et al. 2006. Association of inadequate feed intake during lactation with removal of sows from the breeding herd. *J. Swine Health Prod.*, 14, 296–301.

Araújo, É. B. C., et al. 2009. Reproductive performance of sows submitted to intrauterine insemination. *Revista Brasileira de Zootecnia*, 38, 1460–1467.

Arend, L. S. 2022. *External and Internal Factors Influencing Ovarian Follicle Development in Weaned Sows and and the Use of Hormaonal Technology to Improve Subsequent Fertility.* Ph.D., University of Illinois.

Arend, L. S., et al. 2019. Effects of feeding melatonin during proestrus and early gestation to gilts and parity 1 sows to minimize effects of seasonal infertility. *J. Anim. Sci.*, 97, 4635–4646.

Arend, L. S., et al. 2023. Effects of nursing a large litter and ovarian response to gonadotropins at weaning on subsequent fertility in first parity sows. *J. Anim. Sci.*, 101.

Ashworth, C. J., et al. 2001. Causes and consequences of fetal growth retardation in pigs. In: Geisert, R. D., et al. (eds.) *Control of Pig Reproduction VI.* Cambridge, UK: Society for Reproduction and Fertility.

Athorn, R. Z., et al. 2013. Effect of feeding level on luteal function and progesterone concentration in the vena cava during early pregnancy in gilts. *Reprod. Fertil. Dev.*, 25, 531–538.

Auvigne, V., P. et al. 2010. Seasonal infertility in sows: a five year field study to analyze the relative roles of heat stress and photoperiod. *Theriogenology*, 74, 60–66.

Baldwin, D. M. and Stabenfeldt, G. H. 1975. Endocrine changes in the pig during late pregnancy, parturition and lactation. *Biol. Reprod.*, 12, 508–515.

Bauer, R., et al. 1998. Body weight distribution and organ size in newborn swine (sus scrofa domestica) – A study describing an animal model for asymmetrical intrauterine growth retardation. *Exp. Toxicol. Pathol.*, 50, 59–65.

Baxter, E. M., et al. 2020. Managing the litter from hyperprolific sows. In: Farmer, C. (ed.) *The Suckling and Weaned Piglet.* Wageningen, Netherlands: Wageningen Academic Publishers.

Beaulieu, A. D., et al. 2010. Impact of piglet birth weight, birth order, and litter size on subsequent growth performance, carcass quality, muscle composition, and eating quality of pork1. *J. Anim. Sci.*, 88, 2767–2778.

Belstra, B. A., et al. 2020. Effects of the number of sperm and site of uterine semen deposition on conception rate and the number of embryos in weaned sows receiving a single fixed-time insemination. *J. Anim. Sci.*, 98, 1–10.

Bergman, P., et al. 2018. Sow removal in commercial herds: patterns and animal level factors in Finland. *Prev. Vet. Med.*, 159, 30–39.

Boripun, R., et al. 2021. Analysis of culling reasons during the breeding cycle and lifetime performance: the strategy to remove crossbred Landrace and Large White sows under tropical climate. *Vet. World*, 14, 3170–3174.

Bortolozzo, F. P., et al. 2015. New artificial insemination technologies for Swine. *Reprod. Domest. Anim.*, 50, 80–84.

Bracken, C. J., et al. 2003. Factors affecting follicular populations on Day 3 postweaning and interval to ovulation in a commercial sow herd. *Theriogenology*, 60, 11–20.

Britt, J. H., et al. 1983. Characterization of summer infertility of sows in large confinement units. *Theriogenology*, 20, 133–140.

Britt, J. H., et al. 1985. Control of follicular development during and after lactation in sows. In: Foxcroft, G. R., et al. (eds.) *Control of Pig Reproduction II*. Cambridge, UK: Journals of Reproduction and Fertility Ltd.

Broekhuijse, M. L. W. J., et al. 2012. Artificial insemination in pigs: predicting male fertility. *Vet. Q.*, 32, 151–157.

Brussow, K. P., et al. 2009. Studies on fixed-time ovulation induction in the pig. In: Rodriguez-Martinez, H., et al. (eds.) *Control of Pig Reproduction VIII*. Nottingham, UK: Nottingham University Press.

Calderón Díaz, J. A., et al. 2015. Age at puberty, ovulation rate, and uterine length of developing gilts fed two lysine and three metabolizable energy concentrations from 100 to 260 d of age. *J. Anim. Sci.*, 93, 3521–3527.

Chen, T. Y., et al. 2012. Undernutrition during early follicle development has irreversible effects on ovulation rate and embryos. *Reprod. Fertil. Dev.*, 24, 886–892.

Cheng, H., et al. 2001. Prostaglandin F2α added to extended boar semen at processing elicits in vitro myometrial contractility after 72 hours of storage. *Theriogenology*, 55, 1901–1906.

Claus, R. 1990. Physiological role of seminal plasma components in the reproductive tract of the female pig. *J. Reprod. Fertil. Suppl.*, 40, 117–131.

Costermans, N. G. J., et al. 2020. Consequences of negative energy balance on follicular development and oocyte quality in primiparous sows. *Biol. Reprod.*, 102, 388–398.

Craig, J. R., et al. 2017. Poorer lifetime growth performance of gilt progeny compared with sow progeny is largely due to weight differences at birth and reduced growth in the preweaning period, and is not improved by progeny segregation after weaning1. *J. Anim. Sci.*, 95, 4904–4916.

Cronin, G. M., et al. 1983. The incidence of, and factors associated with, failure to mate by 245 days of age in the gilt. *Anim. Reprod. Sci.*, 5, 199–205.

Da Silva, C. L. A., et al. 2017. Relationship between ovulation rate and embryonic characteristics in gilts at 35 d of pregnancy1. *J. Anim. Sci.*, 95, 3160–3172.

Davis, D. L., et al. 1985. Scheduled breeding of gilts after estrous synchronization with altrenogest. *J. Anim. Sci.*, 60, 599–602.

De Jong, E., et al. 2012. Management factors associated with sow reproductive performance after weaning. *Reprod. Domest. Anim.*, 48, 435–440.

De Rensis, F. and Kirkwood, R. N. 2016. Control of estrus and ovulation: fertility to timed insemination of gilts and sows. *Theriogenology*, 86, 1460–1466.

De Rensis, F., et al. 2012. Prostaglandin F2α and control of reproduction in female swine: a review. *Theriogenology*, 77, 1–11.

De Rensis, F., et al. 2017. Seasonal infertility in gilts and sows: Aetiology, clinical implications and treatments. *Theriogenology*, 96, 111–117.

Dos Santos, J. M., et al. 2004. Early-weaned sows: altrenogest therapy, estrus, ovulation, and reproductive performance. *Anim. Reprod. Sci.*, 84, 407–413.

Duziński, K., et al. 2014. The use of oxytocin in liquid semen doses to reduce seasonal fluctuations in the reproductive performance of sows and improve litter parameters–a 2-year study. *Theriogenology*, 81, 780–786.

Dyck, G. W. 1988. The effect of housing facilities and boar exposure after weaning on the incidence of postlactational anestrus in primiparous sows Can. *J. Anim. Sci.*, 68, 983–985.

Dziuk, P. J. 1996. Factors that influence the proportion of offspring sired by a male following heterospermic insemination. *Anim. Reprod. Sci.*, 43, 65–88.

Eissen, J. J., et al. 2003. The importance of a high feed intake during lactation of primiparous sows nursing large litters. *J. Anim. Sci.*, 81, 594–603.

Eliasson, L., et al. 1991. Relationships between puberty and production traits in the gilt. 1. Age at puberty. *Anim. Reprod. Sci.*, 25, 143–154.

Ellendorff, F., et al. 1979. Endocrinology of parturition in the pig. *Anim. Reprod. Sci.*, 2, 323–334.

Engblom, L., et al. 2007. Sow removal in Swedish commercial herds. *Livest. Sci.*, 106, 76–86.

Fernandez, L., et al. 2005. Reproductive performance in primiparous sows after postweaning treatment with a progestagen. *J. Swine Health Prod.*, 13, 28–30.

Flowers, W. L. 1998. Insemination programs for swine to increase fertility. *J. Anim. Sci.*, 76, 39–46.

Flowers, W. L. and Esbenshade, K. L. 1993. Optimizing management of natural and artificial matings in swine. In: Foxcroft, G. R., et al. (eds.) *Control of Pig Reproduction IV.* Cambridge, UK: The Journals of Reproduction and Fertility Ltd.

Flowers, W. L., et al. 2013. Boar seminal plasma proteins and thier relevance to reproductive technologies. In: Rodriguez-Martinez, H., et al. (eds.) *Control of Pig Reproduction IX.* Leicestershire, UK: Context Products, Ltd.

Foisnet, A., et al. 2010. Relationships between colostrum production by primiparous sows and sow physiology around parturition1. *J. Anim. Sci.*, 88, 1672–1683.

Foisnet, A., et al. 2011. Farrowing induction induces transient alterations in prolactin concentrations and colostrum composition in primiparous sows. *J. Anim. Sci.*, 89, 3048–59.

Galli, M. C., et al. 2021. The changing face and associated drivers of research on welfare of the gestating sow. Ital. *J. Anim. Sci.*, 20, 2174–2187.

Gama, L. L. and Johnson, R. K. 1993. Changes in ovulation rate, uterine capacity, uterine dimensions, and parity effects with selection for litter size in swine. *J. Anim. Sci.*, 71, 608–617.

Garcia, A. and J. J. Mcglone 2022. Animal welfare and the acknowledgment of cultural differences. *Animals (Basel)*, 12, 474, 1–11.

Geisert, R. D., et al. 2015. Implantation and establishment of pregnancy in the pig. In: Geisert, R. D. and Bazer, F. W. (eds.) *Regulation of Implantation and Establishment of Pregnancy in Mammals: Tribute to 45 Year Anniversary of Roger V. Short's "Maternal Recognition of Pregnancy".* Switzerland: Springer International Publishing.

Gerritsen, R., et al. 2005. Effects of (artificial) boar stimuli on uterine activity in estrous sows. *Theriogenology*, 64, 1518–1525.

Gerritsen, R., et al. 2014. Cystic ovaries in intermittently-suckled sows: follicle growth and endocrine profiles. *Reprod. Fertil. Dev.*, 26, 462–468.

Gianluppi, R. D. F., et al. 2020. Effects of different amounts and type of diet during weaning-to-estrus interval on reproductive performance of primiparous and multiparous sows. *Animal*, 14, 1906–1915.

Givens, M. D. and Marley, M. S. 2008. Infectious causes of embryonic and fetal mortality. *Theriogenology*, 70, 270–285.

Gonçalves, M. A. D., et al. 2016. Fact sheet - Impact of increased feed intake during late gestation on reproductive performance of gilts and sows. *J. Swine Health Prod.*, 24, 264–266.

Guo, J. Y., et al. 2019. Effect of suckling intensity of primiparous sows on production performance during current and subsequent parities1. *J. Anim. Sci.*, 97, 4845–4854.

Hazeleger, W., et al. 2005. The effect of feeding strategy during the pre-follicular phase on subsequent follicular development in the pig. *Dom. Anim. Endocrinol.*, 29, 362–370.

Hemsworth, P. H. and Hansen, C. 1990. The effects of continuous boar contact on the oestrus detection rate of weaned sows. *Appl. Anim. Behav. Sci.*, 28, 281–285.

Hernández-Caravaca, I., et al. 2012. Reproductive performance and backflow study in cervical and post-cervical artificial insemination in sows. *Anim. Reprod. Sci.*, 136, 14–22.

Houska, L. 2009. The relationship between culling rate, herd structure and production efficiency in a pig nucleus herd. *Czech J. Anim. Sci.*, 54, 365–375.

Hughes, P. E., et al. 1990. Mechanisms mediating the stimulatory effects of the boar on gilt reproduction. In: Cole, D. J. A., et al. (eds.) *Control of Pig Reproduction III*. Cambridge, UK: Journals of Reproduction and Fertility Ltd.

Hughes, P. E., et al. 2010. Relationships among gilt and sow live weight, P2 backfat depth, and culling rates. *J. Swine Health Prod.*, 18(6), 301–305.

Jindal, R., et al. 1996. Effect of nutrition on embryonal mortality in gilts: association with progesterone. *J. Anim. Sci.*, 74, 620–624.

Kauffold, J., et al. 2008. Effects of feeding Level during lactation on FSH and LH secretion patterns, and follicular development in primiparous sows. *Reprod. Domest. Anim.*, 43, 234–238.

Kemp, B. and Soede, N. M. 1996. Relationship of weaning-to-estrus interval to timing of ovulation and fertilization in sows. *J. Anim. Sci.*, 74, 944–949.

Kemp, B. and Soede, N. M. 2012. Should weaning be the start of the reproductive cycle in hyper-prolific sows? A physiological view. *Reprod. Domest. Anim.*, 47, 320–326. 10.1111/j.1439-0531.2012.02092.x.

Kemp, B., et al. 2005. Effects of boar contact and housing conditions on estrus expression in sows. *Theriogenology*, 63, 643–656.

Ketchum, R. and Rix, M. 2013. Zeroing in on female retention rates. In: *National Hog Farmer*. London, UK: Informa PLC.

Kim, J. S., et al. 2016. Relationship between body weight of primiparous sws during late gestation and subsequent reproductive efficiency over six parities. Asian-Australas. *J. Anim. Sci.*, 29, 768–774.

King, R. H. and Williams, I. H. 1984. The influence of ovulation rate on subsequent litter size in sows. *Theriogenology*, 21, 677–680.

Kirkwood, R. N., et al. 2008. Practical application of seminal plasma. *Theriogenology*, 70, 1364–1367.

Knight, J. W., et al. 1977. Conceptus development in intact and unilaterally hysterectomized-ovariectomized gilts: Interrelations among hormonal status, placental development, fetal fluids and fetal growth. *J. Anim. Sci.*, 44, 620–637.

Knox, R., et al. 2014. Seasonal effects of photoperiod and heat stress on fertility in the swine breeding herd. In: Bortolozzo, F. P., et al. (eds.) *XII SINSUI, May 2019 Porto Alegre, Rio Grande do Sul*. Brazil: UFRGS GRAFICA, 17–30. https://www.sinsui.com.br/?menu=pagina&CODPAGINA=453

Knox, R. V. 2005. Recruitment and selection of ovarian follicles for determination of ovulation rate in the pig. *Dom. Anim. Endocrinol.*, 29, 385–397.

Knox, R. V. 2015. Recent advancements in the hormonal stimulation of ovulation in swine. *Vet. Med. Research and Reports*, 6, 309–320.

Knox, R. V. 2016. Artificial insemination in pigs today. *Theriogenology*, 85, 83–93.

Knox, R. V. 2019. Physiology and Endocrinology Symposium: Factors influencing follicle development in gilts and sows and management strategies used to regulate growth for control of estrus and ovulation. *J. Anim. Sci.*, 97, 1433–1445.

Knox, R. V. 2022a. Follicle development in pigs: State of the art. *Mol. Reprod. Dev.*, 1–11.

Knox, R. V. 2022b. Overview of management of reproduction: pigs. *Merck Manual: Veterinary Manual* [Online]. Available: https://www.merckvetmanual.com/management-and-nutrition/management-of-reproduction-pigs/overview-of-management-of-reproduction-pigs.

Knox, R. V. and Probst-Miller, S. 2004. Evaluation of transrectal ultrasonography for use in identifying sources of reproductive in wean sows. *J. Swine Health Prod.*, 12(2), 71–74.

Knox, R. V., et al. 2002. Effect of frequency of boar exposure and adjusted mating times on measures of reproductive performance in weaned sows. *J. Anim. Sci.*, 80, 892–829.

Knox, R. V., et al. 2003. Plasma gonadotropins and ovarian hormones during the estrous cycle in high compared to low ovulation rate gilts. *J. Anim. Sci.*, 81, 249–260.

Knox, R. V., et al. 2004. Effect of housing system and boar exposure on estrus expression in weaned sows. *J. Anim. Sci.*, 82, 3088–3093.

Knox, R. V., et al. 2013. An analysis of survey data by size of the breeding herd for the reproductive management practices of North American sow farms. *J. Anim. Sci.*, 91, 433–445.

Knox, R. V., et al. 2017. Effect of numbers of sperm and timing of a single, post-cervical insemination on the fertility of weaned sows treated with OvuGel. *Theriogenology*, 92, 197–203.

Knox, R. V., et al. 2018. Design and biological effects of a vaginally administered gel containing the GnRH agonist, triptorelin, for synchronizing ovulation in swine. *Theriogenology*, 112, 44–52.

Knox, R. V., et al. 2011. Synchronization of ovulation and fertility in weaned sows treated with intravaginal triptorelin is influenced by timing of administration and follicle size. *Theriogenology*, 75, 308–319.

Knox, R. V., et al. 2021. Effect of physical or fenceline boar exposure and exogenous gonadotropins on puberty induction and subsequent fertility in gilts. *J. Anim. Sci.*, 99, No. 12, 1–11.

Koketsu, Y. 2003. Re-serviced females on commercial swine breeding farms. *J. Vet. Med. Sci.*, 65, 1287–1291.

Koketsu, Y. and Dial, G. D. 1997. Factors influencing the postweaning reproductive performance of sows on commercial farms. *Theriogenology*, 47, 1445–1461.

Koketsu, Y., et al. 1996. Feed intake pattern during lactation and subsequent reproductive performance of sows. *J. Anim. Sci.*, 74, 2875–2884.

Koketsu, Y., et al. 1997. Returns to service after mating and removal of sows for reproductive reasons from commercial swine farms. *Theriogenology*, 47, 1347–1363.

Koketsu, Y. and Iida, R. 2020. Farm data analysis for lifetime performance components of sows and their predictors in breeding herds. *Porc. Health Manag.*, 6(24), 1–12.

Koketsu, Y., et al. 2017. Factors for improving reproductive performance of sows and herd productivity in commercial breeding herds. *Porc. Health Manag.*, 3(1), 1–10.

Kraeling, R. R. and Webel, S. K. 2015. Current strategies for reproductive management of gilts and sows in North America. *J. Anim. Sci. Biotechnol.*, 6(3), 1–14.

Kruse, S., et al. 2011. Analysis of water, feed intake and performance of lactating sows. *Livest. Sci.*, 135, 177–183.

Kummer, R., et al. 2006. Reproductive performance of high growth rate gilts inseminated at an early age. *Anim. Reprod. Sci.*, 96, 47–53.

Kummer, R., et al. 2009. Reproductive performance of gilts with similar age but with different growth rate at the onset of puberty stimulation. *Reprod. Domest. Anim.*, 44, 255–259.

Kunavongkrit, A., et al. 1982. Follicular development in primiparous lactating sows. *Anim. Reprod. Sci.*, 5, 47–56.

Kuster, C. E. and Althouse, G. C. 1999. The fecundity of porcine semen stored for 2 to 6 days in Androhep and X-CELL extenders. *Theriogenology*, 52, 365–376.

Lamberson, W. R. and Safranski, T. J. 2000. A model for economic comparison of swine insemination programs. *Theriogenology*, 54, 799–807.

Lambert, E., et al. 1991. The extent and timing of prenatal loss in gilts. *Theriogenology*, 36, 655–665.

Langendijk, P., et al. 2000. Effects of boar contact and housing conditions on estrus expression in weaned sows. *J. Anim. Sci.*, 78, 871–878.

Leenhouwers, J. I., et al. 1999. Analysis of stillbirth in different lines of pig. *Livest. Prod. Sci.*, 57, 243–253.

Levis, D. G. and Connor, L. 2013. Group housing systems: choices and designs. *Pork Production & Managment: Animal Wellbeing: Sow Housing* [Online], 03643-3/13. Available: https://porkcdn.s3.amazonaws.com/sites/all/files/documents/2013SowHousingWebinars/1%20-%20Group%20Housing%20Systems.Choices%20and%20Designs%20-03643.pdf [Accessed May 2023].

Lewchalermwong, D., et al. 2020. Investigation into the variation in follicular and endocrine responses of prepubertal gilts treated with exogenous gonadotropins. *Anim. Reprod. Sci.*, 223(106622), 1–11.

Lopes, T. P., et al. 2017. Altrenogest treatment before weaning improves litter size in sows. *Reprod. Domest. Anim.*, 52(Suppl 4), 75–77.

Lopes, T. P., et al. 2020. Ovarian follicle growth during lactation determines the reproductive performance of weaned sows. *Animals (Basel)*, 10. https://doi.org/10.3390/ani10061012

Lucia Jr, T., et al. 2000. Lifetime reproductive performance in female pigs having distinct reasons for removal. *Livest. Prod. Sci.*, 63, 213–222.

Lucy, M. C., et al. 2001. Ovarian follicular growth in sows. In: Geisert, R. D., et al. (eds.) *Control of Pig Reproduction VI*. Cambridge, UK: Society for Reproduction and Fertility.

Maes, D., et al. 2011. Artificial insemination in pigs. *Artificial Insemination in Farm Animals* [Online]. Available: http://www.intechopen.com/books/artificial-insemination-in-farm-animals/artificial-insemination-in-pigs

Maes, D., et al. 2016. Impact of group housing of pregnant sows on health. *Porc. Health Manag.*, 2(17), 1–7.

Mallmann, A. L., et al. 2020. Effects of flush feeding strategy before breeding on reproductive performance of modern replacement gilts: impacts on ovulation rate and litter traits. *J. Anim. Sci.*, 98 No. 6, 1–10.

Mallmann, A. L., et al. 2019a. Impact of feed intake during late gestation on piglet birth weight and reproductive performance: a dose-response study performed in gilts. *J. Anim. Sci.*, 97, 1262–1272.

Mallmann, A. L., et al. 2019b. Maternal nutrition during early and late gestation in gilts and sows under commercial conditions: impacts on maternal growth and litter traits1. *J. Anim. Sci.*, 97, 4957–4964.

Mcglone, J. J., et al. 2004. The physical size of gestating sows. *J. Anim. Sci.*, 82, 2421–2427.

Meile, A., et al. 2020. Ultrasonographic examination of postpartum uterine involution in sows. *Anim. Reprod. Sci.*, 219 106540, 1–7.

Miller, D. J. 2018. Review: the epic journey of sperm through the female reproductive tract. *Animal*, 12, s110–s120.

Miller, G., et al. 2003. Pregnancy diagnosis in swine: a comparison of two methods of real-time ultrasound, and characterization of image and labor requirements for positive pregnancy diagnosis. *J. Swine Health Prod.*, 11, 233–239.

Moreira, R. H. R., et al. 2020. Variability of piglet birth weights: A systematic review and meta-analysis. *J. Anim. Physiol. Anim. Nutr. (Berl)*, 104, 657–666.

Nielsen, S. E., et al. 2018. Litter size of Danish crossbred sows increased without changes in sow body dimensions over a thirteen year period. *Livest. Sci.*, 209, 73–76.

Noguchi, M., et al. 2010. Peripheral concentrations of inhibin A, ovarian steroids, and gonadotropins associated with follicular development throughout the estrous cycle of the sow. *Reproduction*, 139, 153–161.

Okazaki, T., et al. 2014. Addition of oxytocin to semen extender improves both sperm transport to the oviduct and conception rates in pigs following AI. *Anim. Sci. J.*, 85, 8–14.

Oliviero, C. 2022. Offspring of hyper prolific sows: Immunity, birthweight, and heterogeneous litters. *Mol. Reprod. Dev.*, 1–5.

Patterson, J., et al. 2020. Associations among individual gilt birth weight, litter birth weight phenotype, and the efficiency of replacement gilt production. *J. Anim. Sci.*, 98(11), 1–13.

Patterson, J. and Foxcroft, G. 2019. Gilt management for fertility and longevity. *Animals (Basel)*, 9 434, 1–15.

Patterson, J., et al. 2016. Validation of the use of exogenous gonadotropins (PG600) to increase the efficiency of gilt development programs without affecting lifetime productivity in the breeding herd1. *J. Anim. Sci.*, 94, 805–815.

Patterson, J. L., et al. 2002a. The effect of lean growth rate on puberty attainment in gilts. *J. Anim. Sci.*, 80, 1299–1310.

Patterson, J. L., et al. 2002b. Impact of boar exposure on puberty attainment and breeding outcomes in gilts. *Theriogenology*, 57, 2015–2025.

Pearce, G. P. and Paterson, A. M. 1992. Physical contact with the boar is required for maximal stimulation of puberty in the gilt because it allows transfer of boar pheromones and not because it induces cortisol release. *Anim. Reprod. Sci.*, 27, 209–224.

Peltoniemi, O. A. T., et al. 1999. Seasonal and management effects on fertility of the sow: a descriptive study. *Anim. Reprod. Sci.*, 55, 47–61.

Peltoniemi, O. A. T., et al. 2009. Uterine insemination with a standard AI dose in a sow pool system. *Reprod. Domest. Anim.*, 44, 414–418.

Peña, F. J., et al. 1998. Treatment of swine summer infertility syndrome by means of oxytocin under field conditions. *Theriogenology*, 49, 829–836.

Pena, F. J., et al. 2000. Intrauterine infusion of PGF2α at insemination enhances reproductive performance of sows during the low fertility season. *Vet. J.*, 159, 259–261.

Perry, J. S. and Rowlands, I. W. 1962. Early pregnancy in the pig. *Reproduction*, 4, 175–188.

Pierozan, C. R., et al. 2020. Herd-level factors associated with piglet weight at weaning, kilograms of piglets weaned per sow per year and sow feed conversion. *Animal*, 14, 1283–1292.

Pigchamp 2022. *Benchmarking USA summary.* Ames, IA: PigCHAMP Inc. Available: https://www.pigchamp.com/benchmarking/benchmarking-summaries3-2022

Pope, W. F. 1994. Embryonic mortality in swine. In: Zavy, M. T. and Geisert, R. D. (eds.) *Embryonic Motrality In Domestic Species.* Ann Arbor, MI: CRC Press.

Quesnel, H. 2009. Nutritional and lactational effects on follicular development in the pig. In: Rodriguez-Martinez, H., et al. (eds.) *Control of Pig Reproduction VIII.* Nottingham, UK: Nottingham University Press.

Quesnel, H., et al. 2010. Post-insemination level of feeding does not influence embryonic survival and growth in highly prolific gilts. *Anim. Reprod. Sci.*, 120, 120–124.

Quesnel, H., et al. 2008. Influence of some sow characteristics on within-litter variation of piglet birth weight. *Animal*, 2, 1842–1849.

Quesnel, H., et al. 2000. Feed restriction in cyclic gilts: gonadotropin-independent effects of follicular growth. *Reprod. Nutr. Dev.*, 40, 405–414.

Reese, D. E., et al. 1982. Influence of energy intake during lactation on the interval from weaning to first estrus in sows. *J. Anim. Sci.*, 55, 590–598.

Riddersholm, K. V., et al. 2021. Identifying risk factors for low piglet birth weight, high within-litter variation and occurrence of intrauterine growth-restricted piglets in hyperprolific sows. *Animals*, 11, 2731.

Riesenbeck, A. 2011. Review on international trade with boar semen. *Reprod. Domest. Anim.*, 46, 1–3.

Roca, J., et al. 2016. Will AI in pigs become more efficient? *Theriogenology*, 86, 187–193.

Rodibaugh, M. 2002. Real time ultrasound. *J. Swine Health Prod.*, 10, 227.

Rodriguez-Zas, S. L., et al. 2006. Impact of biological and economic variables on optimal parity for replacement in swine breed-to-wean herds. *J. Anim Sci.*, 84, 2555–2565.

Roongsitthichai, A., et al. 2014. Influence of growth rate and onset of boar contact on puberty attainment of replacement gilts raised in Thailand. *Trop. Anim. Health Prod.*, 46, 1243–1248.

Ross, J. W., et al. 2017. Physiological mechanisms through which heat stress compromises reproduction in pigs. *Mol. Reprod. Dev.*, 84, 934–945.

Rozeboom, K. J., et al. 2004. The reproductive performance and factors affecting on-farm application of low-dose intrauterine deposit of semen in sows. *J. Anim. Sci.*, 82, 2164–2168.

Rutherford, K., et al. 2013. The welfare implications of large litter size in the domestic pig I: biologica factors. *Anim. Welf.*, 22, 199–218.

Ryan, D. P., et al. 1994. Follicular development during early pregnancy and the estrous cycle of the sow. *Theriogenology*, 42, 623–632.

Salogni, C., et al. 2016. Infectious agents identified in aborted swine fetuses in a high-density breeding area:a three-year study. *J. Vet. Diagn. Invest.*, 28, 550–554.

Sasaki, Y., et al. 2011. Assessment of lifetime economic returns of sows by parity of culled sows in commercial breeding herds. *J. Vet. Epidemiol.*, 16, 37–45.

Sbardella, P. E., et al. 2014. The post-cervical insemination does not impair the reproductive performance of primiparous sows. *Reprod. Domest. Anim.*, 49, 59–64.

Schenkel, A. C., et al. 2010. Body reserve mobilization during lactation in first parity sows and its effect on second litter size. *Livest. Sci.*, 132, 165–172.

Schulze, M., et al. 2019. New trends in production management in European pig AI centers. *Theriogenology*, 137.

Schwarz, T., et al. 2008. Physiological mechanisms of ovarian follicular growth in pigs – A review. *Acta Vet. Hung.*, 56, 369–378.

Schwarz, T., et al. 2021. Effects of individual versus group housing system during the weaning-to-estrus interval on reproductive performance of sows. *Animal*, 15 100122, 1–7.

Sherwood, O. D., et al. 1976. Relaxin concentrations in pig plasma following the administration of prostaglandin F2 alpha during late pregnancy. *Endocrinology*, 98, 875–879.

Soede, N. M., et al. 1994. Periovulatory profiles of oestradiol, LH and progesterone in relation to oestrus and embryo mortality in multiparous sows using transrectal ultrasonography to detect ovulation. *J. Reprod. Fertil.*, 101, 633–641.

Soede, N. M. and Kemp, B. Recent advances in pig reproduction: 2. Higher litter size. Anais do XXIII Congresso Brasileiro de Reprodução Animal (CBRA-2019); Gramado, RS, 15 a 17 de maio de 2019, 2019.

Soede, N. M., et al. 1995. Effects of time of insemination relative to ovulation, as determined by ultrasonography, on fertilization rate and accessory sperm count in sows. *J. Reprod. Fertil.*, 104, 99–106.

Stalder, K. J., et al. 2003. Financial impact of average parity of culled females in a breed-to-wean swine operation using replacement gilt net present value analysis. *J. Swine Health Prod.*, 11, 69–74.

Steverink, D. W., et al. 1999. Duration of estrus in relation to reproduction results in pigs on commercial farms. *J. Anim. Sci.*, 77, 801–809.

Tani, S., et al. 2018. Culling in served females and farrowed sows at consecutive parities in Spanish pig herds. *Porc. Health Manag.*, 4, 3.

Tantasuparuk, W., et al. 2001. Body weight loss during lactation and its influence on weaning-to-service interval and ovulation rate in Landrace and Yorkshire sows in the tropical environment of Thailand. *Anim. Reprod. Sci.*, 65, 273.

Tarocco, C., et al. 2000. Effect of split weaning interval on retrun to estrus and sow fertility. *J. Swine Health Prod.*, 8, 221–223.

Taverne, M. A. and Van Der Weijden, G. C. 2008. Parturition in domestic animals: targets for future research. *Reprod. Domest. Anim.*, 43(Suppl 5), 36–42.

Taverne, M. A. M. 1992. Physiology of parturition. *Anim. Reprod. Sci.*, 28, 433.

Terry, R., et al. 2013. Split weaning increases the incidence of lactation oestrus in boar-exposed sows. *Anim. Reprod. Sci.*, 142, 48–55.

Town, S. C., et al. 2005. Embryonic and fetal development in a commercial dam-line genotype. *Anim. Reprod. Sci.*, 85, 301–316.

Tummaruk, P. 2012. Effects of season, outdoor climate and photo period on age at first observed estrus in Landrace×Yorkshire crossbred gilts in Thailand. *Livest. Sci.*, 144, 163–172.

Tummaruk, P., et al. 2010. Influence of repeat-service and weaning-to-first-service interval on farrowing proportion of gilts and sows. *Prev. Vet. Med.*, 96, 194–200.

Ulguim, R. R., et al. 2014. Female lifetime productivity in a swine integration system using segregated gilt development units. *Tropical Animimal Health and Production*, 46, 697–700.

Ulguim, R. R., et al. 2018. Ovulation and fertility responses for sows receiving once daily boar exposure after weaning and OvuGel® followed by a single fixed time post cervical artificial insemination. *Theriogenology*, 105, 27–33.

Ullrey, D. E., et al. 1965. Growth of the swine fetus. *J. Anim. Sci.*, 24, 711–717.

Van Der Waaij, E. H., et al. 2010. Effect of excessive, hormonally induced intrauterine crowding in the gilt on fetal development on day 40 of pregnancy. *J. Anim. Sci.*, 88, 2611–2619.

Van Leeuwen, J. J. J., et al. 2011a. Effects of altrenogest treatments before and after weaning on follicular development, farrowing rate, and litter size in sows. *J. Anim. Sci.*, 89, 2397–2406.

Van Leeuwen, J. J. J., et al. 2011b. The effect of different postweaning altrenogest treatments of primiparous sows on follicular development, pregnancy rates, and litter sizes. *J. Anim. Sci.*, 89, 397–403.

Vanderhaeghe, C., et al. 2010. Longitudinal field study to assess sow level risk factors associated with stillborn piglets. *Anim. Reprod. Sci.*, 120, 78–83.

Vanderhaeghe, C., et al. 2011. Incidence and prevention of early parturition in sows. *Reprod. Domest. Anim.*, 46, 428–433.

Vanderhaeghe, C., et al. 2013. Non-infectious factors associated with stillbirth in pigs: a review. *Anim. Reprod. Sci.*, 139, 76–88.

Vargas, A. J., et al. 2006. Time of ovulation and reproductive performance over three parities after treatment of primiparous sows with PG600. *Theriogenology*, 66, 2017–2023.

Vargas, A. J., et al. 2009a. Factors associated with return to estrus in first service swine females. *Prev. Vet. Med.*, 89, 75–80.

Vargas, A. J., et al. 2009b. Reproductive performance of swine females re-serviced after return to estrus or abortion. *Anim. Reprod. Sci.*, 113, 305–310.

Waberski, D., et al. 1994. The initial fertilizing capacity of longerm-stored liquid boar semen following pre- and postovulatory insemination. *Theriogenology*, 41, 1367–1377.

Waberski, D., et al. 2019. Application of preserved boar semen for artificial insemination: past, present and future challenges. *Theriogenology*, 137.

Wang, C., et al. 2019. An analysis of culling patterns during the breeding cycle and lifetime production from the aspect of culling reasons for gilts and sows in southwest China. *Animals (Basel)*, 9, 1–9.

Weitze, K. F., et al. 1994. The onset of heat after weaning, heat duration, and ovulation as major factors in AI timing in sows. *Reprod. Dom. Anim.*, 29, 433–443.

Werlang, R. F., et al. 2011. Effects of breeding at the second oestrus or after post-weaning hormonal treatment with altrenogest on subsequent reproductive performance of primiparous sows. *Reprod. Domest. Anim.*, 46, 818–823.

Wrathall, A. E., et al. 1974. A radiographic study of development of the appendicular skeleton in the fetal pig. *Res. Vet. Sci.*, 17, 154–168.

Chapter 11

Accelerating genomic improvement in pigs by using reproductive biotechnologies

Bethany K. Redel and Paula R. Chen, USDA-ARS, Plant Genetics Research Unit, USA;
Bhanu Telugu, Caroline Gomes Lucas, Kiho Lee, Kristin M. Whitworth and Randall S. Prather, University of Missouri, USA

1 Introduction

Livestock production, specifically pig production, is an integral part of the food supply, and sustainable production is important. It is estimated that at least 70% increase in animal protein productivity will be needed to feed the world by 2050 (1). Conventional strategies aimed to improve food productivity often result in incremental improvements and will be impossible to nearly double the productivity by 2050. Genetic engineering can result in animals that perform well even in the face of threats such as disease and improve animal welfare. The repurposing of the CRISPR/Cas9 system to introduce novel alleles has revolutionized our ability to produce targeted modifications in cells and animals. While this technology has proved useful, there are complex events that need to take place in order to generate high-efficiency production of genetically engineered (GE) pigs. A robust system of embryo production is a necessary enabling technology for the genetic engineering of livestock. In

http://dx.doi.org/10.19103/AS.2024.0137.16

the review below, the current status of *in vitro* maturation of pig oocytes, *in vitro* fertilization (IVF) to create zygotes and embryo culture is described. Two main types of genetic modification routes are also described. Embryo-based methods include microinjection of reagents into oocytes or zygotes and sperm-mediated gene transfer, and cell-based methods include the development of embryonic stem cells and somatic cell nuclear transfer (SCNT).

The review concludes with specific examples of genetic engineering that may improve swine production. More than 20% of global animal production losses are linked to animal disease (1). While animals that are resistant to diseases can reduce the costs associated with outbreaks, other possible examples of GE pigs that improve production traits include making the animals more efficient at producing meat, changing the composition of the carcass, eliminating the necessity of castration, providing antimicrobial resistance, and making the animals adjust to the current environment. A real advantage of these genetic engineering technologies is that new genetic variations can be introduced in a single step. When no known variation occurs within a gene in a species, the use of genetic engineering can provide a solution like that of Cluster of Differentiation 163 (*CD163*). Numerous efforts to find natural porcine reproductive and respiratory (PRRS) virus resistance have failed (2, 3), and when genome databases were analyzed, there was no natural variation identified in *CD163* that would have resulted in PRRS resistance (4, 5). Currently, the ONLY solution to the significant costs resulting from PRRS is introducing new genetic variation via genetic engineering. Likely there will be other genetic engineering solutions developed for other intractable situations affecting production agriculture.

2 *In vitro* production of pig embryos

Selective breeding has brought tremendous improvement to livestock production. However, this breeding approach comes with challenges in that (1) variation must be seen within the herd that is associated with a positive trait and (2) improved or novel alleles from other species cannot be introgressed into the herd by breeding. Therefore, to accelerate the production of livestock with improved traits, biotechnology provides a solution. Assisted reproductive technologies, including the *in vitro* production (IVP) of embryos, are critically important to ensure high efficiency of embryo production and subsequent animal production. While the IVP system in humans is generally established with predictable outcomes, the IVP of embryos from agriculturally important species is becoming more prevalent and efficient than it has been in recent years (6, 7). Specifically, the importance of *in vitro* porcine embryo production has been recognized for agricultural applications (8). The IVP of embryos starts with *in vitro* oocyte maturation (IVM) to supply viable oocytes to generate *in vitro*-derived embryos. IVF or SCNT are used to produce embryos, and lastly, the embryos should not lose their viability once placed into *in vitro* culture (IVC). The IVP of porcine embryos can be used to accelerate genetic improvement by introducing superior genetics, providing a convenient method of transportation of genetics internationally, reducing potential importation of animal diseases to a farm during re-population, or producing GE animals (9). Although much research has been done to improve IVP systems, it remains a challenge to produce porcine embryos that are able to produce viable young at high efficiency, especially when using oocytes derived from pre-pubertal ovaries. For instance, oocytes derived from pre-pubertal animals are often used for IVP because of their availability at local abattoirs; however, embryos generated by using these oocytes are in general less competent to produce piglets than oocytes from sexually mature females. Researchers aiming to enhance the quality of oocytes derived from pre-pubertal animals identified methods to improve the efficiency of IVP. Most methods used to produce genome-engineered animals, which are discussed later, require the use of the IVP system before the embryos can be transferred to a recipient. Therefore, developing the most optimal culture environments that produce competent embryos is vital to GE pig production.

Recent advances in oocyte maturation, fertilization procedures, and embryo culture environments have drastically improved the efficiency of *in vitro* embryo production and the number of piglets born (10). A few of the methods and advancements to the system are highlighted here.

2.1 *In vitro* maturation

While each stage of the IVP process is important, the oocyte maturation step may be the most influential as the growth of the oocyte and its developmental competence directly determines whether a successful pregnancy and viable offspring can be achieved. *In vitro* oocyte maturation systems have been established to generate embryos; however, these *in vitro* embryos often lack the potential to develop into viable offspring compared to those that are produced in the *in vivo* environment. The differences are not clearly defined, but the IVM environment is considered to cause improper oocyte nuclear and cytoplasmic maturation, as well as oocytes subject to polyspermy during fertilization (11).

Under normal physiological conditions, immature oocytes at the primordial follicle stage gradually attain developmental competence by the time they grow to the antral follicle stage (12–14). During the growth of the oocyte from the primordial to the antral stage, the oocyte goes through programmed events that lead to the completion of cytoplasmic and nuclear maturation. The exact molecular and signaling mechanisms driving this oocyte maturation process remain incompletely understood. The somatic cells surrounding the oocytes

Published by Burleigh Dodds Science Publishing Limited, 2025.

during the time of growth, called cumulus cells (CCs), play a critical role in the oocyte attaining developmental competency. These cells supply the oocyte with the required nutrients and molecules needed to support subsequent embryo development. The communication between the oocyte and CCs is not a one-way relationship, but rather a bi-directional orchestration of the communication necessary for regulating meiosis and for successful fertility to be obtained (15, 16).

During IVM, the communication between the oocyte and surrounding CCs can be disrupted. Removal of immature oocytes from the follicle can cause the spontaneous resumption of meiosis in a subset of oocytes. Normally, the meiotic arrest is maintained due to high levels of oocyte-secreted cyclic adenosine monophosphate (cAMP), which facilitates to keep the level of meiosis-promoting factor (MPF) high in oocytes (17). Once a cumulus-oocyte complex (COC) is physically removed from the follicle, disruption of gap junctions between the oocyte and CCs may occur leading to a reduced influx of cyclic guanosine monophosphate (cGMP), an inhibitor of cAMP hydrolysis (18, 19). The reduced cGMP levels in oocytes prematurely removed from follicles can cause resumption of meiosis and cause the oocytes to be nuclear mature but not cytoplasmically mature thereby compromising development (20). As a result, researchers have treated COCs that were physically removed from the follicular environment with phosphodiesterase inhibitors to inhibit the hydrolyzation of cyclic nucleotides cGMP and cAMP. This inhibition can lead to a decrease in the number of COCs that spontaneously resume meiosis. Milrinone, a phosphodiesterase 3A inhibitor, treatment of porcine COCs improved both nuclear and oocyte competence (21).

Key oocyte mediators are epidermal growth factor (EGF), follicle-stimulating hormone (FSH), and luteinizing hormone (LH) (22, 23) and play a role on maturation and cumulus cell expansion. Together, these signaling molecules activate a downstream signaling pathway, mitogen activated protein kinase-3 and -1 (MAPK3/1) pathway by phosphorylation (24–26). Aspirated oocytes may not respond to LH when they are retrieved from small to medium-sized antral follicles during IVM. For this reason, FSH is often supplemented in the maturation medium to stimulate LH receptors on CCs (24–30). While progress has been made to enhance oocyte maturation, the maternal environment is composed of many unknown factors that cannot be fully established *in vitro*.

A cocktail of cytokines including fibroblast growth factor 2 (FGF2), leukemia inhibitory factor (LIF), and insulin-like factor 1 (IGF1), were found to have a synergistic effect on oocyte maturation (10). This trio of cytokines, termed 'FLI', promoted CC expansion, activation of MAPK pathway within CCs, and accelerated the severance of transzonal projections between CCs and the oocyte. Surprisingly, the supplementation of FLI to maturation medium produced two-fold more blastocysts and doubled the mean litter size of piglets

Published by Burleigh Dodds Science Publishing Limited, 2025.

derived from embryo transfer of these FLI-matured pre-pubertal gilt oocytes (10). Recently, by using an RNA sequencing approach, researchers identified three more cytokines: C-X-C motif chemokine ligand 12 (CXCL12), vascular endothelial growth factor (VEGFA), and Wingless-type MMTV integration site family member 5A (WNT5A), that together improved nuclear maturation, MAPK activation, and severance of the transzonal projections in oocytes matured in this cytokine supplemented medium (31). Together, research targeting improved oocyte maturation has had a huge impact on the success of porcine embryo production.

2.2 *In vitro* fertilization

A persistent challenge in the IVP of pig embryos is the IVF procedure. In pigs, the IVF procedure is not as successful as compared to other species, such as cows and humans. A higher incidence of polyspermic fertilization often occurs when using lower quality IVM oocytes and semen at the time of fertilization (32). Embryos that contain more than two pronuclei may develop to the blastocyst stage but often fail to develop to piglets (33, 34). Natural prevention of polyspermic fertilization occurs in oocytes after fertilization when conditions are appropriate. Since recreating the exact conditions, oocyte quality, and sperm quality as found *in vivo* is not possible, identifying alternative methods to produce monospermic fertilized embryos at a high efficiency has been the priority. Research focusing on the IVF medium components, sperm concentration, sperm/oocyte incubation time, and oocyte source have been analyzed with minimal improvements identified (35–41). Testing boar semen at different concentrations to determine monospermic fertilization rates has had an impact on our porcine IVF success. Currently, with our IVM protocol that is supplemented with an FLI medium, we see average monospermic fertilization rates around 50–80% and is often higher than what is described elsewhere in the literature (42).

2.3 Embryo culture

The embryo culture environment is the last critical step in producing competent IVP embryos. For effective *in vitro* embryo production, different media formulations such as modified Chatot, Ziomek, Bavister (CZB) medium (43), Beltsville embryo culture medium (BECM) (44), modified Whitten's medium (45), porcine zygote medium (PZM) (46), and North Carolina State University (NCSU)-23 and NCSU-37 media (47) have been developed to culture porcine embryos to the blastocyst stage. These media are designed based on the understanding of embryo metabolism, differing in energy and amino acid sources. Currently, NCSU-23 and PZM have been the most commonly used basal formulations (42, 48). NCSU-23 medium contains glucose (5.5 mM) and glutamine (1 mM), as energy substrates, and taurine (7 mM) and hypotaurine (5 mM) as amino acid sources, but pyruvate and lactate are absent (47). Certain NCSU-23 variants supplement with pyruvate (0.5 mM) and lactate (5 mM) for the first 48 h which are then replaced by glucose (5.5 mM) for the remaining period of culture (49). NCSU-37 was supplemented with sorbitol (12 mM) but lacks taurine and hypotaurine (47). PZM variations including PZM-3, a chemically semi-defined medium, PZM-4, a fully defined medium, and PZM-5, an undefined medium, are based on the composition of pig oviductal fluid with amino acids supplementation and glucose replacement by pyruvate (0.2 mM) and lactate (2 mM) (46, 50). Development to the blastocyst stage was higher in the PZM-3 medium when compared to PZM-4 (51). PZM-3 also improved parthenogenetic embryo development to the blastocyst stage when compared to PZM-4 and PZM-5 (50). While the availability of a fully defined medium, specifically PZM-4, offers a valuable resource to study embryo metabolism, there are contradicting results on outcomes of blastocyst formation *in vitro* (46, 51). Both PZM-3 and NCSU-23 media are proven to produce live piglets, but some reports have demonstrated that culture with PZM-3 compared to NCSU-23, improved IVF and SCNT-derived blastocyst development and increased the mean number of cells in *in vitro*-derived blastocysts (46, 52, 53).

While these base medium formulations have proved successful at producing competent embryos, producing blastocyst-stage embryos at high efficiency is still not routine. To improve the percentage of embryos that develop to the blastocyst stage, research has focused on ways to supplement these base formulations. Medium supplementation with different small molecules has become an important route to improve the efficiency of *in vitro* embryo production and accelerate the production of genetically modified livestock models (42, 54). Deep sequencing analysis of *in vivo*-derived (IVV) and *in vitro*-cultured (IVC) porcine blastocysts identified changes in the abundance of transcripts involved in amino acid transport and metabolism (55). Porcine embryos cultured in PZM-3 medium supplemented with arginine, now named MU1 medium, decreased arginine transporter *SLC7A1* message levels to that seen in IVV embryos. Supplementation of arginine also increased the percentage of embryos that developed to the blastocyst stage, as well as the total cell number in the embryos (56). Multiple improvements to MU1 have since taken place. Supplementation of MU1 medium with 5 μM of 5-(4-Chloro-phenyl)-3-phenyl-pent-2-enoic acid (PS48), referred as MU2 medium, is able to replace bovine serum albumin (BSA) present in the basal PZM-3 formulation resulting in a chemically defined medium. MU2 media improved *in vitro* embryo development through PS48 stimulation of phosphoinositide 3 kinase (PI3K) and subsequent increase in phosphorylation of v-akt murine thymoma viral oncogene homolog (AKT) (57). Afterward, MU2 containing BSA was modified by replacing 1 mM of glutamine with 3.75 mM GlutaMAX, an

L-alanyl-L-glutamine dipeptide. This medium, known as MU3 medium, enhanced development to blastocyst stage and caused a decrease of glutamine-related transcript abundance to levels comparable to the IVV embryos and improved mitochondrial function (58). Recently, the MU4 medium was developed by the removal of hypotaurine, which is an expensive antioxidant compound. When culture was conducted in 5% O_2, the absence or the presence of hypotaurine did not affect blastocyst development nor total cell number (59).

Other factors such as oxygen tension as well as temperature and osmolality have been shown to have an effect on the outcomes of *in vitro* embryo culture. García-Martínez and coworkers (60) demonstrated that the use of the same O_2 conditions as those found in the pig reproductive tract (7% O_2 during IVF and embryo culture) was beneficial for embryo production *in vitro*. The cleavage percentage and the number of cells in porcine IVF embryos were higher under 7% O_2 than those cultured under 20% O_2. Variations in temperature during embryo culture produced detrimental effects during porcine IVC. Currently, the physiological temperature of 38.5°C is the most common for embryo incubation. When the incubator temperature was higher than 43°C during embryo culture, a decrease in porcine embryo development was observed, but a molecular indicator of heat stress, heat shock protein 70, was not altered (61). Similarly, heat shock, at 42°C for 9 h in embryos at the 1- or 2-cell stage, induced developmental perturbations when compared to the non-heat shock controls (62). Interestingly, heat shock right after fertilization showed beneficial effects with an increase in cleavage percentages and a tendency to improve blastocyst percentage (63). The osmolality of embryo culture medium typically is around 280–290 mOsm (46, 47). Media with values below 270 mOsm or above 300 mOsm can adversely impact development during culture (64). However, contrasting reports demonstrated that when culture media were at higher osmolality levels (≤320 mOsm), no detrimental effect on embryonic development was detected (65). In further support of this, researchers also identified an increase in the total cell number of blastocyst embryos cultured with glycine-supplemented medium at 275 mOsm (66). Embryo vitrification and embryo transfer are also important methodologies for the production of GE pigs. Tu et al. describe the advancement of these technologies and the challenges that still exist (67).

3 Genetic engineering: embryo-based genetic engineering

Genetic engineering of zygotes is principally achieved by delivery of editors, targeting reagents and/or transgenes and via microinjection and sperm-mediated gene transfer. Lately, electroporation is being increasingly used for delivering reagents.

3.1 Microinjection of zygotes

In a landmark publication, Gordon et al. reported the first generation of a transgenic mouse by microinjecting a DNA construct containing SV40 virus and herpes simplex virus thymidine kinase genes into the pronucleus of the donor zygotes followed by implantation into pseudo-pregnant recipient animals (68). Soon after, transgenic mice with functional transgene (rat and human growth hormone genes) were generated by utilizing the microinjection approach (69). Following the proof-of-concept experiments in mice, the first transgenic pig was generated by zygotic microinjection (70). Following the initial breakthrough in generating transgenic pigs, progress in this field has been slow due to technical difficulties in utilizing pig embryos. Specifically, (a) poor visibility of pronuclei (71, 72), which requires either centrifugation of the zygotes to coalesce the lipid droplets and expose the pronuclei or cytoplasmic injections; (b) poor efficiencies in generating transgenic animals, with fewer than 15% of the offspring carrying the transgene (73); (c) lack of control over the number and location(s) of integrated transgenes, and consequently, the level of transgene expression; (d) positional variegation; (e) insertional mutagenesis; and (f) a high degree of mosaicism, to name a few of the challenges. These limitations necessitate the generation of multiple transgenic founder animals to screen for optimal transgene copy number integration, expression, and transmission to the next generation. Alternative methodologies, such as lentiviral transduction, transposon-mediated, or sperm-mediated genetic transfer were developed to modify zygotes (74). Specifically, (1) microinjection of transgene coated with bacterial recombinase (RecA) (75); (2) co-injection of *piggyBac* transposons and transposase combinations directly into the ooplasm/cytoplasm of the zygote (76–78); (3) injecting the lentiviruses into the perivitelline space of the zygote (79–81); and (4) sperm mediated gene transfer (briefly discussed below) (82, 83).

The availability and successful adoption of designer site-specific nucleases (SSN), including meganucleases, ZFNs, TALENs, and CRISPR/Cas has been a game-changer for genetic engineering efforts in pigs (Fig. 1). The SSN introduce a double-stranded break (DSB) at the target site, which is repaired predominantly by an error-prone non-homologous end joining DNA repair pathway resulting in small insertions, deletions and consequently alterations in the open reading frame. At a low frequency (and efficiency), the DSBs can be resolved by a template-assisted homology-directed repair (HDR) pathway, which results in precise donor-driven alterations at the cut site. Among the genome editors, the CRISPR/Cas system offers several advantages and has emerged as a clear front-runner for genetic engineering efforts. The CRISPR reagents are relatively inexpensive, are easily accessible, can be delivered directly into the oocytes (84) or zygotes (85) via microinjection or electroporation, and often

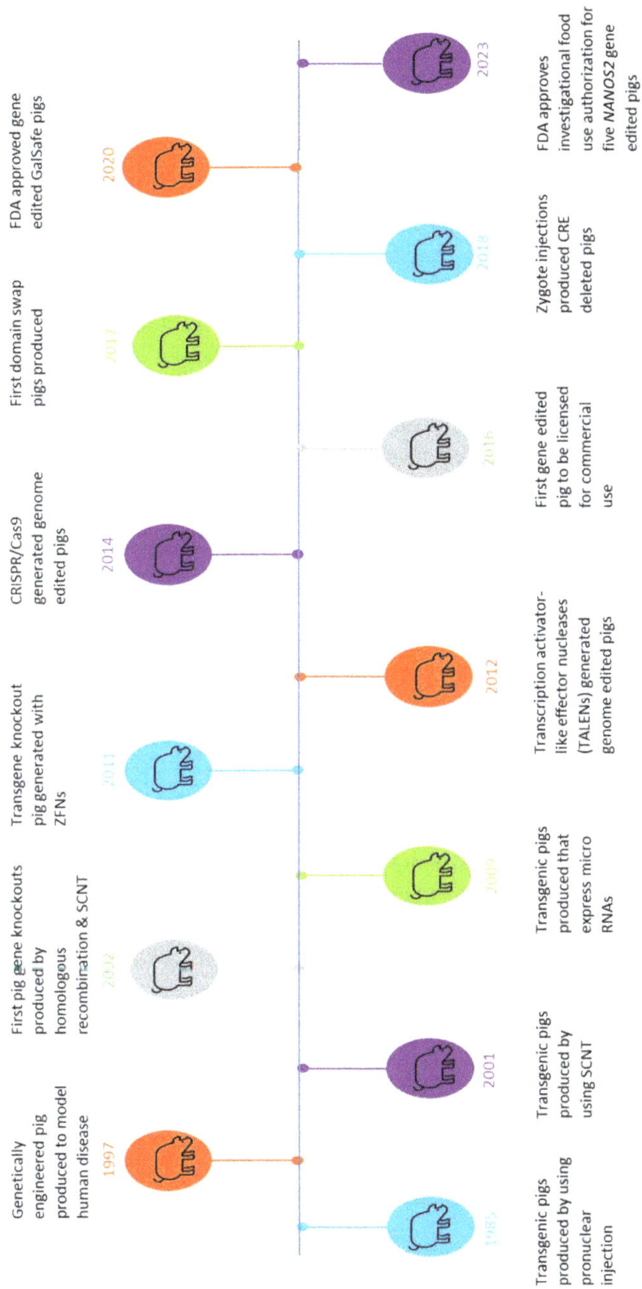

Figure 1 Milestones in the productio 1 of genetically engineered pigs.

result in high-efficiency editing at the target site. A distinct advantage of editing in zygotes over somatic cells is the relative ease of access of the editors to the target gene of interest. This may in part be due to the relatively relaxed state of chromatin prior to the first cleavage division. Added to this, the pregnancy rates are quite high using both *in vitro* (9) and *in vivo* fertilized zygotes (86) in the desired genetics. A major limitation of this approach, however, is the high rates of mosaicism, with the resulting offspring potentially carrying distinct somatic vs germline modifications. This makes the task of screening and predicting heritable mutations in a large animal model like pigs somewhat tricky. The mosaicism rates are especially pronounced in HDR-based targeting (87, 88) in zygotes. Several approaches are currently being tested or in use to mitigate mosaicism and off-targeting concerns emanating from zygotic injections including but not limited to (1) rational design of CRISPR guides and targeting reagents, (2) chemical alteration of targeting oligos/plasmids, (3) altering concentrations of the reagents and injection times, and (4) the use of CRISPR inhibitors, to name a few. However, every target allele requires independent validation, and the practitioners are encouraged to perform validation of the reagents for the target of interest prior to generating offspring.

3.2 Sperm-mediated transgenesis

One technique for producing GE pigs is the sperm-mediated gene transfer (SMGT) technique which is based on the intrinsic ability of spermatozoa to spontaneously deliver exogenous DNA to the oocyte (82, 89, 90). The first indication of spermatozoa having the capacity to hold foreign DNA was suggested 45 years ago when rabbit sperm cells could take up [3H]-thymidine-labeled purified SV40 DNA (91). Nevertheless, this finding did not receive attention in the scientific community and was reevaluated approximately 20 years later and utilized to introduce exogenous DNA into the genome of other species. In 1989, researchers illustrated that exogenous DNA could be detected in sea urchin spermatozoa and during embryonic development (92). In the same year, Lavitrano and colleagues (82) described that mouse spermatozoa were able to internalize DNA into their sperm head and could produce transgenic offspring at a frequency of about 30% after insemination. However, in the same year, Brinster and colleagues, reported that scientists had attempted to repeat Lavitrano's study and none of the laboratories were able to produce transgenic mice by SMGT (93). Wang et al. (94) used SMGT to produce transgenic rabbits and observed in their F1 offspring that one of six animals expressed the transgene. In the pig, transgenic expression of the human decay accelerating factor (hDAF) was detected in their progeny at a rate of 20–50% (95). Additionally, Weber and colleagues used the SMGT technology to create multi gene transgenic pigs expressing three fluorescent proteins. In this study, they identified multiple sites of integration of the genes as well as different integration numbers among cells from the same animal (96).

The sperm-DNA binding process is not a random event and relies on the presence of DNA-binding proteins (DBP) on the sperm cell surface (97). DNA/DBP complex formation is mediated by the major histocompatibility complex (MHC) class II molecules and by the CD4 antigen (90, 98). Natural factors present in the seminal fluid, such as the inhibitor factor (IF-1) and DNases can block DNA-binding proteins and cause the degradation of exogenous DNA, respectively (90, 98, 99). Over the years, the optimization of DNA uptake protocols has shown to be one of the key steps to increase the efficiency of this technique and to overcome the lack of repeatability observed when using spermatozoa as a vector (100–103). Technical advances in SMGT have been proposed to improve the ability to pass the transgene onto its offspring. Liposomes (99–106), adenovirus and retrovirus vectors (107–109), electroporation (110–117), restriction enzyme-mediated integration (REMI) (118–120), nanomaterial mediated delivery (99–123), as well as the combination of SMGT with intracytoplasmic sperm injection (ICSI)-mediated gene transfer (71–128) were some strategies developed to increase the DNA uptake by sperm cells.

SMGT could minimize the costs and the complexity of transgenic animal production; however, more research is needed to make this technique reproducible among different laboratories. Although we described the success of SMGT in other species, this technique has not been considered a reliable tool for genetic manipulation. Overall, the approach is time-consuming and inconsistent. Notwithstanding the drawbacks, zygotic injections remained the only option for genetic manipulation for two decades and have been employed with varying success in pigs (70, 79, 80).

4 Genetic engineering: cell-mediated approaches

4.1 Stem cells and chimeras

The establishment of pluripotent embryonic stem cells (ESC) from the inner cell mass of the early blastocyst stage embryo represents a major milestone for genetic engineering efforts (129). In mice, the ESC can be routinely and stably established in culture by using pre-implantation blastocysts from permissive strains. The ESC maintain pluripotency for an extended period in culture, permit iterative and compound genetic modification(s), and do not undergo senescence. The latter attribute facilitates clonal expansion and selection for the intended genetic modifications over multiple generations. The resulting stem cells when introduced into a recipient embryo, efficiently contribute to somatic and germline lineage, and even a whole organism when the placenta is derived from tetraploid donor embryos (130). This ability to efficiently contribute to the germline in the resulting chimera or a whole individual effectively allows the genetic modifications to be passed on to the next generation. Needless to say, the ESC has ushered in a new age for genetic engineering. In another landmark development, Takahashi and Yamanaka reprogrammed somatic cells to induced pluripotent stem cells (iPSC) by upregulating a core of four reprogramming genes (131, 132). The iPSC mimicked ESC in many ways, including the ability to give rise to germline chimeras (133, 134). Much of this excitement and promise surrounding ESC and iPSC, however, has yet to be translated to other animals besides permissive mouse strains. This is especially true for pigs. Although progress has been made, authenticated and chimera-competent porcine stem cells are still lacking (131, 132). With further refinements in the culture techniques that are currently being pursued, we expect this perceived bottleneck will be overcome in the coming years.

4.2 Somatic cell nuclear transfer

The use of a somatic cell as a nuclear donor for generating genetically modified offspring predates early genetic engineering efforts, with Hans Spemann first theorizing SCNT in 1938. Briggs and King successfully performed nuclear cloning in the frog (135). Four decades later, Campbell et al. succeeded in generating a cloned mammal, a sheep named Dolly (136). Success in pigs quickly followed, and to date, SCNT remains the cornerstone for genetic engineering efforts in pigs (137, 138). The SCNT procedure involves a series of technically challenging steps, with each of them having a potentially significant impact on cloning outcomes. These include: (a) genetic modification of donor cells, (b) enucleation (either aspiration or bisection) of metaphase chromosomes from MII-arrested oocytes; (c) injection of a donor cell into the perivitelline space of the enucleated oocyte; (d) fusion and activation of the reconstructed pseudo-zygote; (e) embryo culture and transfer into synchronized surrogate recipients; and (f) birth of healthy offspring fee of cloning abnormalities (139). As such, the procedure is technically challenging, expensive, and is less amenable for widescale adoption. Technical challenges aside, a major bottleneck for SCNT in pigs is the use of fetal fibroblasts for genetic modification and as nuclear donors. The fetal fibroblasts divide slowly in culture (long cell cycle interval), have a limited lifespan in culture, undergo senescence following clonal expansion and often do not survive cryopreservation. This severely limits generation of the compound and multiplex targeting of the somatic cells, and often requires the rederivation of 'refreshed' fibroblast lines for subsequent genetic modification. Additionally, the longer culture and associated drug selection to screen for the intended genetic modification impacts the clonability of cells. It is not uncommon that gene targeting in somatic cells even though resulting in an intended genetic modification, may not result in a viable offspring following

SCNT, or worse off result in offspring with overt imprinting defects manifested as lameness, respiratory issues, immunodeficiency, obesity, and early death due to incomplete reprogramming of the donor cell (140, 141). Due to the inherent inefficiencies and abnormalities associated with SCNT, the founder (G0) generation is often precluded from phenotyping. Nonetheless, in the absence of credible alternatives, SCNT remained a go-to method for genetic engineering in pigs. The advent of genome editors and advances in stem cell biology is expected to improve the efficiency of GE pig production.

Embryonic fibroblasts are a reliable alternative for genetic engineering. Recently, Park et al. (142) successfully established primary cells from embryonic outgrowths. The authors also demonstrated the establishment of primary embryonic fibroblast (EF)-like cells from both wild type and edited embryos, and alternatively performed genetic engineering in the EF cells. The advantage of this new paradigm is that it combines high-efficiency editing in zygotes with the establishment of primary and clonal lines from the edited embryos for screening intended genetic modification prior to generating edited offspring and overcoming mosaicism concerns. Additionally, the establishment of EFs also serves as a 'fitness' test for edited embryos with the embryonic outgrowths providing a readout of the implantation potential of the embryos and consequently pregnancy outcomes. As expected, the authors have demonstrated high SCNT efficiencies with the use of EFs and primary embryonic cells as nuclear donors (142).

In summary, embryo-based genetic modification allows for genetic modification from elite founders, which can be introduced into the breeding pipeline. To overcome mosaicism concerns emanating from embryo injections, primary embryonic cells can be established in culture, prescreened for genetic modification, and can be used as nuclear donors for SCNT to generate a clonal population of animals. When dealing with rare genetics or less prolific breeds (Yucatan, Ossabaw, other minipigs) where zygotic genetic modification is less practical, SCNT remains the only credible option, and the establishment of pluripotent cells or embryonic primary lines are expected to serve as an alternative. Regardless of the source (embryos or cells) for genetic engineering, it is clear that the genome editors have provided much-needed fillip for widescale adoption and deployment of genetic engineering in pigs for agricultural and biomedical research.

5 Gene editing to improve swine production: disease resistance

Application of the CRISPR/Cas9 system allowed for the production of pigs with viral resistance. Inactivation of CD163 by using the genome editing technology generated *CD163* knockout pigs that are fully resistant to both Type 1

(European) and Type 2 (North American) porcine reproductive and respiratory syndrome virus (PRRSVs) (143–146). The same strategy was used to produce gene-edited pigs resistant to other viruses, including the amino peptidase N (*ANPEP/APN/CD13*) edited pigs, which were fully resistant to the coronavirus transmissible gastroenteritis virus (TGEV) (147), and the anthrax toxin receptor 1 (*ANTXR1*) edited pigs that were found to be resistant to Senecavirus A (SVA) (148). A transmembrane serine protease (*TMPRSS2*) knockout pig was created to be resistant to a wide variety of viruses that includes influenza (149, 150). The models above do not introduce a transgene into the genome but were created by precision editing to include a base pair insertion or deletion (INDEL) in the targeted gene of interest. Other models have been created that used a transgene to confer viral resistance include the porcine radical SAM domain-containing 2 (*pRSAD2*) knock-in pigs that are resistance to Classical Swine Fever Virus (CSFV) and pseudorabies (PRV) (151) and the altered RELA proto-oncogene, NF-kB subunit (*RELA*) model that shows decreased severity to resistance to African Swine Fever Virus by altering 3 amino acids (152). Another transgene model utilizes a transgene so that pigs express recombinant human lysozyme (hLZ) in milk. This resulted in decreased mortality and morbidity to scours in growing piglets (153).

As mentioned earlier, one of the most economically and reproductively important viral-resistant models is the PRRSV-resistant *CD163* knockout model. Porcine Reproductive and Respiratory Syndrome is caused by a positive-stranded RNA Arterivirus that results in multi-million-dollar economic losses worldwide each day. Both type 1 or type 2 (154, 155) now appear to have worldwide distribution, and there is a highly pathogenic type 2 PRRSV (hpPRRSV) prevalent in Asia (156) that all use the same entry mediator, CD163. PRRSV replicates in pulmonary alveolar macrophages (PAMs) and results in prolonged viremia, respiratory distress, lethargy, and a mortality rate of 12–15% in young pigs. PRRS also results in reproductive failure by inducing late term abortion, early farrowing, decreased litter size and increased number of mummies in sows. PRRS also negatively impacts boars by reducing libido, inducing high fevers and reducing sperm count (157). Vaccinations have not proven useful as non-neutralizing antibodies enhance viral replication in alveolar macrophages (158). Currently the only way to guarantee that pigs do not become infected is via a GE pig (159).

Diseases such as PRRS result in significant costs. One calculatable cost is the economic cost to producers and consumers. There are not only economic costs, as mentioned above, to diseases like PRRS, but the decrease in food production is a threat to food security. When animals get sick, underperform, die, or are euthanized the food supply is interrupted. There is a sustainability and environmental cost. The impact of PRRS generates animal welfare costs as well because many of the infected animals need to be euthanized. The economic and animal welfare costs result in psychological and emotional costs to producers. Seeing the animals sick and having to euthanize them results in psychological and emotional stress in their lives. The economic costs also result in psychological and emotional distress as many swine operations are family-owned farms.

Due to the huge economic and animal well-being losses caused by PRRS, years of research determined that the PRRS virus was entering the cytoplasm of alveolar macrophages by either binding (160) the N-terminal region of sialic acid binding Ig like lectin 1 (SIGLEC1) or being uncoated by CD163 (160) resulting in infectivity (161–163). To address the question of SIGLEC1's role in PRRSV infection, *SIGLEC1* knockout pigs were created by homologous recombination and were challenged with type 2 PRRSV (164), but interestingly the pigs were not resistant and became infected with PRRS. The focus then shifted to the second candidate entry mediator, CD163. The first PRRSV resistant pigs were created at the University of Missouri by introducing INDELs into exon 7 of *CD163* resulting in complete loss of function of the protein. *CD163* knockout pigs were challenged with either a type 1 or type 2 PRRSV, and there was no evidence of infection as measured by fever, coughing, viremia, antibody response or lung pathology (143, 165). Pregnant *CD163* knockout sows were also shown to protect in utero fetuses with an intact expression of CD163 after PRRSV inoculation of the sow (166). Pigs were also edited to replace or remove exon 7 that encodes the viral binding domain keeping the remaining CD163 protein remained intact (167, 168). The pigs were fully resistant to both type 1 and type 2 PRRS. PRRS resistance provides an absolute solution to PRRS infection and will hopefully be a tool that swine producers have access to after Food and Drug Administration approval.

In 2013, a porcine epidemic diarrhea virus (PEDV) outbreak in the United States resulted in the death of nearly 7 million pigs, an estimated 10% loss in U.S. pig production for that year (169). It was determined that the PEDV outbreak was caused by coronaviruses which are enveloped, single-stranded, positive sense RNA viruses, placed in the order, *Nidovirales*. There are multiple coronaviruses that infect pigs that include the alphacoronaviruses, transmissible gastroenteritis coronavirus (TGEV) and PEDV. TGEV and PEDV infection of immunologically naïve newborn piglets results in losses approaching 100% mortality caused by mal-absorptive diarrhea and dehydration after the destruction of infected gut enterocytes (170, 171). TGEV is typically less destructive in pigs that have been previously infected with porcine respiratory coronavirus (PRCV). Pigs typically recover well from PRCV infection and produce neutralizing antibodies that also neutralize TGEV. This cross-reactivity results in a less severe infection in piglets exposed to TGEV (172). In 2014, another coronavirus, deltacoronavirus (PDCoV), was isolated from five infected farms in Ohio and rapidly spread throughout the United States (173). Porcine aminopeptidase N (APN, ANPEP, CD13) was characterized and hypothesized to be an entry mediator for pig coronaviruses. *ANPEP* knockout pigs were developed at the University of Missouri with the CRISPR/Cas9 system (174) and were challenged with both PEDV, TGEV and PDCoV. *ANPEP* knockout pigs were fully resistant to TGEV infection but retained susceptibility to infection with PEDV and PDCoV (175). Tu et al. generated pigs with a modified CMP-N-glycolylneuraminic acid hydrozylase (CMAH) gene (176). These pigs were hypothesized to be resistant to PEDV infection, however, the modified pigs still showed clinical symptoms but to a lesser extent (176).

Senecavirus A (SVA), a non-enveloped, positive-sense, single-stranded RNA virus belonging to the genus Senecavirus in the family *Picornaviridae*, was first reported in the United States in 2010, and by 2015, there were over 230 cases causing increased concern because the clinical signs are similar to foot and mouth disease virus (FMDV). Anthrax toxin receptor 1 (ANTXR1) was identified to be the cellular receptor for SVA as it is necessary for permissivity after infection of cell lines with SVA (177). Gene edited *ANTXR1* knockout pigs were created to investigate the role of ANTXR1 in SVA infection (148). ANTXR1 knockout pigs did not get infected with SVA or showed clinical symptoms or circulating viremia (148).

Other models have also been created to study disease resistance and receptor binding ability including the transmembrane protease, serine S1, member 2 (*TMPRSS2*) knockout pigs which were created to determine the role of TMPRSS2 protease in the pathogenesis of swine influenza (149–180). The CRISPR/Cas9 system was used to create *TMPRSS2* knockout pigs and then challenged with H1N1 and H3N2 subtype influenza virus (150). These GE pigs exhibited a significant reduction of virus shedding in the upper respiratory tract and lower titers in the lower respiratory tract in comparison to wild type pigs (150). Another GE pig model was created with a site-specific knock-in of Radical S-Adenosyl Methionine Domain Containing 2 (*RSAD2*) gene by using CRISPR/Cas9. The resulting transgenic pigs expressing *RSAD2* resulted in reduced CSFV and PRV infectivity (151, 153, 181).

Knowledge of viral entry mediators continues to increase as does the ability to design pigs to prevent infectivity. One virus, the African swine fever virus (ASFV), is quite frightening and entry of this virus into the United States and Western Europe would decimate pork production in those areas. ASFV causes a lethal, hemorrhagic disease in domestic swine. ASFV is a large, enveloped double-stranded DNA virus in the family *Asfarviridae*. Warthogs act as a host to the virus which causes a non-clinical, persistent infection, but does not typically kill warthogs compared to infection of domestic swine which results in a 100% mortality rate (182). One of the differences between the warthog and domestic pigs is a three amino acid difference in the RELA protein, a subunit of the NF-KB transcription factor that plays a key role in regulating immune response after infection. Gene-edited domestic pigs were created with either 2 or 3 of these amino acid changes and were challenged with ASFV. Pigs with all 3 warthog amino acids had a delayed onset of clinical signs and less viral DNA in blood and nasal samples, but unfortunately there was not resistance to ASFV (152, 183). One candidate molecule as the entry mediator of ASRV is CD163; however, *CD163* knockout pigs were also not resistant to ASFV (184). To date, an on/off type of entry mediation as observed with PRRSV or TGEV resistance has not been identified for ASFV, but the research continues robustly to identify an effective entry mediator to edit.

6 Gene editing to improve swine production: welfare concerns

6.1 Male aggressiveness and boar taint

In the swine industry, particular traits of males are undesirable, such as aggressiveness and boar taint in the meat. Therefore, piglets are surgically castrated after birth without anesthesia which has become a major welfare issue in several countries. The use of genome editing for reduction or feminization of the male gonads has shown promise as a solution for these issues. The short arm of the Y chromosome (*SRY* gene) has been identified as essential for the masculinization of the gonads in mammals (185), and pigs are known to have two head-to-head copies of the *SRY* gene (186). Deletion of the entire 'high mobility group' (HMG) domain of both *SRY* loci led to the production of three genetically modified male piglets that had complete external female genitalia and development of ovaries, oviducts, and uteri (187). However, by nine months of age, which is past the onset of puberty, the *SRY* knockout pigs did not present signs of heat, and internal examination revealed underdeveloped reproductive tracts and no follicular development in the ovaries compared to age-matched controls. Another target for controlling the size of male gonads in pigs is the kisspeptin-1 (*KISS1*) gene which encodes for a peptide that stimulates the release of gonadotropin-releasing hormone (GnRH) and subsequent gonadotropins (188). Disruption of porcine *KISS1* in males resulted in decreased testicular weight, reduced serum testosterone and LH, and abnormal development of seminiferous tubules (189). The lower production of testosterone would potentially indicate a decrease in androstenone and skatole which are responsible for boar taint. Moreover, *KISS1* knockout females demonstrated hypogonadism, no follicular development, and reduced serum follicular stimulating hormone (FSH) (189). Incorporation of these gene edited pigs into commercial herds would be difficult because all males are infertile, and each generation would need to be modified.

6.2 Thermoregulation

Neonatal piglets are unable to efficiently regulate their body temperature and are highly susceptible to cold exposure. This may be partially explained by the fact that pigs lack a functional uncoupling protein 1 (UCP1) that uncouples proton pumping across the inner mitochondrial membrane from ATP synthesis to generate heat due to a genetic event that deleted exons 3 to 5 (190). Adipose-specific knock-in (KI) of mouse *Ucp1* was achieved by CRISPR/Cas9-mediated homologous recombination into the endogenous porcine *UCP1* locus, and KI piglets at 1 and 6 months of age demonstrated an improved ability to withstand cold exposure compared to wild type controls (191). Moreover, KI pigs had decreased adiposity and increased lipolysis without differences in body weight or physical activity. Further characterization revealed larger mitochondria with highly developed cristae and upregulation of several mitochondrial-related transcripts in the white adipose tissue of the KI pigs (192). Certain bacteria, such as *Oscillospira* and *Coprococcus*, were increased in the feces of KI pigs and the concentration of these bacteria had a negative relationship with backfat thickness, indicating that expression of *UCP1* can alter gut microbiota and reduce fat deposition (193).

6.3 Growth traits

Myostatin (MSTN) is a well-known negative regulator of muscle growth, and natural mutations result in a double muscling phenotype in cattle (194). When *MSTN*-KO pigs were produced through SCNT, a 15% increase in birthweight was observed compared to wild-type littermates as well as more muscling due to increased myofiber number (195). Enlarged tongues were attributed to the lack of MSTN; however, this is a common phenotype seen in pigs produced by SCNT. Additionally, *MSTN*-KO pigs were shown to have leaner carcasses and increased expression of markers for beige adipose tissue, such as uncoupling protein 3 (UCP3) (196). Several negative effects were observed in *MSTN*-KO Landrace, Duroc, and Large White pigs, including abnormal legs and mortality shortly after birth; however, *MSTN*-KO in certain Chinese native breeds did not show the same adverse phenotypes, pointing to striking differences in genetic background (197, 198). Similar to cattle, mutations in *MSTN* also impact the reproductive capabilities of pigs as $MSTN^{-/+}$ sows had delayed age to puberty, increased number of services per pregnancy, and decreased litter sizes compared to wild-type sows (199).

Insulin-like growth factor 2 (IGF2) is a highly characterized growth factor that promotes growth during fetal and postnatal stages. Disruption of the binding site for a negative regulatory element that binds to intron 3 of *IGF2*, known as zinc finger BED-type containing 6 (ZBED6), increases the expression of *IGF2* in the skeletal muscle of pigs (200). After targeting of the ZBED6 binding site with CRISPR/Cas9 mutagenesis, edited pigs were shown to have increased growth rates and muscle development compared to wild-type pigs (201, 202), demonstrating a positive effect on production traits after editing a noncoding region. Likewise, knockout of *ZBED6* resulted in pigs with increased lean mass and upregulation of IGF2 expression at the mRNA and protein levels in skeletal muscle (203). However, weights of internal organs, such as the heart and liver, were increased in *ZBED6*-KO pigs which could have unknown effects on health and metabolism.

6.4 Surrogate sires

Males that lack endogenous germline can be used as recipients for sperm-producing stem cells of boars that possess superior genetics or desired traits. Specific genes that control the formation of the germline have been targeted by using different gene editing systems. For example, porcine nanos C2HC-type zinc finger 2 (*NANOS2*) was disrupted by using the CRISPR/Cas9 system which led to ablation of the germline in males, but female *NANOS2*-KO pigs were fertile (86). Allogenic transplantation of spermatogonial stem cells from wild-type boars into the rete testis of *NANOS2*-KO boars resulted in the appearance of sperm in the ejaculate of one boar at approximately 100 days post-transplantation (204). However, two of the six *NANOS2*-KO boars developed osteopathic degeneration which could be a result of NANOS2 deficiency or a potential off-target effect of the CRISPR/Cas9 system. Recently, deleted in azoospermia-like (*DAZL*)-KO pigs were generated by using TALENs which is not sex-specific for germline ablation (205). Successful transplantation of spermatogonial stem cells into *DAZL*-KO males was observed in 16 out of 18 individuals, albeit sperm concentrations were lower than wild-type controls which may limit their functionality in a production setting.

7 Conclusion and future trends

Significant advancements in reproductive technologies now allow the utilization of *in vitro-produced* embryos to rapidly improve the genetic value of pigs. Combined with recently developed genome editing systems, new technologies are available to present solutions to many challenges that are facing livestock producers and consumers today. Genome editing offers unpresented opportunities to introduce and merge validated genetic traits into a line of animals without prolonged breeding steps and incorporate novel traits into animals. While these tools are available, the current regulatory policies are not established to support the use of these technologies to produce genome-edited animals for production purposes. Other countries have recognized the value of the new technology and are already moving forward with the approval of food production from genome-edited animals (206). Constructing and streamlining the regulatory process that can accommodate the efficient application of the technology will assist producers globally to generate more resilient livestock to ensure a sustainable food supply.

8 Where to look for further information

Numerous other reviews highlight specific advancements in gene editing technology, as well as its application to enhancing livestock health and production.

- eBook chapter: Perisse I. V., Fan Z., Singina G. N., White K. L., and Polejaeva I. A. (2021) Improvements in Gene Editing Technology Boost Its Applications in Livestock. Frontiers in Genetics 11. [In English] doi:10.3389/fgene.2020.6146882.
- eBook: Lin J. C. and Van Eenennaam A. L. (2021) Electroporation-Mediated Genome Editing of Livestock Zygotes. Frontiers in Genetics 12. [In English] doi:10.3389/fgene.2021.6484823.
- Tait-Burkard, C., Doeschl-Wilson, A., McGrew, M. J. et al. Livestock 2.0 – genome editing for fitter, healthier, and more productive farmed animals. Genome Biol 19, 204 (2018). https://doi.org/10.1186/s13059-018-1583-14.
- Chen, P. J., Liu, D. R. Prime editing for precise and highly versatile genome manipulation. Nat Rev Genet 24, 161–177 (2023). https://doi.org/10.1038/s41576-022-00541-1.

9 References

1. World Organization for Animal Health 2024 [Available from: https://www.woah.org/en/what-we-do/animal-health-and-welfare/.
2. You, X., et al. Role of genetic factors in different swine breeds exhibiting varying levels of resistance/susceptibility to PRRSV. *Virus Research*. 2023;326:199057.
3. Su, C.M., et al. Recent advances in PRRS virus receptors and the targeting of receptor-ligand for control. *Vaccines (Basel)*. 2021;9(4):935.
4. Schroyen, M., et al., Bioinformatic analyses in early host response to Porcine Reproductive and Respiratory Syndrome virus (PRRSV) reveals pathway differences between pigs with alternate genotypes for a major host response QTL. BMC Genomics, 2016. 17(1): p. 196.
5. Johnsson, M., et al., Sequence variation, evolutionary constraint, and selection at the CD163 gene in pigs. Genet Sel Evol, 2018. 50(1): p. 69.
6. Mueller, M.L. and Van Eenennaam, A.L. Synergistic power of genomic selection, assisted reproductive technologies, and gene editing to drive genetic improvement of cattle. *CABI Agriculture and Bioscience*. 2022;3(1):13.

7. Paramio, M.T. and Izquierdo, D. Current status of in vitro embryo production in sheep and goats. *Reproduction in Domestic Animals*. 2014;49(Suppl 4):37–48.
8. Whitworth, K.M., et al. Improvements in pig agriculture through gene editing. *CABI Agriculture and Bioscience*. 2022;3(1):41.
9. Fowler, K.E., et al. The production of pig preimplantation embryos in vitro: current progress and future prospects. *Reproductive Biology*. 2018;18(3):203–211.
10. Yuan, Y., et al. Quadrupling efficiency in production of genetically modified pigs through improved oocyte maturation. *Proceedings of the National Academy of Sciences of the United States of America*. 2017;114(29):E5796–E804.
11. Sun, Q.Y. and Nagai, T. Molecular mechanisms underlying pig oocyte maturation and fertilization. *Journal of Reproductive Development*. 2003;49(5):347–359.
12. Sirard, M.A., et al. Contribution of the oocyte to embryo quality. *Theriogenology*. 2006;65(1):126–136.
13. Gilchrist, R.B., et al. Oocyte-secreted factors: regulators of cumulus cell function and oocyte quality. *Human Reproduction Update*. 2008;14(2):159–177.
14. Reader, K.L., et al. The role of oocyte organelles in determining developmental competence. *Biology (Basel)*. 2017;6(3):35.
15. Eppig, J.J. Oocyte control of ovarian follicular development and function in mammals. *Reproduction*. 2001;122(6):829–838.
16. Russell, D.L., et al. Bidirectional communication between cumulus cells and the oocyte: old hands and new players? *Theriogenology*. 2016;86(1):62–68.
17. Mehlmann, L.M., et al. The Gs-linked receptor GPR3 maintains meiotic arrest in mammalian oocytes. *Science*. 2004;306(5703):1947–1950.
18. Norris, R.P., et al. Cyclic GMP from the surrounding somatic cells regulates cyclic AMP and meiosis in the mouse oocyte. *Development*. 2009;136(11):1869–1878.
19. Vaccari, S., et al. Cyclic GMP signaling is involved in the luteinizing hormone-dependent meiotic maturation of mouse oocytes. *Biology of Reproduction*. 2009;81(3):595–604.
20. Sela-Abramovich, S., et al. Disruption of gap junctional communication within the ovarian follicle induces oocyte maturation. *Endocrinology*. 2006;147(5):2280–2286.
21. Roy, P.K., et al. Enhancing oocyte competence with milrinone as a phosphodiesterase 3A inhibitor to improve the development of porcine cloned embryos. *Frontiers in Cell Developmental Biology*. 2021;9:647616.
22. Su, Y.-Q., et al. Oocyte-dependent activation of mitogen-activated protein kinase (ERK1/2) in cumulus cells is required for the maturation of the mouse oocyte–cumulus cell complex. *Developmental Biology*. 2003;263(1):126–138.
23. Fan, H.Y., et al. MAPK3/1 (ERK1/2) in ovarian granulosa cells are essential for female fertility. *Science*. 2009;324(5929):938–941.
24. Shimada, M., et al. Luteinizing hormone receptor formation in cumulus cells surrounding porcine oocytes and its role during meiotic maturation of porcine oocytes. *Biology of Reproduction*. 2003;68(4):1142–1149.
25. Shimada, M., et al. Paracrine and autocrine regulation of epidermal growth factor-like factors in cumulus oocyte complexes and granulosa cells: key roles for prostaglandin synthase 2 and progesterone receptor. *Molecular Endocrinology*. 2006;20(6):1352–1365.
26. Hsieh, M., et al. Luteinizing hormone-dependent activation of the epidermal growth factor network is essential for ovulation. *Molecular and Cellular Biology*. 2007;27(5):1914–1924.

27. Prochazka, R., et al. Epidermal growth factor-receptor tyrosine kinase activity regulates expansion of porcine oocyte-cumulus cell complexes in vitro. *Biology of Reproduction*. 2003;68(3):797–803.
28. Yamashita, Y., et al. Positive feedback loop between prostaglandin E2 and EGF-like factors is essential for sustainable activation of MAPK3/1 in cumulus cells during in vitro maturation of porcine cumulus oocyte complexes. *Biology of Reproduction*. 2011;85(5):1073–1082.
29. Yuan, Y. and Krisher, R.L. In vitro maturation (IVM) of porcine oocytes. *Methods in Molecular Biology*. 2012;825:183–198.
30. Agung, B., et al. In vitro fertilization and development of porcine oocytes matured in follicular fluid. *Journal of Reproductive Development*. 2013;59(2):103–106.
31. Liu, X., et al. Maternal cytokines CXCL12, VEGFA, and WNT5A promote porcine oocyte maturation via MAPK activation and canonical WNT inhibition. *Frontiers in Cell Developmental Biology*. 2020;8:12.
32. Suzuki, H., et al. In vitro fertilization and polyspermy in the pig: factors affecting fertilization rates and cytoskeletal reorganization of the oocyte. *Microscopy Research Techniques*. 2003;61(4):327–334.
33. Funahashi, H. Polyspermic penetration in porcine IVM-IVF systems. *Reproductive, Fertility and Development*. 2003;15(3):167–177.
34. Somfai, T., et al. In vitro development of polyspermic porcine oocytes: relationship between early fragmentation and excessive number of penetrating spermatozoa. *Animal Reproduction Science*. 2008;107(1–2):131–147.
35. Abeydeera, L.R. and Day, B.N. Fertilization and subsequent development in vitro of pig oocytes inseminated in a modified tris-buffered medium with frozen-thawed ejaculated spermatozoa. *Biology of Reproduction*. 1997;57(4):729–734.
36. Abeydeera, L.R. and Day, B.N. In vitro penetration of pig oocytes in a modified Tris-buffered medium: effect of BSA, caffeine and calcium. *Theriogenology*. 1997;48(4):537–544.
37. Rath, D., et al. In vitro production of sexed embryos for gender preselection: high-speed sorting of X-chromosome-bearing sperm to produce pigs after embryo transfer. *Journal of Animal Science*. 1999;77(12):3346–3352.
38. Coy, P., et al. Sperm concentration influences fertilization and male pronuclear formation in vitro in pigs. *Theriogenology*. 1993;40(3):539–546.
39. Gil, M.A., et al. Brief coincubation of gametes in porcine in vitro fertilization: role of sperm:oocyte ratio and post-coincubation medium. *Theriogenology*. 2007;67(3):620–626.
40. Coy, P., et al. In vitro fertilization of pig oocytes after different coincubation intervals. *Theriogenology*. 1993;39(5):1201–1208.
41. Romar, R., et al. In vitro fertilization in pigs: new molecules and protocols to consider in the forthcoming years. *Theriogenology*. 2016;85(1):125–134.
42. Chen, P.R., et al. Challenges and considerations during in vitro production of porcine embryos. *Cells*. 2021;10(10):2770.
43. Pollard, J.W., et al. Comparison of development of pig zygotes and embryos in simple and complex culture media. *Journal of Reproduction and Fertility*. 1995;103(2):331–337.
44. Dobrinsky, J.R., et al. Development of a culture medium (BECM-3) for porcine embryos: effects of bovine serum albumin and fetal bovine serum on embryo development. *Biology of Reproduction*. 1996;55(5):1069–1074.

Published by Burleigh Dodds Science Publishing Limited, 2025.

45. Beckmann, L.S. and Day, B.N. Effects of media NaCl concentration and osmolarity on the culture of early-stage porcine embryos and the viability of embryos cultured in a selected superior medium. *Theriogenology*. 1993;39(3):611–622.
46. Yoshioka, K., et al. Birth of piglets derived from porcine zygotes cultured in a chemically defined medium. *Biology of Reproduction*. 2002;66(1):112–119.
47. Petters, R.M. and Wells, K.D. Culture of pig embryos. *Journal of Reproduction and Fertility*. 1993;48:61–73.
48. Lucas, C.G., et al. Applications of omics and nanotechnology to improve pig embryo production in vitro. *Molecular Reproduction and Development*. 2019;86(11):1531–1547.
49. Canovas, S., et al. DNA methylation and gene expression changes derived from assisted reproductive technologies can be decreased by reproductive fluids. *Elife*. 2017;6.
50. Cao, Z., et al. Effects of chemically defined medium on early development of porcine embryos derived from parthenogenetic activation and cloning. *Zygote*. 2012;20(3):229–236.
51. Spate, L.D., et al. Replacement of bovine serum albumin with N-methyl-D-aspartic acid and homocysteine improves development, but not live birth. *Molecular Reproduction and Development*. 2012;79(5):310.
52. Im, G.S., et al. In vitro development of preimplantation porcine nuclear transfer embryos cultured in different media and gas atmospheres. *Theriogenology*. 2004;61(6):1125–1135.
53. Nánássy, L., et al. Effects of activation methods and culture conditions on development of parthenogenetic porcine embryos. *Animal Reproduction Science*. 2008;104(2):264–274.
54. Lucas, C.G., et al. Effects of RAD51-stimulatory compound 1 (RS-1) and its vehicle, DMSO, on pig embryo culture. *Reprod Toxicol*. 2021;105:44–52.
55. Bauer, B.K., et al. Transcriptional profiling by deep sequencing identifies differences in mRNA transcript abundance in in vivo-derived versus in vitro-cultured porcine blastocyst stage embryos. *Biology of Reproduction*. 2010;83(5):791–798.
56. Redel, B.K., et al. Arginine increases development of in vitro-produced porcine embryos and affects the protein arginine methyltransferase?dimethylarginine dimethylaminohydrolase?nitric oxide axis. *Reproduction, Fertility and Development*. 2015. https://doi.org/10.1071/rd14293
57. Spate, L.D., et al. PS48 can replace bovine serum albumin in pig embryo culture medium, and improve in vitro embryo development by phosphorylating AKT. *Molecular Reproduction and Development*. 2015;82(4):315–320.
58. Chen, P.R., et al. Glutamine supplementation enhances development of in vitro-produced porcine embryos and increases leucine consumption from the medium. *Biology of Reproduction*. 2018;99(5):938–948.
59. Chen, P.R., et al. Removal of hypotaurine from porcine embryo culture medium does not impair development of in vitro-fertilized or somatic cell nuclear transfer-derived embryos at low oxygen tension. *Molecular Reproduction and Development*. 2020;87(7):773–782.
60. García-Martínez, S., et al. Mimicking physiological O2 tension in the female reproductive tract improves assisted reproduction outcomes in pig. *Molecular Human Reproduction*. 2018;24(5):260–270.

61. Kojima, T., et al. Effect of heat stress on development in vitro and in vivo and on synthesis of heat shock proteins in porcine embryos. *Molecular Reproduction and Development*. 1996;43(4):452–457.
62. Isom, S.C., et al. Heat shock of porcine zygotes immediately after oocyte activation increases viability. *Molecular Reproduction and Development*. 2009;76(6):548–554.
63. Isom, S.C., et al. Enhanced developmental potential of heat-shocked porcine parthenogenetic embryos is related to accelerated mitogen-activated protein kinase dephosphorylation. *Reproduction, Fertility, and Development*. 2009;21(7):892–900.
64. Li, R., et al. Concentration and composition of free amino acids and osmolalities of porcine oviductal and uterine fluid and their effects on development of porcine IVF embryos. *Molecular Reproduction and Development*. 2007;74(9):1228–1235.
65. Hwang, I.S., et al. Osmolarity at early culture stage affects development and expression of apoptosis related genes (Bax-alpha and Bcl-xl) in pre-implantation porcine NT embryos. *Molecular Reproduction and Development*. 2008;75(3):464–471.
66. Redel, B.K., et al. Glycine supplementation in vitro enhances porcine preimplantation embryo cell number and decreases apoptosis but does not lead to live births. *Molecular Reproduction and Development*. 2016;83(3):246–258.
67. Tu, C.F., et al. Invited review: reproductive technologies needed for the generation of precise gene-edited pigs in the pathways from laboratory to farm. *Animal Biosciences*. 2023;36(2):339–349.
68. Gordon, J.W. and Ruddle, F.H. Integration and stable germ line transmission of genes injected into mouse pronuclei. *Science*. 1981;214(4526):1244–1246.
69. Palmiter, R.D., et al. Dramatic growth of mice that develop from eggs microinjected with metallothionein-growth hormone fusion genes. *Nature*. 1982;300(5893):611–615.
70. Hammer, R.E., et al. Production of transgenic rabbits, sheep and pigs by microinjection. *Nature*. 1985;315(6021):680–683.
71. Perry, A.C., et al. Mammalian transgenesis by intracytoplasmic sperm injection. *Science*. 1999;284(5417):1180–1183.
72. Salamone, D., et al. Transgenesis in farm animals. *Animal Reproduction*. 2018;9(4):772–776.
73. Palmiter, R.D., et al. Differential regulation of metallothionein-thymidine kinase fusion genes in transgenic mice and their offspring. *Cell*. 1982;29(2):701–710.
74. Talluri, T. and Telugu, B.P. Advances and applications of transgenesis in farm animals. *Frontier Technologies in Bovine Reproduction*. 2022:255–288.
75. Maga, E.A., et al. Increased efficiency of transgenic livestock production. *Transgenic Research*. 2003;12:485–496.
76. Garrels, W., et al. Germline transgenic pigs by Sleeping Beauty transposition in porcine zygotes and targeted integration in the pig genome. *PLoS One*. 2011;6(8):e23573.
77. Li, Z., et al. Generation of transgenic pigs by cytoplasmic injection of piggyBac transposase-based pm GENIE-3 plasmids. *Biology of Reproduction*. 2014;90(5):93, 1–10.
78. Garrels, W., et al. Identification and re-addressing of a transcriptionally permissive locus in the porcine genome. *Transgenic Research*. 2016;25:63–70.
79. Hofmann, A., et al. Efficient transgenesis in farm animals by lentiviral vectors. *EMBO Reports*. 2003;4(11):1054–1058.

80. Whitelaw, C.B.A., et al. Efficient generation of transgenic pigs using equine infectious anaemia virus (EIAV) derived vector. *FEBS Letters*. 2004;571(1–3):233–236.
81. Cabot, R., et al. Transgenic pigs produced using in vitro matured oocytes infected with a retroviral vector. *Animal Biotechnology*. 2001;12(2):205–214.
82. Lavitrano, M., et al. Sperm cells as vectors for introducing foreign DNA into eggs: genetic transformation of mice. *Cell*. 1989;57(5):717–723.
83. Lavitrano, M., et al. Sperm-mediated gene transfer. *Reproduction, Fertility and Development*. 2005;18(2):19–23.
84. Su, X., et al. Production of non-mosaic genome edited porcine embryos by injection of CRISPR/Cas9 into germinal vesicle oocytes. *Journal of Genetics and Genomics*. 2019;46(7):335–342.
85. Hai, T., et al. One-step generation of knockout pigs by zygote injection of CRISPR/Cas system. *Cell Research*. 2014;24(3):372–375.
86. Park, K.-E., et al. Generation of germline ablated male pigs by CRISPR/Cas9 editing of the NANOS2 gene. *Scientific Reports*. 2017;7(1):40176.
87. Park, K.-E., et al. Targeted gene knockin in porcine somatic cells using CRISPR/Cas ribonucleoproteins. *International Journal of Molecular Sciences*. 2016;17(6):810.
88. Park, K.-E., et al. One-step homology mediated CRISPR-Cas editing in zygotes for generating genome edited cattle. *The CRISPR Journal*. 2020;3(6):523–534.
89. Francolini, M., et al. Evidence for nuclear internalization of exogenous DNA into mammalian sperm cells. *Molecular Reproduction and Development*. 1993;34(2):133–139.
90. Zani, M., et al. The mechanism of binding of exogenous DNA to sperm cells: factors controlling the DNA uptake. *Experimental Cell Research*. 1995;217(1):57–64.
91. Brackett, B.G., et al. Uptake of heterologous genome by mammalian spermatozoa and its transfer to ova through fertilization. *Proceedings of the National Academy of Sciences of the United States of America*. 1971;68(2):353–357.
92. Arezzo, F. Sea urchin sperm as a vector of foreign genetic information. *Cell Biology International Reports*. 1989;13(4):391–404.
93. Brinster, R.L., et al. No simple solution for making transgenic mice. *Cell*. 1989;59(2):239–241.
94. Wang, H.J., et al. Expression of porcine growth hormone gene in transgenic rabbits as reported by green fluorescent protein. *Animal Biotechnology*. 2001;12(2):101–110.
95. Lavitrano, M., et al. Efficient production by sperm-mediated gene transfer of human decay accelerating factor (hDAF) transgenic pigs for xenotransplantation. *Proceedings of the National Academy of Sciences*. 2002;99(22):14230–14235.
96. Webster, N.L., et al. Multi-transgenic pigs expressing three fluorescent proteins produced with high efficiency by sperm mediated gene transfer. *Molecular Reproduction and Development*. 2005;72(1):68–76.
97. Lavitrano, M., et al. Sperm-mediated genetic modifications. *Methods in Molecular Biology*. 2020;2066:125–132.
98. Magnano, A.R., et al. Sperm/DNA interaction: integration of foreign DNA sequences in the mouse sperm genome. *Journal of Reproductive Immunology*. 1998;41(1–2):187–196.
99. Campos, V.F., et al. NanoSMGT: transgene transmission into bovine embryos using halloysite clay nanotubes or nanopolymer to improve transfection efficiency. *Theriogenology*. 2011;76(8):1552–1560.

100. Campos, V.F., et al. Exogenous DNA uptake by South American catfish (Rhamdia quelen) spermatozoa after seminal plasma removal. *Animal Reproduction Science*. 2011;126(1–2):136–141.
101. Collares, T., et al. Transgene transmission in chickens by sperm-mediated gene transfer after seminal plasma removal and exogenous DNA treated with dimethylsulfoxide or N,N-dimethylacetamide. *Journal of Biosciences*. 2011;36(4):613–620.
102. Cong, L., et al. Reducing blood glucose level in TIDM mice by orally administering the silk glands of transgenic hIGF-I silkworms. *Biochemical and Biophysical Research Communications*. 2011;410(4):721–725.
103. Lavitrano, M., et al. Sperm mediated gene transfer in pig: selection of donor boars and optimization of DNA uptake. *Molecular Reproduction and Development*. 2003;64(3):284–291.
104. Bachiller, D., et al. Liposome-mediated DNA uptake by sperm cells. *Molecular Reproduction and Development*. 1991;30(3):194–200.
105. Lai, L., et al. Development of porcine embryos and offspring after intracytoplasmic sperm injection with liposome transfected or non-transfected sperm into in vitro matured oocytes. *Zygote*. 2001;9(4):339–346.
106. Rubessa, M., et al. SLIM microscopy allows for visualization of DNA-containing liposomes designed for sperm-mediated gene transfer in cattle. *Molecular Biology Reports*. 2019;46(1):695–703.
107. Farre, L., et al. Adenovirus-mediated introduction of DNA into pig sperm and offspring. *Molecular Reproduction and Development*. 1999;53(2):149–158.
108. Nagano, M., et al. Retrovirus-mediated gene delivery into male germ line stem cells. *FEBS Letters*. 2000;475(1):7–10.
109. Zi, X.D., et al. The effect of retroviral vector on uptake of human lactoferrin DNA by Yak (Bos Grunniens) spermatozoa and their fertilizability in vitro. *Animal Biotechnology*. 2009;20(4):247–251.
110. Cavalcanti, P.V., et al. Cell viability of bovine spermatozoa subjected to DNA electroporation and DNAse I treatment. *Theriogenology*. 2016;85(7):1312–1322.
111. Collares, T., et al. Transgene transmission in South American catfish (Rhamdia quelen) larvae by sperm-mediated gene transfer. *Journal of Biosciences*. 2010;35(1):39–47.
112. Daneluz, L.O., et al. Efficiency and cell viability implications using tip type electroporation in zebrafish sperm cells. *Molecular Biology Reports*. 2020;47(8):5879–5887.
113. Dos Santos da Silva, L., et al. Capillary electroporation affects the expression of miRNA-122-5p from bull sperm cells. *Gene*. 2021;768:145286.
114. Gagné, M.B., et al. Electroporation of bovine spermatozoa to carry foreign DNA in oocytes. *Molecular Reproduction and Development*. 1991;29(1):6–15.
115. Horan, R., et al. Effects of electropermeabilization on the association of foreign DNA with pig sperm. *Archives of Andrology*. 1992;28(2):105–114.
116. Kumar Pramod, R., et al. Transgenic expression of green fluorescent protein in caprine embryos produced through electroporation-aided sperm-mediated gene transfer. *Gene*. 2016;576(1 Pt 3):505–511.
117. Simões, R., et al. Sperm-mediated gene transfer: effect on bovine in vitro embryo production. *Zygote*. 2013;21(4):325–329.
118. Harel-Markowitz, E., et al., editors. Use of sperm plasmid DNA lipofection combined with REMI (restriction enzyme-mediated insertion) for production of transgenic chickens expressing eGFP (enhanced green fluorescent protein) or human

follicle-stimulating hormone. *Biology of Reproduction*; 2009;80(5):1046–1052. doi: 10.1095/biolreprod.108.070375. Epub 2009 Jan 21. PMID: 19164177
119. Kroll KL, Amaya E. Transgenic Xenopus embryos from sperm nuclear transplantations reveal FGF signaling requirements during gastrulation. *Development*. 1996;122(10):3173–3183.
120. Shemesh, M., et al. Gene integration into bovine sperm genome and its expression in transgenic offspring. *Molecular Reproduction and Development*. 2000;56(2 Suppl):306–308.
121. Barkalina, N., et al. Nanotechnology in reproductive medicine: emerging applications of nanomaterials. *Nanomedicine*. 2014;10(5):921–938.
122. Barkalina, N., et al. Functionalization of mesoporous silica nanoparticles with a cell-penetrating peptide to target mammalian sperm in vitro. *Nanomedicine (London)*. 2015;10(10):1539–1553.
123. Wang, Y., et al. Production of transgenic mice through sperm-mediated gene transfer using magnetic nano-carriers. *Journal of Biomedical Nanotechnology*. 2017;13(12):1673–1681.
124. Bevacqua, R.J., et al. High rates of bovine blastocyst development after ICSI-mediated gene transfer assisted by chemical activation. *Theriogenology*. 2010;74(6):922–931.
125. Li, Q., et al. Production of transgenic rabbit embryos through intracytoplasmic sperm injection. *Zygote*. 2010;18(4):301–307.
126. Pereyra-Bonnet, F., et al. A unique method to produce transgenic embryos in ovine, porcine, feline, bovine and equine species. *Reproduction, Fertility and Development*. 2008;20(7):741–749.
127. Sánchez-Villalba, E., et al. Improved expression of green fluorescent protein in cattle embryos produced by ICSI-mediated gene transfer with spermatozoa treated with streptolysin-O. *Animal Reproduction Science*. 2018;196:130–137.
128. Wu, Y., et al. Factors affecting the efficiency of producing porcine embryos expressing enhanced green fluorescence protein by ICSI-mediated gene transfer method. *Animal Reproduction Science*. 2009;113(1–4):156–166.
129. Evans, M.J. and Kaufman, M.H. Establishment in culture of pluripotential cells from mouse embryos. *Nature*. 1981;292(5819):154–156.
130. Imai, H., et al. Tetraploid embryonic stem cells maintain pluripotency and differentiation potency into three germ layers. *PLoS One*. 2015;10(6):e0130585.
131. Malaver-Ortega, L.F., et al. The state of the art for pluripotent stem cells derivation in domestic ungulates. *Theriogenology*. 2012;78(8):1749–1762.
132. Park, K. E. and Telugu, B.P.V. Role of stem cells in large animal genetic engineering in the TALENs–CRISPR era. *Reproduction, Fertility and Development*. 2014;26(1):65–73.
133. Hamanaka, S., et al. Generation of germline-competent rat induced pluripotent stem cells. *PLoS One*. 2011;6(7):e22008.
134. Okita, K., et al. Generation of germline-competent induced pluripotent stem cells. *Nature*. 2007;448(7151):313–317.
135. Briggs, R. and King, T.J. Transplantation of living nuclei from blastula cells into enucleated frogs' eggs. *Proceedings of the National Academy of Sciences*. 1952;38(5):455–463.
136. Campbell, K.H., et al. Sheep cloned by nuclear transfer from a cultured cell line. *Nature*. 1996;380(6569):64–66.
137. Grisham, J. Pigs cloned for first time. *Nature Biotechnology*. 2000;18(4):365–366.

138. Onishi, A., et al. Pig cloning by microinjection of fetal fibroblast nuclei. *Science*. 2000;289(5482):1188–1190.
139. Polejaeva, I.A. 25th anniversary of cloning by somatic cell nuclear transfer: generation of genetically engineered livestock using somatic cell nuclear transfer. *Reproduction*. 2021;162(1):F11–F22.
140. Loi, P., et al. A new, dynamic era for somatic cell nuclear transfer? *Trends in Biotechnology*. 2016;34(10):791–797.
141. Ogura, A., et al. Recent advancements in cloning by somatic cell nuclear transfer. *Philosophical Transactions of the Royal Society B: Biological Sciences*. 2013;368(1609):20110329.
142. Park, C.H., et al. Establishment, characterization, and validation of novel porcine embryonic fibroblasts as a potential source for genetic modification. *Frontiers in Cell and Developmental Biology*. 2022;10:1059710.
143. Whitworth, K.M., et al. Gene-edited pigs are protected from porcine reproductive and respiratory syndrome virus. *Nature Biotechnology*. 2016;34(1):20–22.
144. Wells, K.D., et al. Replacement of porcine CD163 scavenger receptor cysteine-rich domain 5 with a CD163-like homolog confers resistance of pigs to genotype 1 but not genotype 2 porcine reproductive and respiratory syndrome virus. *Journal of Virology*. 2017;91(2).
145. Tanihara, F., et al. Generation of CD163-edited pig via electroporation of the CRISPR/Cas9 system into porcine in vitro-fertilized zygotes. *Animal Biotechnology*. 2021;32(2):147–154.
146. Hung, S.W., et al. Activated macrophages of CD 163 gene edited pigs generated by direct cytoplasmic microinjection with CRISPR gRNA/Cas9 mRNA are resistant to PRRS virus assault. *Animal Biotechnology*. 2023;34(9):4196–4209.
147. Whitworth, K.M., et al. Resistance to coronavirus infection in amino peptidase N-deficient pigs. *Transgenic Research*. 2018. https://doi.org/10.1007/s11248-018-0100-3
148. Chen, P.R., et al. Disruption of anthrax toxin receptor 1 in pigs leads to a rare disease phenotype and protection from senecavirus A infection. *Scientific Reports*. 2022;12(1):5009.
149. Whitworth, K.M., et al. Zygote injection of CRISPR/Cas9 RNA successfully modifies the target gene without delaying blastocyst development or altering the sex ratio in pigs. *Transgenic Research*. 2017;26(1):97–107.
150. Kwon, T., et al. Gene editing of pigs to control influenza A virus infections. *bioRxiv*. 2024:2024.01.15.575771.
151. Xie, Z, et al. Generation of pRSAD2 gene knock-in pig via CRISPR/Cas9 technology. *Antiviral Research*. 2020;174:104696.
152. McCleary, S., et al. Substitution of warthog NF-κB motifs into RELA of domestic pigs is not sufficient to confer resilience to African swine fever virus. *Scientific Reports*. 2020;10(1):8951.
153. Lu, D., et al. High-level recombinant human lysozyme expressed in milk of transgenic pigs can inhibit the growth of Escherichia coli in the duodenum and influence intestinal morphology of sucking pigs. *PLoS One*. 2014;9(2):e89130.
154. Brar, M.S., et al. Evolutionary diversification of type 2 porcine reproductive and respiratory syndrome virus. *Journal of General Virology*. 2015;96(Pt 7):1570–1580.
155. Stadejek, T., et al. Molecular evolution of PRRSV in Europe: current state of play. *Veterinary Microbiology*. 2013;165(1–2):21–28.

156. Shi, M., et al. Molecular epidemiology of PRRSV: a phylogenetic perspective. *Virus Research*. 2010;154(1–2):7–17.
157. Schulze, M., et al. Effects on boar semen quality after infection with porcine reproductive and respiratory syndrome virus: a case report. *Acta Veterinaria Scandinavica*. 2013;55:16.
158. Yoon, K.-J., et al. Antibody-dependent enhancement (ADE) of porcine reproductive and respiratory syndrome virus (PRRSV) infection in pigs. *Viral Immunology*. 1996;9(1):51–63.
159. Yuan, H., et al. Current status of genetically modified pigs that are resistant to virus infection. *Viruses*. 2022;14(2):417.
160. Van Breedam, W., et al. Porcine reproductive and respiratory syndrome virus entry into the porcine macrophage. *Journal of General Virology*. 2010;91(Pt 7):1659–1667.
161. Van Gorp, H., et al. Scavenger receptor CD163, a Jack-of-all-trades and potential target for cell-directed therapy. *Molecular Immunology*. 2010;47(7–8):1650–1660.
162. Van Gorp, H., et al. Sialoadhesin and CD163 join forces during entry of the porcine reproductive and respiratory syndrome virus. *Journal of General Virology*. 2008;89(Pt 12):2943–2953.
163. Calvert, J.G., et al. CD163 expression confers susceptibility to porcine reproductive and respiratory syndrome viruses. *Journal of Virology*. 2007;81(14):7371–7379.
164. Prather, R.S., et al. An intact sialoadhesin (Sn/SIGLEC1/CD169) is not required for attachment/internalization of the porcine reproductive and respiratory syndrome virus. *Journal of Virology*. 2013;87(17):9538–9546.
165. Wells, K.D., et al. Pigs with a CD163 SRCR domain 5 replaced with human CD163-like homolog SRCR domain 8 are resistant to genotype 1 but not genotype 2 porcine reproductive and respiratory syndrome virus. *Journal of Virology*. 2016;91(2). https://doi.org/10.1128/jvi.01521-16
166. Prather, R.S., et al. Knockout of maternal CD163 protects fetuses from infection with porcine reproductive and respiratory syndrome virus (PRRSV). *Scientific Reports*. 2017;7(1):13371.
167. Burkard, C., et al. Precision engineering for PRRSV resistance in pigs: macrophages from genome edited pigs lacking CD163 SRCR5 domain are fully resistant to both PRRSV genotypes while maintaining biological function. *PLoS Pathogens*. 2017;13(2):e1006206.
168. Burkard, C., et al. Pigs lacking the scavenger receptor cysteine-rich domain 5 of CD163 are resistant to porcine reproductive and respiratory syndrome virus 1 infection. *Journal of Virology*. 2018;92(16).
169. Stevenson, G.W., et al. Emergence of Porcine epidemic diarrhea virus in the United States: clinical signs, lesions, and viral genomic sequences. *Journal of Veterinary Diagnostic Investigation*. 2013;25(5):649–654.
170. Madson, D.M., et al. Characterization of porcine epidemic diarrhea virus Isolate US/Iowa/18984/2013 infection in 1-day-old cesarean-derived colostrum-deprived piglets. *Veterinary Pathology*. 2016;53(1):44–52.
171. Saif, L.J.P.M., et al. Coronaviruses. In: Zimmerman, J.J., et al. editor. *Diseases of Swine*. 10 Hoboken, NJ: Wiley-Blackwell, 2019, ed2012. p. 501–524.
172. Kim, L., et al. Molecular characterization and pathogenesis of transmissible gastroenteritis coronavirus (TGEV) and porcine respiratory coronavirus (PRCV) field isolates co-circulating in a swine herd. *Archives of Virology*. 2000;145(6):1133–1147.

173. Wang, L., et al. Detection and genetic characterization of deltacoronavirus in pigs, Ohio, USA, 2014. *Emerging Infectious Diseases*. 2014;20(7):1227–1230.
174. Whitworth, K.M., et al. Resistance to coronavirus infection in amino peptidase N-deficient pigs. *Transgenic Research*. 2019;28(1):21–32.
175. Stoian, A., et al. The use of cells from ANPEP knockout pigs to evaluate the role of aminopeptidase N (APN) as a receptor for porcine deltacoronavirus (PDCoV). *Virology*. 2020;541:136–140.
176. Tu, C.F., et al. Lessening of porcine epidemic diarrhoea virus susceptibility in piglets after editing of the CMP-N-glycolylneuraminic acid hydroxylase gene with CRISPR/Cas9 to nullify N-glycolylneuraminic acid expression. *PLoS One*. 2019;14(5):e0217236.
177. Miles, L.A., et al. Anthrax toxin receptor 1 is the cellular receptor for Seneca Valley virus. *The Journal of Clinical Investigation*. 2017;127(8):2957–2967.
178. Hatesuer, B., et al. Tmprss2 is essential for influenza H1N1 virus pathogenesis in mice. *PLoS Pathogens*. 2013;9(12):e1003774.
179. Tarnow, C., et al. TMPRSS2 is a host factor that is essential for pneumotropism and pathogenicity of H7N9 influenza A virus in mice. *Journal of Virology*. 2014;88(9):4744–4751.
180. Fleming, T.P., et al. The embryo and its future. *Biology of Reproduction*. 2004;71(4):1046–1054.
181. Huang, G., et al. Lysozyme improves gut performance and protects against enterotoxigenic Escherichia coli infection in neonatal piglets. *Veterinary Research*. 2018;49(1):20.
182. Blome, S., et al. Pathogenesis of African swine fever in domestic pigs and European wild boar. *Virus Research*. 2013;173(1):122–130.
183. Lillico, S.G., et al. Mammalian interspecies substitution of immune modulatory alleles by genome editing. *Scientific Reports*. 2016;6:21645.
184. Popescu, L., et al. Genetically edited pigs lacking CD163 show no resistance following infection with the African swine fever virus isolate, Georgia 2007/1. *Virology*. 2017;501:102–106.
185. Gubbay, J., et al. A gene mapping to the sex-determining region of the mouse Y chromosome is a member of a novel family of embryonically expressed genes. *Nature*. 1990;346(6281):245–250.
186. Skinner, B.M., et al. The pig X and Y chromosomes: structure, sequence, and evolution. *Genome Research*. 2016;26(1):130–139.
187. Kurtz, S., et al. Knockout of the HMG domain of the porcine SRY gene causes sex reversal in gene-edited pigs. *Proceedings of the National Academy of Sciences*. 2021;118(2):e2008743118.
188. Lents, C.A., et al. Central and peripheral administration of kisspeptin activates gonadotropin but not somatotropin secretion in prepubertal gilts. Reproduction. 2008;135:879–887.
189. Flórez, J.M., et al. CRISPR/Cas9-editing of KISS1 to generate pigs with hypogonadotropic hypogonadism as a castration free trait. *Frontiers in Genetics*. 2023;13:1078991.
190. Berg, F., et al. The uncoupling protein 1 gene (UCP1) is disrupted in the pig lineage: a genetic explanation for poor thermoregulation in piglets. *PLoS Geneicst*. 2006;2(8):e129-e.
191. Zheng, Q., et al. Reconstitution of UCP1 using CRISPR/Cas9 in the white adipose tissue of pigs decreases fat deposition and improves thermogenic capacity.

Proceedings of the National Academy of Sciences of the United States of America. 2017;114(45):E9474–E9482.

192. Pan, J., et al. Adipose lipidomics and RNA-Seq analysis revealed the enhanced mitochondrial function in UCP1 knock-in pigs. *Biochimica et ciophysica acta (BBA)-Molecular and Cell Biology of Lipids.* 2019;1864(10):1375–1383.
193. Pan, J., et al. Fecal microbiota was reshaped in UCP1 knock-in pigs via the Adipose-Liver-Gut Axis and contributed to less fat deposition. *Microbiology Spectrum.* 2023;11(1):e03540-22.
194. Kambadur, R., et al. Mutations in myostatin (GDF8) in double-muscled Belgian Blue and Piedmontese cattle. *Genome Research.* 1997;7(9):910–915.
195. Wang, K., et al. Efficient generation of myostatin mutations in pigs using the CRISPR/Cas9 system. *Scientific Reports.* 2015;5(1):16623.
196. Xuan, M.-F., et al. Skeletal muscle-secreted myokine interleukin-6 induces white adipose tissue conversion into beige adipose tissue in myostatin gene knockout pigs. *Domestic Animal Endocrinology.* 2022;78:106679.
197. Kang. Q., et al., editors. *Improving pig genetic resistance and muscle production through molecular biology.* Proceedings, 10th world congress of genetics applied to livestock production; 2014.
198. Wang, K., et al. CRISPR/Cas9-mediated knockout of myostatin in Chinese indigenous Erhualian pigs. *Transgenic Research.* 2017;26:799–805.
199. Han, S.Z., et al. Reproduction traits of heterozygous myostatin knockout sows crossbred with homozygous myostatin knockout boars. *Reproduction in Domestic Animals.* 2021;56(1):26–33.
200. Markljung, E., et al. ZBED6, a novel transcription factor derived from a domesticated DNA transposon regulates IGF2 expression and muscle growth. *PLoS Biology.* 2009;7(12):e1000256.
201. Xiang, G., et al. Editing porcine IGF2 regulatory element improved meat production in Chinese Bama pigs. *Cellular and Molecular Life Sciences.* 2018;75:4619–4628.
202. Liu, X., et al. Disruption of the ZBED6 binding site in intron 3 of IGF2 by CRISPR/Cas9 leads to enhanced muscle development in Liang Guang Small Spotted pigs. *Transgenic Research.* 2019;28(1):141–150.
203. Wang, D., et al. Porcine ZBED6 regulates growth of skeletal muscle and internal organs via multiple targets. *PLoS Genetics.* 2021;17(10):e1009862.
204. Ciccarelli, M., et al. Donor-derived spermatogenesis following stem cell transplantation in sterile NANOS2 knockout males. *Proceedings of the National Academy of Sciences.* 2020;117(39):24195–24204.
205. Lara, N.L., et al. DAZL knockout pigs as recipients for spermatogonial stem cell transplantation. *Cells.* 2023;12(21):2582.
206. Wray-Cahen, D., et al. Advancing genome editing to improve the sustainability and resiliency of animal agriculture. *CABI Agriculture and Bioscience.* 2022; 3(1):1-21.

Chapter 12

Neonatal piglet nutritional supplementation to increase growth and survival: a review

Kara Stewart, Purdue University, USA

1 Introduction

Over the past few decades, the swine industry has genetically selected highly prolific animals with economically desirable carcass traits. The average total number of piglets born per litter increased from 11.5 to 15.1 in the United States from 2004 to 2021 (PigChamp USA). Along with this increase has come an increase in variation in piglet birthweights within litters in addition to the development of a light birthweight phenotype sow (Foxcroft et al., 2009; DeVillers et al., 2007; Milligan et al., 2002). These light birthweight piglets, as well as piglets born into litters with high birthweight variation (Milligan et al., 2002), are more susceptible to death by crushing and disease which has driven preweaning mortality up from 12.4% in 2004 to 16.7% in 2022 (PigChamp USA). Panzardi et al. (2013) reported that 55% of all deaths occur in the first 3 days of life from piglets that weigh less than 1.3 kg. It is important to examine birthweight not only as a risk factor for mortality but also to identify opportunities to improve survival.

Table 1 shows data from a commercial farm with a total mortality rate of 13.5% (K. Stewart, unpublished). Piglets that are less than 1 kg at birth are at the highest risk of mortality (29%) but only represent approximately 11% of

http://dx.doi.org/10.19103/AS.2024.0137.17

Table 1 Piglet survival by birthweight

Birth weight category[a]	Birth weight (kg)	NBA[b]	% of total born	PWM[c], %	Number dead piglets	Percent of total mortality
Light weight: 10.9% of piglets born 29% of total mortality	<1.0	417	10.9	36.2	151	29
Medium weight: 55% of piglets born 50% of total mortality	1.0–1.2	517	13.5	16.1	83	16
	1.2–1.4	748	19.5	12.2	91	17.6
	1.4–1.6	832	21.7	10.1	84	16.2
Heavy weight: 34.5% of piglets born 21% of total mortality	1.6–1.8	721	18.8	8.7	63	12.2
	1.8–2.0	409	10.7	6.8	28	5.4
	2.0–2.2	154	4	7.8	12	2.3
	>2.2	41	1.1	14.6	6	1.2

[a]Breakpoints for categories based on literature where <1.0 kg is considered potential for Intrauterine Growth Retardation in piglets(IUGR), then evaluated 0.6 kg of body weight as a group, where the very few animals that were >2.2 kg were included in the heavy weight piglet category.
[b]NBA = number of piglets born alive.
[c]PWM = pre-weaning mortality in piglets.
Data unpublished from K. Stewart from commercial farm trial with The Maschhoffs utilizing 3839 piglets with overall mortality 13.5%.

all piglets born. Piglets weighing between 1.0 kg and 1.6 kg at birth represent the greatest percentage of all piglets born and account for 50% of all mortality. This group of piglets likely represents the group of piglets with the greatest opportunity to increase survival.

Several review papers have been published describing risk factors for preweaning mortality in piglets (Lay et al., 2002; Muns et al., 2016; Baxter and Edwards, 2018; Panzardi et al., 2013), so this topic will not be discussed in detail here. However, in these reviews, it is evident that the cause of preweaning mortality is multifactorial with the increased litter sizes negatively affecting piglet survivability beginning *in utero* from poor placental function to the birthing process by increasing stress on sows and asphyxiation in piglets and to the neonatal period with reduced colostrum intake and poor thermoregulation of newborns. In 2011, Devillers et al. published a study describing risk factors for preweaning mortality as low birth weight, consuming less than 200 g of colostrum in the first 24 h of life, lower weight gain in the first 24 h, lower rectal temperature, and lower plasma IgG concentrations.

Several of these observations are likely to be correlated with each other and most are related to the level of colostrum intake by the piglet, a driver of piglet survival (Le Devidich et al., 2005; Quesnel et al., 2012). At birth, the energy reserves in the piglet in the form of glycogen and body fat are less than what the piglet needs for maintenance for the first day of life (Mellor and Cockburn, 1986; Farmer et al., 2006) and so they rely on the energy provided from colostrum for metabolism and growth. Piglets that consume little colostrum are less active and unable to thermoregulate, increasing susceptibility to crushing by the sow, which is most prevalent in the first 3 days of life (Lay et al., 2002). Inadequate colostrum intake could also result in decreased passive immunity for the piglet and increase their risk of death due to disease, which typically occurs in mid-to-late lactation and early nursery phases of development.

The importance of colostrum continues for the piglet beyond the lactation period as it sets the developmental trajectory for piglets into adulthood (Vallet et al., 2016). Piglets that consume more than 200 g of colostrum in the first 24 h of life have a higher likelihood of survival until weaning (Devillers et al., 2011; Moreira et al., 2017; K. Stewart unpublished) (Table 2). This level of colostrum intake is based off a formula from Devillers et al. (2004) using bottle-fed piglets to create a predictive equation for colostrum intake based on body weight gain in the first 24 h. Moreira et al. (2017) found that piglets between 1.0 kg and 1.3 kg at birth that consumed more than 200 g of colostrum had a morality rate of only 3.8% compared to 31% in the piglets of the same birth weight range that consumed 200 g or less of colostrum. Nonetheless, this minimum colostrum intake target of 200 g as estimated to minimize mortality may not be enough for a piglet to grow and thrive. Estimations of 250 g have been shown to improve weight gain preweaning and post-weaning (Devillers et al., 2004; Le Devidich et al., 2005; Devillers et al., 2011).

Gilts that have preweaning ADG >0.22 kg/day attain puberty at younger ages and have increased longevity in the herd (Knauer, 2016; Vallet et al., 2016). To achieve this rate of growth during the preweaning period, gilts may need to consume >250–300 g of colostrum in the first 24 h of life (K. Stewart, unpublished, Table 2), although this intake is confounded with piglet body weight. A sow's colostrum yield does not increase with the birth of additional piglets in our hyperprolific sows today (Devillers et al., 2007; Quesnel, 2011).

Table 2 Colostrum intake and piglet growth and survival from 648 piglets born at Purdue University in 2020-2021

Colostrum intake[a], g	% of piglets	Average BW, kg	Average Daily Gain (ADG) birth to wean, kg/day	PWM, %
150–200	2.8	0.70	0.187	39
200–250	11.9	0.92	0.204	7.8
250–300	21.8	1.16	0.222	6.4
300–350	30.0	1.4	0.245	5.6
350–400	22.3	1.63	0.264	6.9
400–450	8.6	1.9	0.269	7.1
>450	2	2.1	0.296	7.7

[a]Colostrum intake was calculated based on body weight changes in the first 24 h of life using formulas published in.

Therefore, as litter sizes increase, the number of nursing piglets exceeding the number of functional teats becomes more likely. This results in a reduction of ~42 g of colostrum intake per piglet for each additional piglet born into the litter (Devillers et al., 2005). Colostrum yield from sows can vary from 1.5 kg to 6 kg during the first 24 h after the birth of the first piglet (reviewed by Quesnel et al. (2012)) and individual piglet's colostrum intake can vary from 210 g/kg to 370 g/kg birth weight (reviewed by Farmer et al. (2006)).

Several researchers have attempted to supplement piglets with porcine colostrum and other colostrum substitutes to increase preweaning growth and survival in piglets. While others have reviewed how colostrum is produced and the importance of colostrum to piglets (Quesnel et al., 2012; 2015; Farmer et al., 2006), this chapter will focus on a review of supplementation to young piglets prior to weaning on growth, immune status, and survival.

2 Piglet supplementation: sow colostrum

Sow colostrum is a complex, early milk composed of lipids, proteins, carbohydrates, vitamins, and minerals, as well as growth factors, hormones, immune cells, and bioactive molecules as reviewed by Hurley (2015) and McGrath et al. (2016). Colostrum is a readily digestible food source for the newborn piglet providing energy for thermoregulation, nutrition for final maturation of body organs and growth, and immunoglobulins (Ig) for passive immunity for the piglet, especially IgG. Colostrum is so complex in make-up having impacts on multiple body systems, making it hard to replicate with supplements. Therefore, some research has investigated direct supplementation of sow colostrum to piglets.

The piglets that typically consume the least amount of colostrum are the light birth weight (LBW) piglets (Moreira et al., 2017; K. Stewart, unpublished). Therefore, many studies that have provided supplementation of sow colostrum and other colostrum substitutes have focused on treating LBW piglets. According to literature, LBW piglets are defined as those with maximal birth weights between 1.1 kg and 1.35 kg. It should be noted that these piglets are heavier than those defined as IUGR (Bauer et al., 1998), which typically weigh less than 1 kg at birth and have additional impairments in organ development (Wu et al., 2006).

Supplementation of 15–30 mL of pooled sow colostrum to LBW piglets within the first few hours following birth increased the efficiency of IgG absorption and early growth with no impact on survival until weaning (Campbell et al., 2012; Muns et al., 2014, 2015; Pinsumrit et al., 2004; Miguel et al., 2021). Muns et al. (2015) evaluated a 15 mL dose of sows' colostrum administered to LBW piglets around the time of birth and found that mortality was decreased only for litters from primiparous sows. This benefit in primiparous sows could be due to them having lower colostrum yield (Devillers et al., 2007) and lower colostrum quality (Nuntapaitoon et al., 2019; Vallet and Miles, 2017). Miguel and colleagues (2021) fed 30 mL of sows' colostrum in two feedings after birth on a commercial farm and found a decrease in antibiotic usage and diarrhea-associated mortality.

3 Piglet supplementation: protein and energy supplements

Alternatives to colostrum supplementation that have been evaluated for piglets typically consist of either protein, energy, or Ig supplements. Energy and protein supplements are expected to provide piglets with sufficient energy to thermoregulate after birth thereby enabling them to access the sow's teats to drink colostrum on their own. Evaluation of this body of literature is challenging as most studies are performed on relatively small sample sizes and do not have sufficient statistical power to evaluate mortality. Additionally, studies evaluate supplements provided to piglets at various doses and frequencies, with little consistency, and do not all evaluate the same outcomes. Most studies provide a supplement in one to two doses directly fed to piglets sometime within the first 12 h after birth.

Most hypotheses predict a change in rectal temperature, body weight gain, IgG absorption, and/or survival during the lactation period. Piglets that do not consume adequate colostrum have lower rectal temperatures through 24 h, suggesting their inability to thermoregulate (Suarez-Trujillo et al., 2020). Nonetheless, energy or protein supplementation studies have not been successful in increasing rectal temperature in piglets (Schmitt et al., 2019). Effects on piglet weight gain during lactation from energy supplements appear equivocal where most studies show no effect on weight gain (Schmitt et al., 2019; Moreira et al., 2017; Muns et al., 2017, 2015). One study that provided free access to milk replacer beginning on d1 of lactation showed improved weight gains (Azain et al., 1996).

Despite limited data showing improvement in thermoregulation and growth, several studies have found an improvement in piglet survival during lactation. Casellas et al. (2005) supplemented medium-long-chain triglycerides daily for 3 days to piglets weighing less than 1.25 kg at birth and found a 1.9-fold reduction in the risk of death. Muns et al. (2017) supplemented light birthweight piglets with a protein-based product (LianoColostro, Ardol BV, The Netherlands) in two doses after birth and found lower mortality in supplemented piglets at d1 but no differences at weaning. Again, only 1882 piglets were used in this trial, and only those weighing <1.3 kg at birth were supplemented, limiting the statistical power needed to properly evaluate mortality.

4 Piglet supplementation: immunoglobulins

While low birth weight is a large driver of mortality in piglets, the transfer of Ig from the dam to the offspring after birth (as estimated by the immunocrit assay) accounts for four times the variation in piglet survival compared to birth weight (Vallet et al., 2013). Due to the epitheliochorial placenta in swine, transfer of large immune cells from the dam to the fetus while *in utero* is lacking resulting in piglets born with an immature immune system and very low levels of IgG (Franz et al., 1982; Klobasa et al., 1981). Therefore, intake of sow's colostrum by piglets is necessary to provide passive immunity and begin building the piglet's immune system, which typically can begin producing its own IgG's at about 7 days of age (Drew and Owen, 1988).

Ig are absorbed in the small intestine of piglets as intact proteins prior to gut closure which occurs at about 24 h in pigs (Svendsen et al., 2005) but may be shorter depending on colostrum consumption by the neonate. It is important to note that IgG and macromolecule absorption by the piglet is mediated by the piglet's diet, or in this case, the colostrum composition and quantity consumed (Carlsson et al., 1980; Werhahn et al., 1981; Jensen et al., 2001). Sow's colostrum contains many bioactive components that regulate immune function in the piglet such as nutrient binders, enzymes, growth factors, lymphocytes, macrophages, and neutrophils. Some of these colostrual components also aid in the maturation of the gastrointestinal tracts in piglets (Xu et al., 2002).

Studies have attempted to supplement piglets with Ig from porcine and bovine sources to increase circuiting Ig in piglets. Research has suggested that supplementing directly with porcine colostrum can increase circulating IgG in LBW piglets compared to suckled counterparts (Muns et al., 2014). This increase was more evident in litters with heterogeneous body weights of piglets but was not enough to improve mortality during lactation likely due to small sample size of only 46 litters. In 2012, Campbell and colleagues evaluated orally feeding porcine plasma as an Ig source to piglets in combination with three different energy sources (undisclosed details) and compared serum IgG to piglets-fed sow colostrum. The efficiency for IgG absorption was greatest in sow colostrum at 28.5%, whereas the porcine plasma ranged from 25.5% to 27.6%, indicating potential for porcine plasma to serve as a supplement to increase circulating IgG in piglets.

5 Piglet supplementation: bovine colostrum

Bovine colostrum has been studied as an alternative source of Igs to newborn piglets; however, the ability of the piglets to absorb and utilize bovine Igs remains unclear. Similar to porcine colostrum, bovine colostrum contains digestible nutrients, bioactive molecules, an abundance of Igs (de Lange et al., 2010), and whey proteins with immune-modulatory effects known during the post-weaning period (Boudry et al., 2008). Bovine colostrum has been shown to have positive effects on growth and immune function in post-weaned pigs (Boundary and Thewis, 2009; Rasmussen et al., 2016), with the impacts of bovine colostrum on the neonate piglet less studied.

Early studies attempted to rear piglets in a colostrum-free environment where milk replacers and various supplements were tested to increase survival of artificially reared pigs. In 1998, Gomez and colleagues fed piglets either sow's colostrum, milk replacer, bovine colostrum (obtained from a local dairy) or a basal diet supplemented with porcine Ig for 2 days after birth followed by a common diet for the subsequent 17 days. At d2, total serum protein was greatest in piglets consuming sow's colostrum, followed by piglets supplemented with porcine Ig and those fed bovine colostrum, and lowest in piglets fed milk replacer; however, these effects were not seen beyond d2. This increase in total proteins in circulation in bovine colostrum-fed piglets at d2 suggests similar macromolecule absorption. However, these pigs had an 80% survival rate compared to 30% in the colostrum-free piglets, so bovine-derived macromolecules provide benefit to piglet survival. The piglets fed porcine Ig only had a slight increase in circulating porcine IgG and no increase in piglets fed bovine colosturm reiterating that bioactive components of sow colostrum likely increase absorption of IgG in newborn piglets.

These results are similar to earlier studies where survival of artificially reared piglets by supplementing purified bovine or porcine Ig was increased over piglets fed milk replacer alone (McCallum et al., 1977; Drew and Owen, 1988). Early studies in 1981 fed purified porcine IgG with and without bovine colostrum (fresh, liquid) and found a 50–70% increase in circulating IgG in piglets supplemented with bovine colostrum, suggesting that bovine colostrum can increase absorption of porcine IgG's similar to porcine colostrum bioactive molecules (Werhahn et al., 1981). It does appear as though piglets are able to select for absorption of porcine-derived macromolecules (Drew and Owens, 1988).

Jensen and colleagues (2001) fed piglets bovine colostrum (fresh, liquid) supplemented with porcine plasma at a ratio of 3:1 and found the two IgG's equally in circulation, suggesting preferential absorption of porcine IgG's. This suggests that IgG absorption in piglets is specific; however, the mechanism by which this occurs is not known. For artificially reared piglets, it appears that bovine colostrum, perhaps in combination with porcine Ig, was a suitable alternative to porcine colostrum.

There is limited data on the effects of supplementing small doses of bovine colostrum, instead of as the piglet's entire diet, around the time of birth on newborn piglet growth and survival. In theory, supplementing bovine colostrum would increase the total Ig content of the piglet's diet on the first day of life while the piglet still consumes its mother's colostrum, ensuring adequate gut maturation and providing the bioactive molecules needed for macromolecule absorption. However, the body of literature in this area is quite limited and it is not clear if there is a particular ratio of bovine:porcine colostrum that is most effective at providing adequate immune protection and intestinal maturation for piglets. Even if this ratio was known, it would likely be challenging to achieve due to the variation in piglet colostrum intake from nursing its dam.

Muns and colleagues (2017) supplemented bovine-derived, dried Ig (ColoBoost, Raizup Nutrition Care, France) in 2 mL and 5 mL doses administered 4 h and 12 h after birth to LBW piglets that were also nursing sows and found it increased circulating porcine IgG 5 days later compared to unsupplemented controls and decreased mortality in LBW piglets (6.3% vs 18.3%), suggesting that bovine colostrual products can increase porcine IgG absorption. There is evidence that supplementing the dam with various vitamins that have antioxidant effects can increase IgG absorption in young piglets (reviewed by Rooke and Bland (2002)), but these effects on piglet growth and survival are not known. Further research is warranted into the role that bovine colostrum can have at modulating the absorption of porcine IgG in newborn piglets.

6 Bovine colostrum and human health

Many cultures around the world have used bovine colostrum to boost immunity in young and elderly people, to aid in tissue healing, to treat bacterial infections, and to aid in recovery following abdominal surgery. However, many of these uses are anecdotal and lacking in controlled trials. A recent series of review papers have been published to present the known findings from research evaluating potential roles for bovine colostrum in human health (Chandwe and Kelly, 2021; Davison, 2021; Ghosh and Iacucci, 2021; Playford and Weiser, 2021). These papers discuss research supplementing bovine colostrum to athletes, infants, elderly, and humans with gastrointestinal disease. While randomized controlled trials are still required, the current evidence suggests that components of bovine colostrum can function as immune modulators and as antimicrobials in the gut.

Due to colostrum's bioactive components of immune cells, antimicrobials, and growth factors, it has been studied as a supplement to human athletes in attempts to improve performance. It must be mentioned that there is tremendous variability in results from human studies due to variation in bovine colostrum bioactivity, quantity of colostrum supplemented, and the lack of nutritionally matched control groups. While moderate exercise improves immune function in humans, athletes undergo intensive workouts and training that provide physical and emotional stress to the athlete. One could argue that the rapid growth rates of modern-day pigs are similar to athletes in terms of muscle protein accretion and, therefore, make them susceptible to experiencing stress. Davison (2021) suggests that colostrum supplementation may have few benefits for altering body composition but has the potential for larger impacts on gut permeability and immune function, especially during intense training periods in human athletes. Athletes and pigs experience gut inflammation in response to stress and a loss of integrity of gut epithelial cell tight junctions, leading to diarrhea and the potential for endotoxemia. In human studies, evidence has shown a benefit of bovine colostrum supplementation on the maintenance of gut integrity in athletes (Davison, 2021), which may be improved even more in combination with other supplements such as zinc carnosine but has not been studied in pigs. Limited data has also shown that bovine colostrum supplementation reduced intestinal ulcers as a result of non-steroidal anti-inflammatory drug use (Davison, 2021).

Bovine colostrum has also been evaluated in infants and the elderly. A neutraceutical cocktail including colostrum-derived lactoferrin can increase natural killer cell activity in elderly, sedentary subjects with decreased immune function (Naito et al., 2014). Elderly humans tend to have suppressed immune function similar to a late-gestation, peri-parturient sow. Babies, human, and pigs have a developing immune system and there is a growing body of evidence supporting the role of breast milk in organ maturation and immune support in humans, so assessing the use of bovine colostrum in this population is warranted.

A review by Gosh and Iacucci (2021) describes the role of bovine colostrum supplementation to infants and indicates a reduction in gastrointestinal disease likely related to alterations in gut microbiota. The role of bovine colostrum in modifying gut microbiota and increasing immune function has great potential for studying its role in preventing infections, specifically those in the intestinal tract. Playford et al. (2021) conducted *in vitro* assays to assess whether bovine colostrum can aid in reducing the negative impacts of bacterial overgrowth in the small intestine. They found the bovine colostrum to have beneficial effects in maintaining intestinal mucosal cell integrity under bacterial stress. This idea has been supported by *in vivo* trials in humans from developing countries where bacterial-induced diarrhea is common (Kwatra et al., 2021; Gaensbauer et al., 2017), as well as patients with non-steroidal anti-inflammatory drug-induced gastrointestinal injury (Playford et al., 2001).

In the swine industry, there are many stages in development when a pig is susceptible to gastrointestinal challenges. The weaned pig is the most likely animal to suffer a breakdown in gastrointestinal integrity as they experience the stress of weaning from their dam in combination with a switch from liquid milk diet to solid feed. A newborn piglet has an underdeveloped gastrointestinal tract that must be matured by components of the sow's milk. Additionally, finishing hogs, mature sows, and mature boars can experience a loss of gastrointestinal integrity during periods of disease or stress. Therefore, research is needed to assess the role that bovine colostrum supplementation to pigs can play in maintaining gut integrity.

7 Conclusion

The current literature suggests that nutritional supplementation after birth with porcine plasma or bovine colostrum in combination with dam's colostrum has the greatest potential for increasing circulating IgG's and increasing survival in lightweight piglets. However, the labor to supplement individual piglets is substantial and literature defining the quantity and frequency of nutritional supplementation to piglets after birth to provide meaningful improvements in piglet survival is lacking.

8 References

Azain, M., et al., Effect of supplemental pig milk replacer on litter performance: seasonal variation in response. *Journal of Animal Science*, 1996. 74(9): p. 2195–2202.

Bauer, R., et al., Body weight distribution and organ size in newborn swine (*sus scrofa domestica)* – A study describing an animal model for asymmetrical intrauterine growth retardation. *Experimental and Toxicologic Pathology*, 1998. 50(1): p. 59–65.

Baxter, E.M. and Edwards, S.A., Piglet mortality and morbidity: inevitable or unacceptable?, in *Advances in pig welfare*. 2018, Elsevier. p. 73–100.

Boudry, C., et al., Effect of a bovine colostrum whey supplementation on growth performance, faecal Escherichia coli population and systemic immune response of piglets at weaning. *Animal*, 2008. 2(5): p. 730–737.

Boudry, Christelle, and André Thewis. "Bovine colostrum as a natural growth factor for newly-weaned piglets: a review." Bulletin of University of Agricultural Sciences and Veterinary Medicine Cluj-Napoca. Animal Science and Biotechnologies 66.1-2 (2009).

Campbell, J., et al., Evaluation of immunoglobulin G absorption from colostrum supplements gavaged to newborn piglets. *Journal of Animal Science*, 2012. 90(suppl_4): p. 299–301.

Carlsson, L., et al., Intestinal absorption of proteins by the neonatal piglet fed on sow's colostrum with either natural or experimentally eliminated trypsin-inhibiting activity. *Neonatology*, 1980. 38(5–6): p. 309–320.

Casellas, J., et al., Effect of medium-and long-chain triglyceride supplementation on small newborn-pig survival. *Preventive Veterinary Medicine*, 2005. 67(2–3): p. 213–221.

Chandwe, K. and Kelly, P., Colostrum therapy for human gastrointestinal health and disease. *Nutrients*, 2021. 13(6): p. 1956.

Lay, D.C. et al., Preweaning survival in swine. *Journal of Animal Science*, 2002. 80(E. Suppl. 1): p. E74–E86.

Davison, G., The use of bovine colostrum in sport and exercise. *Nutrients*, 2021. 13(6): p. 1789.
De Lange, C., et al., Strategic use of feed ingredients and feed additives to stimulate gut health and development in young pigs. *Livestock Science*, 2010. 134(1–3): p. 124–134.
Devillers, N., et al., Origin and consequences of the variability of colostrum production by the sows and of its intake by the piglets. *Journées de la Recherche Porcine en France*, 2005. 37: p. 435–442.
Devillers, N., et al., Variability of colostrum yield and colostrum intake in pigs. *Animal*, 2007. 1(7): p. 1033–1041.
Devillers, N., et al., Influence of colostrum intake on piglet survival and immunity. *Animal*, 2011. 5(10): p. 1605–1612.
Devillers, N., et al., Estimation of colostrum intake in the neonatal pig. *Animal Science*, 2004. 78(2): p. 305–313.
Drew, M. and Owen, B., The provision of passive immunity to colostrum-deprived piglets by bovine or porcine serum immunoglobulins. *Canadian Journal of Animal Science*, 1988. 68(4): p. 1277–1284.
Farmer, C., et al., Colostrum production in swine: from the mammary glands to the piglets. Pig News and Information, 2006. 27(1).
Foxcroft, G., et al., Prenatal programming of postnatal development in the pig. *Control of Pig Reproduction VIII*, 2009. 66: p. 213–231.
Franz, J., et al., Immunoglobulin secretion in fetal swine organs in in vitro systems. *Veterinarni Medicina*, 1982. 27(1): p. 11–18.
Gaensbauer, J.T., et al., Efficacy of a bovine colostrum and egg-based intervention in acute childhood diarrhoea in Guatemala: a randomised, double-blind, placebo-controlled trial. *BMJ Global Health*, 2017. 2(4): p. e000452.
Ghosh, S. and Iacucci, M., Diverse immune effects of bovine colostrum and benefits in human health and disease. *Nutrients*, 2021. 13(11): 3798.
Gomez, G., et al., Effect of immunoglobulin source on survival, growth, and hematological and immunological variables in pigs. *Journal of Animal Science*, 1998. 76(1): p. 1–7.
Hurley, W., Composition of sow colostrum and milk. *The Gestating and Lactating Sow*, 2015: p. 193–230. https://doi.org/10.3920/978-90-8686-803-2_9
Jensen, A.R., et al., Development of intestinal immunoglobulin absorption and enzyme activities in neonatal pigs is diet dependent. *The Journal of Nutrition*, 2001. 131(12): p. 3259–3265.
Klobasa, F., et al., Regulation of humoral immunity in the piglet by immunoglobulins of maternal origin. *Research in Veterinary Science*, 1981. 31(2): p. 195–206.
Knauer, M.T., *Effects of preweaning factors on sow lifetime productivity*. 2016, National Pork Board: National Pork Board Research Report.
Kwatra, B., et al. (2021) Colostrum as a Therapeutic: A Meta Review. Current Journal of Applied Science and Technology, 40 (34). pp. 1-29. ISSN 2457-1024Text3726-Article Text-6841-2-10-20220914.pdf - Published VersionDownload (500kB)Official URL: https://doi.org/10.9734/cjast/2021/v40i3531570
Le Dividich, J., et al., Nutritional and immunological importance of colostrum for the new-born pig. *The Journal of Agricultural Science*, 2005. 143(6): p. 469–485.
McCallum, I., et al., Survival of colostrum-deprived neonatal piglets fed gamma-globulins. *Canadian Journal of Animal Science*, 1977. 57(1): p. 151–158.

McGrath, B.A., et al., Composition and properties of bovine colostrum: a review. *Dairy Science & Technology*, 2016. 96(2): p. 133–158.

Mellor, D. and F. Cockburn, A comparison of energy metabolism in the new-born infant, piglet and lamb. *Quarterly Journal of Experimental Physiology: Translation and Integration*, 1986. 71(3): p. 361–379.

Miguel, J., et al., Supplementing colostrum from multiparous sows: effects on performance and health in piglets from gilts in farm conditions. *Animals*, 2021. 11(9): p. 2563.

Milligan, B.N., et al., Neonatal-piglet weight variation and its relation to pre-weaning mortality and weight gain on commercial farms. *Preventive Veterinary Medicine*, 2002. 56(2): p. 119–127.

Moreira, L., et al., Effects of colostrum, and protein and energy supplementation on survival and performance of low-birth-weight piglets. *Livestock Science*, 2017. 202: p. 188–193.

Muns, R., et al., Gasa, effect of different management techniques to enhance colostrum intake on piglets' growth and mortality. *Animal Welfare*, 2015. 24: p. 185–192.

Muns, R., et al., Non-infectious causes of pre-weaning mortality in piglets. *Livestock Science*, 2016. 184: p. 46–57.

Muns, R., et al., Effect of oral supplementation with different energy boosters in newborn piglets on pre-weaning mortality, growth and serological levels of IGF-I and IgG. *Journal of Animal Science*, 2017. 95(1): p. 353–360.

Muns, R., et al., Effect of cross-fostering and oral supplementation with colostrums on performance of newborn piglets. *Journal of Animal Science*, 2014. 92(3): p. 1193–1199.

Naito, Y., et al., Gut-targeted immunonutrition boosting natural killer cell activity using Saccharomyces boulardii lysates in immuno-compromised healthy elderly subjects. *Rejuvenation Research*, 2014. 17(2): p. 184–187.

Nuntapaitoon, M., et al., Impact of parity and housing conditions on concentration of immunoglobulin G in sow colostrum. *Tropical Animal Health and Production*, 2019. 51(5): p. 1239–1246.

Panzardi, A., et al., Newborn piglet traits associated with survival and growth performance until weaning. *Preventive Veterinary Medicine*, 2013. 110(2): p. 206–213.

Pinsumrit, K., et al., *Survival rates, growth rates, weaning weights and the impact of a colostrum supplement in low birthweight piglets*. Wetchasan Sattawaphaet, 2004.

Playford, R.J., et al., Co-administration of the health food supplement, bovine colostrum, reduces the acute non-steroidal anti-inflammatory drug-induced increase in intestinal permeability. *Clinical Science*, 2001. 100(6): p. 627–633.

Playford, R.J. and Weiser, M.J., Bovine colostrum: its constituents and uses. *Nutrients*, 2021. 13(1): p. 265.

Quesnel, H., Colostrum production by sows: variability of colostrum yield and immunoglobulin G concentrations. *Animal*, 2011. 5(10): p. 1546–1553.

Quesnel, H., et al., Colostrum intake: influence on piglet performance and factors of variation. *Livestock Science*, 2012. 146(2–3): p. 105–114.

Quesnel, H., et al. Colostrum and milk production, in C. Farmer (ed.) *The Gestating and Lactating Sow*, 2015. Wageningen Academic Publishers.

Rasmussen, S.O., et al., Bovine colostrum improves neonatal growth, digestive function, and gut immunity relative to donor human milk and infant formula in preterm pigs. *American Journal of Physiology-Gastrointestinal and Liver Physiology*, 2016. 311(3): p. G480–G491.

Rooke, J. and I. Bland, The acquisition of passive immunity in the new-born piglet. *Livestock Production Science*, 2002. 78(1): p. 13–23.

Schmitt, O., et al., A single dose of fat-based energy supplement to light birth weight pigs shortly after birth does not increase their survival and growth. *Animals*, 2019. 9(5): p. 227.

Suarez-Trujillo, A., et al. A standardized model to study the effects of varying 24-h colostrum dose on postnatal growth and development. *Translational Animal Science*, 2020. 4(4): txaa212.

Svendsen, J., ēt al., Intestinal macromolecular transmission in newborn pigs: Implications for management of neonatal pig survival and health. *Livestock Production Science*, 2005. 97(2–3): p. 183–191.

Vallet, J., et al., Litter-of-origin trait effects on gilt development. *Journal of Animal Science*, 2016. 94(1): p. 96–105.

Vallet, J., et al., A simple novel measure of passive transfer of maternal immunoglobulin is predictive of preweaning mortality in piglets. *The Veterinary Journal*, 2013. 195(1): p. 91–97.

Vallet, J.L. and Miles, J.R., The effect of farrowing induction on colostrum and piglet serum immunocrits is dependent on parity. *Journal of Animal Science*, 2017. 95(2): p. 688–696.

Werhahn, E., et al., Investigation of some factors which influence the absorption of IgG by the neonatal piglet. *Veterinary Immunology and Immunopathology*, 1981. 2(1): p. 35–51.

Wu, G., et al., Board-invited review: intrauterine growth retardation: implications for the animal sciences. *Journal of Animal Science*, 2006. 84(9): p. 2316–2137.

Xu, R., et al., Bioactive compounds in porcine colostrum and milk and their effects on intestinal development in neonatal pigs. *Biology of Growing Animals*, 2002. 1: p. 169–192.

Chapter 13

Nutritional strategies to optimize reproductive performance of the highly prolific sow

Laura L. Greiner and Caitlyn Wileman, Iowa State University, USA

1 Introduction

Over the last 25 years, sow reproduction rates have increased steadily, with the average total piglets born per sow increasing by approximately 1.1% per year in the United States to an average of just over 11 surviving piglets per litter, as reported by Langemeier (2023), based on United States Department of Agriculture - National Agricultural Statistics Service (USDA-NASS) monthly, quarterly, and annual reports (USDA-NASS 2024). Other research suggests annual increases of up to 2.4% from 1996 to 2021 (Strathe and Bruun 2023) and in Denmark, the number of total born piglets increased from 12.9 in 2000 to 19.6 piglets per litter in 2020 (Danish Pig Research Centre 2001–2020). At

http://dx.doi.org/10.19103/AS.2024.0137.18

the same time, sow mortality rates remain high at 13–14% with annual culling rates averaging 50% of breeding herds, with many sows replaced before their third parity (the third time of giving birth) (Stalder et al. 2003; Paiva et al. 2023; Canavate 2023). The nutrient demands of the sow have changed significantly during this timeframe to account for larger litter sizes and longer lactation intervals, affecting issues such as reproductive efficiency as well as general health.

In order for the sow to maintain a long reproductive life, proper nutrition must start when she is a developing gilt. While studies suggest that the performance of the sow may be impacted while the sow is a fetus (Corson et al. 2009; Foxcroft et al. 2009), this chapter will only focus on nutrition needs at the time the female piglet is weaned from her dam and throughout her reproductive cycle. This chapter covers various aspects of developing gilt and sow nutrition, looking at nutritional requirements during gilt development, gestation and lactation, including optimizing nutritional strategies to support the sow's reproductive health and longevity. A case study discussing increased feed levels at day 90 of gestation is provided to show practical application of research in particular feed programs.

2 Protein requirements: developing gilts, gestation, and lactation

Gilt and sow diets require adequate intake of amino acids for growth, reproduction, and milk production. Some amino acids are identified as essential, meaning that the amino acid cannot be synthesized by the body and must be provided through oral intake. Standard diets are based on lysine (the most limiting of these essential amino acids) as the base value, with other amino acids typically expressed as a percentage of the lysine content.

2.1 Developing gilts

Protein is a critical component of gilt diets during the development stage. The aim is to provide adequate protein levels without leading to too rapid or excessive fat deposition which can cause mobility problems. To achieve this, it is essential to monitor the growth rate of replacement gilts to the point of sexual maturity, defined as two successful estrus events with the second estrus being a breeding estrus at 136 kg body weight (Williams et al. 2005). A balanced protein level in a gilt diet typically ranges from 12% to 13% crude protein (CP). Precise amounts will depend on factors such as body weight, age, and overall feed allowance and may need amending according to regular monitoring of body condition (e.g. to target a back fat level of >12 mm at first service).

Diet formulation should meet standardized ileal digestible (SID) amino acid ratios relative to lysine, including other synthetic amino acids (such as methionine and cysteine) to avoid excess nitrogen excretion. Using diet restriction to slow the growth of gilt to avoid osteochondrosis has been discussed in the literature, as some research suggests a 20% increase in incidence for every 100 g increase in daily gain (Koning et al. 2013). However, many factors beyond nutrition can contribute to osteochondrosis such as genetics and concussive incidents while moving animals (Koning et al. 2014a, 2014b; Quinn et al. 2015). Furthermore, if the growth rate is reduced too much, gilts can experience a delay in estrus (Miller et al. 2011; Calderon Diaz et al. 2015; Athorn et al. 2017).

2.2 Gestation

Protein is an essential nutrient during gestation, supporting both the sow's maintenance requirements and fetal development. It is recommended to provide a balanced protein level in the gestating sow's diet, typically ranging from 12% to 15% CP. Thomas et al. (2018a) fed 11 g, 13.5 g, 16 g, and 18.5 g of digestible lysine per day throughout the gestation period and reported a linear relationship between sow body weight and lysine level without impacting litter birth weights. Appropriate levels of fat (including fat supplementation where required) have been found to be generally beneficial for sow reproductive and litter growth performance (Wang et al. 2022). However, excessively high protein levels should be avoided since they can lead to higher feed costs and excess nitrogen excretion without providing significant additional value.

The specific amino acids required during gestation, especially during the different stages, are still a matter of ongoing research. Lysine and threonine are currently used for direct estimates of the efficiency of amino acid intake utilization for amino acid retention in gestating sows. A recent study reported maximal lysine utilization efficiency in gestating gilts was met at 7.2, 9.1, and 13.5 g SID lysine/day in early-, mid-, and late-gestation (Ramirez-Camba et al. 2020). Another study reported SID lysine requirements at 14 g/day in early gestation (D0–80), and 21 g/day in late gestation (D81–144) (Shi et al. 2016). Based on these studies, SID lysine levels increase as gestation progresses. Optimum lysine efficiency depends on factors such as levels of other amino acids and daily feed allowances.

2.3 Lactation

Lysine requirements of the lactating sow are determined based on feed intake and estimated daily litter growth rate. The National Research Council Nutrient Requirements of Swine (NRC, 2012) modeled the estimated (SID lysine requirement as [(daily milk nitrogen output × 6.3 × 0.0701 – maternal body protein mobilization × 0.0674 / 0.868) / 0.75] × 1.1197. Other research has determined that total lysine intake in sows can be estimated using equations based on daily litter growth rate (kg) for both primiparous [72.68 – (6.04*(3.55 – daily litter growth rate))] and multiparous females [92.03 – (11.9*(4.24 – daily litter growth rate))] (Greiner et al. 2020). These data are further supported by Hojgaard et al. (2019) and, more recently, Theil et al. (2023) in which a series of lactation studies with highly prolific sows demonstrated that a litter with a 3.0 kg/day growth rate required 55 g SID Lys/d with sows utilizing Lys at a level of 88% efficiency, though the study suggested hyper-prolific sows might require up to 57g SID Lys/d (Theil et al. 2023).

Other amino acids should be balanced with lysine. The tryptophan:lysine ratio for the lactating sow appears to be between 17 and 17.6 (Greiner et al. 2017; Libal et al. 1997). Data are conflicting concerning the valine:lysine ratio. Richert et al. (1996) and (1997) determined that the requirement was greater than 100. However, Greiner et al. (2019) and Gaines et al. (2006) determined that the valine:lysine ratio was approximately 73–86%, no higher than the NRC (2012). Other work suggested that feeding levels between 76% and 122% had no effect on performance (Carter et al. 2000; Strathe et al. 2016).

Overall, estimations of branched-chain amino acid (BCAA) ratios in lactation have been conflicting. Differences in sow body weight loss or gain during lactation may help explain the conflicting data associated with valine. Based on earlier work on cellular uptake of valine (e.g. Jackson et al. 2000), a review by Holen et al. (2022a) also found that branched-chain amino acids compete for intestinal uptake and mammary utilization, which means different branched-chain amino acids affect each other and create varied results. For example, work conducted by Humphrey et al. (2023) in nursery pigs demonstrated that valine impacted piglet average daily gain and gain to feed (G:F) ratio, regardless of leucine and isoleucine and that, when valine was fed below NRC (2012) recommendations and leucine was increased, average daily feed intake was decreased. However, a recent study in sows evaluating these relationships (Humphrey et al. 2023 (unpublished)), found no differences in feed intake in sows when valine was fed below the NRC (2012) values and leucine was fed above recommended values. In this study, the sow average daily feed intake was over 7.5 kg and that, while the intake of each branched-chain amino acid was at expected levels, plasma BCAA levels did not correspond to levels consumed, indicating that sows were utilizing BCAAs at different levels. The relationship of valine, isoleucine, and leucine needs to be further researched in evaluations of sow reproduction.

3 Energy requirements: developing gilts, gestation, and lactation

Energy can be provided to the female in the form of carbohydrates or fats. Fats are further broken down into saturated and unsaturated forms. The calculation of energy from an ingredient is measured as gross energy. Common energy values for feedstuffs are generally represented as either metabolizable or net energy, which accounts for energy losses in feces, urine and gas, or feces, urine, gas, and heat increment, respectively.

3.1 Developing gilts

Energy is a fundamental component that must be carefully managed in developing gilt nutrition. Adequate energy levels are necessary to meet the growth and reproductive needs of replacement gilts. Energy-rich ingredients like corn, wheat, barley, and fat supplements such as vegetable oils and animal fats, can be incorporated into the diet. However, it is crucial to consider the growth rate and body condition of gilts to avoid overfeeding energy levels, which can lead to excessive fat deposition and poor lean growth. As gilts progress in their development, their energy intake may be restricted to achieve the desired weight at the time they are selected for breeding. Over-feeding a gilt and breeding her at 170 kg compared to 140 kg of body weight will result in an increased lifetime daily feed intake of 0.2 kg. This is due to an increase in mass that results in a 0.440 MJ ME/kg 0.75 increase in energy requirements (Dourmad et al. 2008). Altering feeding levels from restricted to semi-ad libitum (i.e. the ability to consume as much feed as an animal wants) led to an increase in backfat thickness despite CP levels (Klaaborg et al. 2019; Klindt et al. 1999; Strathe et al. 2019; Thingnes et al. 2015). Therefore, the most efficient way to increase BF is feeding a diet ad libitum with a low CP combined with high energy (Strathe et al. 2019).

3.2 Gestation

Energy is a critical component during gestation, as it contributes to the deposition of maternal lipid and protein. Although Noblet (1985) argued that energy stored in fetuses is mainly stored as protein due to low-fat reserves, dietary energy does contribute to fetal development and growth (Langenduik 2021). It is important to consider energy requirements based on factors such as body condition score (BCS), parity (number of times a sow has given birth), litter size and stage of gestation.

Balancing energy and protein levels is crucial for optimal performance. An insufficient energy or protein balance during early gestation may lead to the sow returning to estrus. Increased energy intake during late gestation can positively impact both fetal growth and maternal weight gain (Rooney et al. 2020). However, excessive energy intake can have a negative impact on farrowing and lactational performance. Theil et al. (2023) have argued that, even though sows have increased energy requirements as gestation progresses, there is a lack of data showing any improvement in sow reproductive responses from phase feeding in gestation. Feyera and Thiel (2017) reported most energy stores in sow body fat (BF) and protein are used to recover from the previous lactation period. In addition, in late gestation, fetal growth makes up part of the protein requirement with energy used primarily for maintenance. Lastly, some fat sources are not fully used for energy, but rather comprise colostrum composition and support reproductive hormones and therefore, the dietary recommendation of linoleic acid levels for gestating sows is 0.10% in standard corn and soybean meal-based diets, the same levels as in lactating sows (NRC 2012).

3.3 Lactation

The energy requirement of the lactating sow is 5–10% greater than that during pregnancy. It has been determined that the maintenance requirement for the lactating sow is 110 kcal/$W^{0.75}$ for lactating sows (NRC 2012). Measurement of the energy requirement of the lactating sow should account for the need to support energy needs along with calculated milk yield (NRC 2012). The addition of dietary fat improves the fat level in colostrum and, based on the fatty acid composition of the diet, can also alter the fatty acid composition of the colostrum (Wensely et al. 2023). It has been suggested that milk fat composition is similar to that of the composition of the fat consumed (Lauridsen and Danielsen 2004). Feeding sows an increased level of added fat from 0% to 6% resulted in an increased level of milk fat, while feeding increased levels of essential fatty acids resulted in increased concentrations of linoleic acid and α-linolenic acid (Rosero et al. 2015).

Linoleic acid intake improved subsequent cycle farrowing rate (%) [$(-1.5 \times 10^{-3} \times$ linoleic acid intake (g/d)$^2) + (0.53 \times$ linoleic acid intake (g/d)$) + (45.2)$] and increased the number of pigs born [$(9.4 \times 10^{-5} \times$ linoleic acid intake (g/d)$^2) + (0.04 \times$ linoleic acid intake (g/d)$) + (10.94)$] (Rosero et al. 2016). As this research suggests, supplemental α-linolenic acid results in a shorter wean to estrus interval and a higher conception rate, suggesting α-linolenic acid intake should be 10 g/d and linoleic acid intake should be 125 g/d. The role of α-linolenic acid in colostrum and milk composition (as well as other variables such as sow reproductive performance and piglet survivability) has been more recently reviewed by Holen et al. (2022b).

4 Fiber requirements: developing gilts, gestation, and lactation

Fiber plays an important role in developing gilts, gestating, and lactating sows (Jo and Kim 2023). Although fiber is a nutrient fraction that cannot be completely broken down by monogastric digestive enzymes, fiber does promote gut health and prevention of constipation. Dietary fiber consists of insoluble and soluble fractions. Common feedstuffs that are high in insoluble fibers include dried distiller grains, soy hulls and wheat middlings. Insoluble fiber generally increases feed transit time and can potentially improve gut health. Soluble fiber tends to be more fermentable by intestinal bacteria which, in turn, produce volatile fatty acids that the pig can utilize for a variety of functions. Fermentable fibers include beet pulp and oat hulls. When feeding high levels of fermentable fiber, dietary considerations such as increased threonine requirements should be considered (Wellington et al. 2018).

4.1 Developing gilts

Fiber plays a crucial role in developing gilt nutrition. It promotes gut health and proper digestion, aiding in the maintenance of a healthy microbiota in the gut (Tian et al. 2020). Fiber sources such as soybean hulls, wheat bran, and dried distillers' grains can be utilized. However, it is important to maintain an adequate balance of fiber levels, as excessive amounts can reduce nutrient digestibility and feed intake. Aiming for a fiber content of around 6% to 8% crude fiber in the diet is generally recommended (NRC 2012).

4.2 Gestation

Fiber is another important component of the gestating sow's diet (Peltoniemi et al. 2010). It is recommended to provide a moderate level (500–600 g/d) of fiber to promote gut health, prevent constipation, and ensure satiety. However, excessive levels of fiber can reduce digestibility or impact feed intake by negatively affecting palatability. Finding the right balance is crucial to ensure optimal nutrition and well-being for gestating sows.

Feeding fiber has the potential to regulate gut microbiota, alleviate immune responses and improve overall performance. Costermans et al. (2019) reported improved oocyte quality, increased early embryo survival and more vigorous litters. This improvement in performance is primarily due to the increase of beneficial microbes, including lactobacilli, which promote digestion and barrier function against pathogens (Sawicki et al. 2017). West et al. (1999) reported no interaction between heat stress and dietary fiber type but suggested that total energy intake has a greater effect on milk yield during a heat stress event. Huang et al. (2020) found that adding 5% resistant starch enhanced postprandial satiety and lowered stillborn rates. The feeding of 34.38% compared to 12.08% dietary fiber throughout gestation resulted in increased lactation feed intake (Li et al. 2023).

4.3 Lactation

Constipation of the female can be of concern both during parturition and lactation. Constipation during farrowing can increase stillborn rates and retained piglets (i.e. piglets retained within their own placenta). Feyera et al. (2017) documented that feeding 350 g of dietary fiber/day and 700 g of dietary fiber/day from days 102 to 108 and 108 until farrowing, respectively, reduced stillborn piglets and total piglet mortality. The reduction in stillborn rates and improved farrowing interval is related to slow energy release associated with fermentable fiber which supports low-energy sows that cannot effectively produce uterine contractions to deliver the fetus (Feyera et al. 2018). Additionally, constipation during lactation may decrease sow comfort, reducing her milking ability and feed intake. The feeding of low-fermentable fiber during lactation increases defecation through increasing fecal bulk. Other fibers that increase water-holding capacity such as sugar beet pulp can also reduce constipation by increasing water intake and moisture content (Oliviero et al. 2009, 2010).

Feeding fiber in gestation has been demonstrated to improve lactation feed intake (Li et al. 2023). Furthermore, the feeding of different types of fiber may influence milk composition. For example, the feeding of wheat bran in gestation and lactation did not result in changes in IgA composition in milk compared to when sugar beet pulp is added to the diet (Shang et al. 2019)

5 Vitamin and mineral requirements

Vitamin and minerals are essential for growth, development, and reproductive performance. Specific requirements for vitamins and minerals are listed in NRC (2012). While all phases of pig growth will require supplementation of vitamins and minerals, specific vitamins and minerals should be considered for the reproductive female. For example, vitamins that support reproduction such as folic acid, biotin, and pyroxidine should be supplemented into the diet. Kornegay (1985) found that the addition of biotin in sow diets improved litter size, conception rates, and wean to estrus interval. Pettigrew et al. (1996) reported that feeding riboflavin levels of 60 mg/day, 110 mg/day, or 160 mg/day of riboflavin compared to 10 mg/day at the beginning of breeding resulted in an increase in the percentage of females that farrowed. Further evaluation of riboflavin in lactation demonstrated that, when sows were fed 16.3 mg/day of riboflavin, the erythrocyte glutathione reductase activity coefficient was optimized (Frank et al. 1988). The NRC (2012) recommends feeding 22.37 mg of riboflavin per day to the lactating sow for optimal performance.

Minerals are divided into macrominerals and microminerals. Macrominerals such as sodium, chloride, calcium, magnesium, potassium, and phosphorus are required to maintain basic metabolic functions within the animal. Typically, total calcium to total phosphorus ratios are between 1:1 and 1.2:1 and total calcium to bio-available phosphorus levels are from 1.4:1 to 2.4:1, which increases as the animal matures (Reese and Hill 2010). Available phosphorus levels should be fed above those that are commonly fed for market hogs and should increase from gilt development through gestation. Bio-available phosphorus levels in lactating sows should be fed based on sow feed intake to ensure that the sow is meeting standard recommendations which are calculated based upon sow age, body weight, and predicted daily weight gain of the piglet (NRC 2012).

Microminerals include zinc, copper, iron, iodine, and selenium. Microminerals can be fed in both inorganic and organic forms. Organic minerals are typically bound to amino acid or amino acid complexes, while inorganic minerals are found in mineral salts or oxides and may need additional steps to be absorbed and utilized in the body. The use of organic minerals (selenium, copper, zinc, and manganese) has been shown to improve conception rates, litter birth weights, immune function, and improved sow longevity within a herd (Peters and Mahan 2008; Tinkel et al. 2018; Wedekind et al. 2015).

Minerals such as calcium, phosphorus, zinc, and selenium are vital for the maintenance of the skeletal system and various physiological functions (Hays 1976; Peo 1976, 1991; Kornegay 1985; Crenshaw 2001). The source of minerals is also important, with organic forms known to have better absorption and utilization compared to inorganic forms (Sampath et al. 2023). Including phytase in the diet enhances the bioavailability of phytate phosphorus and improves utilization of calcium, iron, zinc, and protein (Nasi 1990; Simons et al. 1990; Jongbloed et al. 1992; Pallauf et al. 1992a, 1992b, Cromwell et al. 1993b, 1995; Lei et al. 1993b; Ketaren et al. 1993; Mroz et al. 1994; Kemme et al. 1995; Biehl and Baker 1996).

In addition to vitamins and minerals, other nutrients may need to be considered within the sow diet. One such nutrient is choline. While choline can be produced endogenously through *de novo* synthesis, the concentration of choline available may not always be sufficient to support growth and performance, depending on environmental/dietary conditions. Human data would support the view that additional supplementation of choline through dietary support is needed (Food and Nutrition Board 1998). Studies conducted with gestation and lactating sows would indicate that feeding levels at or slightly above that of the NRC (2012) recommendations optimize birth weight, piglets born alive, piglet mortality, and piglet weaning weight (Getty and Dilger 2015; Seerley et al. 1981; Zhong et al. 2021).

6 Feed and water quality

Regardless of the phase of production, diet composition, and nutrient specifications should be routinely evaluated. Feed should be routinely analyzed for levels of mycotoxins. Mycotoxins such as zearalenone and fumonisin can reduce reproductive performance by causing false heat through estrogenic-like chemicals or reducing immune function, respectively (Bryden 2012; Zinedine et al. 2007). Zearalonene has been documented to interfere with puberty expression in gilts, increased stillborn rates during farrowing, and delayed wean to estrus intervals in recently weaned sows (Diekman and Green 1992; Kanora and Maes 2009).

Water quality should be routinely measured. Sulfates, nitrates, nitrites, and total dissolved solids, pH, and coliforms can lead to issues associated with diarrhea and cyanosis. Recommended maximum allowances are listed in Table 1. Water acidifiers may be used as needed based on the enteric challenges present within a farm. However, caution should be used as acidifiers may cause the disruption of waterline biofilms, resulting in clogged water delivery systems. Water containing high levels of minerals can reduce copper and zinc absorption and increase oxidative stress (Klevay 2001). In addition, water can have high levels of sodium, which means dietary inclusion of sodium may need to be adjusted to prevent sodium toxicosis.

7 Nutrient requirements for developing gilts

Though the trend is to control feed intake (Miller et al. 2011), recent research suggests that developing high-lean gilts should be allowed *ad libitum* feed access, with limiting feeding more appropriate for low- and medium-lean gilts during development (Belkova and Rozkot 2022). The addition of microminerals and macrominerals may be needed to ensure proper bone development. Gilts that will be fed utilizing electronic feeding stations may require training after the gilt selection has occurred. Once gilts have reached puberty, reducing feed intake to manage BCSs is essential.

Table 1 Water quality parameters and related maximum tolerable levels for swine[a]

	Maximum tolerable levels
Sulfates	1000 ppm
Nitrates and nitrites	100 ppm
Total dissolved solids	3000 ppm
pH	7
Coliforms	50 CFU/ml

[a]Table adapted from NRC (2012) and Meegat et al. (2019).

One focus of replacement gilt longevity is to maintain growth to reach 115–140 kg at puberty and 135 kg and 160 kg at breeding (Bortolozzo et al. 2009). Gilts can easily surpass the target weight range at puberty if they are on *ad libitum* feeding systems, indicating the need for limiting feeding systems for gilt development closer to puberty. Slowing down growth in developing gilts can be achieved by restricting energy intake by 25% or by reducing Lys to energy ratios (Lents et al. 2020; Miller at al. 2023). While both of these methods can effectively reduce growth, gilts may overcome the energy restriction deficit in time when utilizing high levels of fiber with increased feed intake. Overweight gilts during their first service may experience locomotion issues which will result in removal from the herd (Filha et al. 2010).

8 Nutrient requirements during gestation

The focus of gestation nutrition programs should be to get the sow into body condition before day 60 of gestation. Prior to day 60, the energy requirements of fetuses are low and, therefore, a majority of dietary energy will be focused toward the mother (Ji et al. 2005; Thomas et al. 2018b).

Effective body condition management is crucial in optimizing the reproductive performance of highly prolific sows. Feeding to condition early in gestation is essential to ensure that sows are at their desired BCS before farrowing. Monitoring and adjusting feed intake during early gestation is needed to prevent excess weight gain, as over-conditioning during this period can lead to issues during farrowing and lactation. Over-conditioned sows develop high levels of fat in the mammary gland which can reduce colostrum and milk yield in lactation (Faccin et al. 2022). Careful management of the sow's body condition is necessary to maintain a healthy and optimal environment for successful reproduction.

Monitoring the body condition of individual sows is essential for effective management. It is important to regularly assess and monitor the BCS of individual sows with a standard scoring system like the sow caliper (Knauer and Baitinger 2015). Monitoring BCS with hand pressure or tools like the sow caliper measures an estimate of the fat stores in the sow. Condition scores can range between 1 and 5, with a score of 3 being the target BCS. The scoring values reflect the ease or difficulty of detecting bones at various pressure points on the sow. Implementing a regular assessment of sow BCS and monitoring with a standardized scoring system allows for adjustments in feed intake and diet composition to reach and maintain the desired condition. The frequency of condition scoring is dependent on the sow management system. The best practice is to implement BCS scoring at breeding and two additional timepoints between breeding and farrowing.

It is important to consider factors such as parity and stage of gestation when determining BCS targets, as sows prioritize the litter during late gestation and lactation. Sows may have a lower BCS during this time, and it is crucial to provide adequate nutrition to support the nutritional demands of both the sow and the developing piglets. This is typically done by increasing feeding levels above maintenance plus conceptus requirements in early gestation until ideal body condition is achieved by 30 days of gestation. Once ideal BCS is established, then sows may be fed at maintenance plus conceptus requirements based on gestational age. Studies have shown that precision feeding programs, which adjust amino acid and energy levels throughout gestation (as recommended by the NRC (2012) guidelines), can lead to increased sow weight gain but not impact litter performance or offspring weaning to market performance (Hansen et al. 2021; Gaillard and Dourmad 2022). It is also important to consider BCS, parity, previous litter size, and stage of gestation when formulating gestation diets.

A study conducted by Ren et al. (2017) found that, when the gestating female gains weight during gestation above acceptable body condition levels, she loses more weight in lactation due to poor feed intake. The weight loss in lactation results in a poorer litter growth rate (Strathe et al. 2017). Kim et al. (2015) demonstrated that, at 109 days of gestation, increased levels of sow backfat depth resulted in a linear reduction in sow feed intake in lactation. In addition, sows with greater levels of backfat at the end of gestation had longer and more difficult labor, resulting in increased stillborn and low-vitality piglets at birth (Oliviero et al. 2010). Lawlor et al. (2007) reported that increasing feed intake during gestation reduced a sow's appetite during lactation. However, when feed intake was increased by 30% from day 3 to 32 after breeding in first and second parity sows, there was an improvement in sow BW gain and increased litter size. However, the sow experienced lactation losses with a resulting reduction in BCS. These findings suggest that early gestation is the best period for lactation loss recovery (Hoving et al. 2011) and emphasize the importance of maintaining an appropriate body condition throughout gestation to support successful farrowing, lactation, and piglet growth.

9 Nutrient requirements during lactation

Milk production represents approximately 80% of the sow's energy requirements during lactation (NRC 2012). It has been found that sows with high milk production potential have high feed intake and are more prone to high body tissue mobilization (Strathe et al. 2017). If a sow loses more than 12% of its protein mass, reductions in follicular development for the subsequent cycle may follow (Clowes et al. 2003). Therefore, it is imperative that the sow be provided fresh feed multiple times a day or *ad libitum* with free access to water.

Formulating diets for a mature herd can create challenges as the young female consumes lower amounts of food relative to the older female (Aherne 2005). Dietary lysine inclusion levels should be adjusted based on sow feed intake and consideration should be made to providing a separate diet for the primiparous gilt. For example, current US farm records indicate that gilts will consume at least 1 kg of feed less/d. If a sow is consuming 7 kg of feed per day, a gilt would be likely to be consuming roughly 6 kg/d. Formulating a diet with a 1.05% SID Lys level would allow for the sow to consume 73.5 g/d while the gilt would consume 63 g/d. According to Theil et al. (2023), this SID Lys level would be adequate for lactation.

However, if feed intake is reduced by 2 kg/d due to heat stress, then the lactating gilt may potentially no longer be consuming her Lys requirements. Based on work conducted by Greiner et al. (2020), this would result in an increased number of days between weaning and estrus and a loss of body weight. In order to ensure that the young female has consumed the desired lysine level, diets may either be top-dressed with a more concentrated ration for the gilt or a separate diet formulated in higher lysine can be used.

During times of heat stress, sow lactation feed intake can be reduced by up to 40% (Black et al. 1993). Providing wet feed can increase feed intake by 10–15%. Other methods to improve feed intake during periods of heat stress could include changing the time of day in which feed is delivered so that feed is presented during the cooler period of the day (Choi et al. 2019b). The feeding of products that can alleviate heat stress metabolically can also be considered. The use of a methyl donor, betaine, has been shown beneficial in multiple species. Betaine is a derivative of glycine which is synthesized from choline and improves water retention within the cell. The use of betaine in sow lactation diets (0.22% inclusion) has been shown to reduce the body temperature of the sow, improve follicular development and increase the subsequent total of piglets born (Cabezon et al. 2016, 2017; Mendoza et al. 2016).

The fatty acid profile of a diet can also influence heat stress response. Sows fed increasing levels of animal-vegetable blend fat or choice white grease (0–3% inclusion in a diet) had improved subsequence conception and farrowing rates and improved litter sizes. However, only increased choice white grease diets resulted in improved feed efficiency of the lactating sow during naturally occurring heat stress with temperatures averaging 27 ± 3°C, and humidity at 61% ± 10% (Rosero et al. 2012). Additionally, water may be added to dry feed to improve feed intake by approximately 1 kg/d (Genest and D'Allaire 1995).

10 Nutrient requirements during the transition period

The transition period of the sow from approximately day 108 of gestation until lactation is a critical period. Sows fed either 4.1 kg/d or 3.0 kg/d during this timeframe had improved farrowing performance, stillborn rates, and colostrum yield (Feyera et al. 2021). These diets supplied 38.8 MJ ME and 23.9 SID Lys/d or 53.0 MJ ME and 32.7 SID Lys. Brunn et al. (2023) also demonstrated that feeding sows 4.0–4.1 kg/d the last week of gestation resulted in improved subsequent lactation performance. Increasing feed from gestation to lactation can occur either at the time of entry into the farrowing room or after farrowing (Graham et al. 2014).

Allowing sows to consume feed *ad libitum* prior to farrowing did not change stillborn rates compared to feeding 2.7 kg once a day but did result in a reduction in body weight loss during lactation. Research indicates the benefits of allowing sows *ad libitum* access to feed during lactation (Theil et al. 2023; Gorr et al. 2024). Gourley et al. (2020) demonstrated that feeding sows once a day resulted in higher fall-behind piglets (those that lag behind the rest of the group in weight gain and general health) compared to feeding four times a day.

Feeding programs post-farrowing vary greatly and should be based on genetic intake patterns. Some breeds may be able to consume large quantities of feed (9 kg) on the first day after farrowing while other breeds may require a step-up program (2 kg, 3 kg, 4 kg, and then *ad libitum*) to prevent sows from having low feed intake during the first week of lactation which can increase culling rates (Koketsu et al. 1996; Rodriguez et al. 2023). Closely monitoring feed intake when evaluating a feeding program should be done to ensure sows are continually increasing feed intake and not refusing feed for multiple days. Allowing sow *ad libitum* access to feed should potentially reduce sow weight loss potential during lactation as weight loss during lactation has been shown to lead to prolonged wean to estrus intervals (Vargas et al. 2009).

11 Nutrient requirements during the post-weaning period

The feeding of the sow post-weaning is less well-researched. Feed intake post-weaning until estrus is important for mediating insulin and insulin-like growth factor (IGF-1) effects on follicle-stimulating hormone (FSH) and lutenizing hormone (LH), which are important for healthy follicular development and high ovulation rates (Langendijk 2021). Recent studies have shown that higher levels of feed during the wean-to-estrus interval did not alter the wean-to-estrus interval (Table 1; Graham et al. 2015). Almedia et al. (2018) also demonstrated that feeding either lactation or gestation feed at either 2.72 or 3.63 kg/day did not change wean to estrus interval or subsequent total born. The limited available research suggests that feeding 2.72–3.63 kg/day during the wean to estrus interval should be sufficient to support follicular development (Table 2).

Table 2 Dietary feeding level during wean to estrus period on sow performance[a]

Item	Feeding level (kg/d)		
	2.72	3.63	5.45
Average daily feed intake (kg)	2.68	3.55	5.18
Wean to estrus interval (d)	5.1	5.0	4.9

[a]Adapted from Graham et al. (2015).

Maintaining a steady and controlled growth rate by making regular adjustments to feed intake levels helps avoid sudden fluctuations in body condition. It is also crucial to adjust the energy content of the feed to meet the sow's specific needs and maintain the desired BCS. However, how fast a newly weaned sow is returned to appropriate body condition varies.

It has been reported that higher feed intake is crucial for embryo survival, especially in the first 72 h of insemination (Jindal et al. 1997). High energy intake post-insemination is beneficial in embryonic survival (Leal et al. 2019). As gestation continues in the first month, higher feed intake is still shown to be beneficial to embryo development (Langendijk 2021). Earlier research by Jindal et al. (1996) demonstrated that high intake in early gestation resulted in lower embryo survivability. This work was conducted in gilts and additional work needs to be conducted to determine if this response only applies to gilts.

12 Case study: increasing feeding levels at day 90 of gestation

Increasing feeding levels (bump feeding) is a practice implemented in late gestation, typically during the last 2–3 weeks before farrowing. Increasing feeding levels during this period is believed to meet the sow's increasing energy and lysine requirements related to fetal growth (Dwyer et al. 1994; Cerisuelo et al, 2009) and mammary development (Weldon et al. 1994; Choi et al. 2019a). However, studies vary in either supporting or opposing phase feeding in late gestation. This section reviews the range of research either supporting or opposing increasing protein levels at day 90 of gestation.

It is recognized that gilts and sows experience an increase in protein requirements in late gestation (Ji et al. 2006; Krogh et al. 2020). Soto et al. (2011) and Shelton et al. (2009) have supported increasing feed from 2.27 to 4.09 kg/d in late gestation, resulting in heavier birth weights (1.31–1.44 kg) for offspring from gilt litters, but no change in birth weights from piglets born from sow litters. It has been suggested that variation in response could be associated with amino acid levels within research trials relative to NRC requirements (Goncalves et al. 2016a).

Recent work suggests an improvement of 30 g per piglet birth weight when sows were fed varying levels of energy in diets (Goncalves et al. 2016b). A 1.2%

reduction in pre-wean mortality was observed when females were fed 20 g Lys per day compared to females fed 10.7 g Lys/d in late gestation (Gonçalves et al. 2016a). These data would support the implementation of a two-phase feeding program during gestation for gilts, with increased levels of lysine from day 90 to improve sow milk yield.

Feyera and Theil (2017) estimated that mammary growth accounted for 16.8% of SID Lys requirements in the last 12 days of gestation. This tends to support the view that dietary lysine recommendations in late gestation (days 90–110) are underestimated and do not account for gilt mammary gland development that occurs in this timeframe. A recent study of feeding gilts 2.65 kg/d with a conventional diet containing 18.6 g/d of SID Lys or 26.0 g/d SID Lys found that the higher level of Lys showed a 44% increase in mammary tissue (Farmer et al. 2022). Other research suggests that gilts require a higher level of amino acids in pre-pubertal development to optimize performance. Feeding growing gilts 2.81 and 2.44 g/Mcal SID Lys/metabolizable energy ratios at 30–60 kg and 60 kg to first mating, respectively, has been shown to improve first parity birth weights and lactation performance (Tuong et al. 2021).

However, others have suggested that there may be no significant benefits associated with bump feeding and that the practice could even reduce sow retention rates over time (Mallaman et al. 2019). Increasing feed intake prior to lactation may result in decreased subsequent lactation feed intake (Mallaman et al. 2018). Feed intake from day 90 of gestation until farrowing is generally increased by 1 kg, which also increases body weight by 7 kg in this timeframe (Gonçalves et al. 2016a). However, one study found no effect on body weight in multi-parity sows fed a control diet (1.8 kg/d of feed (9 g SID Lys/d intake)) or a bump diet (control diet fed at 2.7 kg/d (14 g SID Lys/d intake; Greiner et al. 2016). Other research on gilts identified increased feed intake and weight loss during lactation, but no changes in sow BW during lactation with late gestation due to bump feeding (Shelton et al. 2009). Other studies have observed no changes in total litter birth weights due to bump feeding (Miller et al. 2000; Goncalves et al. 2016b), or improvements in piglet weaning weights from increasing feed in late gestation (Miller et al. 2000; Shelton et al. 2009). A recent international review of multiple studies also supported that there were no differences in birth weights when increasing feed or lysine levels in late gestation (Strathe and Brunn 2023).

Overall, current research suggests that, while bump feeding might support aspects of gilt development such as mammary gland development, this does not automatically translate into benefits such as improved piglet birth weights or overall improvements in sow condition or health. Based on the presented information, it is not recommended to bump feed when sows are of desired body condition at day 90 of gestation. Under-weight sows should receive an increase in feed. In addition, providing 1 kg of additional feed to gilts in late gestation can improve birth weights and should be considered in farms when birth weights are low.

13 Conclusion

Sows continue to increase reproductive capabilities in terms of litter number, size, and milk yield. However, sow longevity within a herd has increasingly become a challenge. The utilization of organic minerals and proper nutrition to reduce weight loss and body composition changes can improve sow longevity. Monitoring growth rates, BCSs, and implementing appropriate diet adjustments are key practices in gilt nutrition. By focussing on providing the right balance of protein, fiber, vitamins, minerals, and energy, replacement gilts can be primed for a successful transition into the sow herd and contribute to enhanced reproductive performance. Tailoring nutritional strategies to ensure optimum body condition during development and gestation are crucial for the reproductive performance of highly prolific sows. Practices such as feeding to condition early, maintaining slow growth, regular feed box calibrations, and monitoring individual sow BCSs are crucial components of successful body condition management strategies. These practices ensure that sows are in optimal condition for farrowing and lactation, maximizing reproductive outcomes and piglet performance. Balancing protein levels, providing adequate fiber, supplementation of vitamins and minerals, and careful consideration of energy requirements contribute to successful gestation outcomes. Data presented within the NRC (2012) on ingredients such as branched-chain amino acids and vitamin and mineral requirements needs to be continually evaluated to assess nutrient requirements as the sow changes. In addition to nutrient requirements, environmental factors such as heat stress and water quality need to be taken into account. There remains a need for continued on all aspects of nutrients to ensure that the reproductive female can be successful.

14 Where to look for further information

Feeding reproducing female swine is a complex topic that encompasses various interconnected factors. Determining their nutritional requirements requires the consideration of multiple resources. Factors such as facilities, environment, parity, feedstuffs, litter size, and health must all be taken into account while aiming to maximize herd longevity. Many studies focus on short-term assessments of sow reproductive performance, making it challenging to predict long-term impacts on longevity. Common resources include the National Academy of Science, Engineering and Medicine Nutrient Requirement of Swine (https://nap.nationalacademies.org/read/13298/chapter/1), Brazilian Tables for Poultry and Swine (https://editorascienza.com.br/pdfs/UFV/poultry_swine_en.pdf), INRAE (https://www.inrae.fr/en) and their software (https://inraporc.inra.fr/inraporc/index_en.html) as well as a multitude of scientific journals and books from around the world. Many of such resources were utilized in the writing of this chapter and further reading of these resources can provide additional information to the reader.

15 References

Aherne, F. 2005. *Feeding the lactating sow*. Pork Information Gateway Factsheet PIG 07-01-05.

Almeida, L.M.D., et al. 2018. Effects of feeding level and diet type during wean-to-estrus interval on reproductive performance of sows. *J Anim Sci* 96(Suppl. 2):92–93. (Abstr.)

Athorn, R.Z., et al. 2017. Early lifetime performance parameters affecting selection and reproductive success in gilts. *Anim Prod Sci* 57:2466–2466.

Belkova, J. and Rozkot, M. 2022. Gilt rearing impacts on sow performance and longevity – a rview. *J Swine Health Prod* 30(1):10–16.

Black, J.L., et al. 1993. Lactation in the sow during heat stress. *Livestock Prod Sci* 35:153–170.

Biehl, R.R. and Baker, D.H. 1996. Efficacy of supplemental 1 α-hydroxycholecalciferol and microbial phytase for young pigs fed phosphorus- or amino acid-deficient corn-soybean meal diets. *J Anim Sci* 74:2960–2966.

Bortolozzo, F.P., et al. 2009. Growth, body state and breeding performance in gilts and primiparous sows. *Control Pig Reprod.* VIII:281–291.

Bruun, T.S., et al. 2023. Feeding level during the last week of gestation can influence performance of sows and their litters in the subsequent lactation. *J Anim Sci* 101:skad349

Bryden, W.L. 2012. Mycotoxin contamination of the feed supply chain: implications for animal productivity and feed security. *Anim Feed Sci Technol* 173:134–158.

Cabezon, F.A., et al. 2016. Effect of betaine supplementation during summer on sow lactation performance and subsequent farrowing performance. *J Anim Sci* 94(Suppl_2):124.

Cabezon, F.A., et al. 2017. Effects of betain and heat stress on lactation and post-weaning reproductive performance on sows. *J Anim Sci* 95(Suppl_2):30.

Calderón Díaz, J.A., et al. 2015. Age at puberty, ovulation rate, and uterine length of developing gilts fed two lysine and three metabolizable energy concentrations from 100 to 260 d of age. *J Anim Sci* 93:3521–3527.

Carter, S.D., et al. 2000. Effects of dietary valine concentration on lactational performance of sows nursing large litters. NCR-42 Committee on Swine Nutrition. *J Anim Sci* 78(11):2879–2884.

Canavate, S. 2023. *Sow robustness: key areas to maximise sow retention*. American Association of Swine Veterinarians Annual Meeting, Aurora (Colorado), USA.

Cerisuelo, A., et al. 2009. Increased sow nutrition during midgestation affects muscle fiber development and meat quality, with no consequences on growth performance. *J Anim Sci* 87(2):729–739.

Choi, Y.H., et al. 2019a. Additional feeding during late gestation improves initial litter weight of lactating sows exposed to high ambient temperature. *Revista brasileira de zootecnia* 48:e20180028.

Choi, Y.H., et al. 2019b. Night feeding in lactating sows is an essential management approach to decrease the deterimental impacts of heat stress. *J Anim Sci Technol* 61(6):333–339.

Clowes, E.J., et al. 2003. Selective protein loss in lactating sows is associated with reduced litter growth and ovarian function. *J Anim Sci* 81(3):753–764.

Costermans, N.G., et al. 2019. Follicular development of sows at weaning in relation to estimated breeding value for within-litter variation in piglet birth weight. *Animal* 13(3):554–563.

Corson, A.M., et al. 2009. Intergenerational effects of birth weight on glucose tolerance and reproductive performance. *Animal* 3(4):579–591.

Crenshaw, T.D. 2001. Calcium, phosphorus, vitamin D, and vitamin K in swine nutrition. pp. 187–212 in *Swine Nutrition*, 2nd Ed., A. J. Lewis and L. L. Southern, eds. Boca Raton, FL: CRC Press.

Cromwell, G.L., et al. 1993. Efficacy of phytase in improving the bioavailability of phosphorus in soybean meal and corn-soybean meal diets for pigs. *J Anim Sci* 71:1831–1840.

Cromwell, G.L., et al. 1995. Efficacy of a recombinant-derived phytase in improving the bioavailability of phosphorus in corn-soybean meal diets for pigs. *J Anim Sci* 73:2000–2008.

Danish Pig Research Centre. 2001–2020. Combination of key figures of production performance in Danish swine herds between 2000 and 2020 [in Danish: Landgennemsnit for produktitvet I svineproduktionen fra 2000 til 2020], SEGES, Danish Pig Research Centre (DPRC).

Diekman, M.A. and Green, M.L. 1992. Mycotoxins and repdocution in domestic livestock. *J Anim Sci* 70(5):1615–1627.

Dourmad, J.Y., et al. 2008. InraPorc: a model and decision support tool for the nutrition of sows. *Anim Feed Sci Technol* 143:372–386.

Dwyer, C.M., et al. 1994. The influence of maternal nutrition on muscle fiber number development in the porcine fetus and on subsequent postnatal growth. *J Anim Sci* 72(4):911–917.

Faccin, J.E.G., et al. 2022. Gilt development to improve offspring performance and survivability. *J Anim Sci*, 100(6):skac128.

Farmer, C., et al. 2022. Dietary supplementation with lysine (protein) stimulates mammary development in late pregnant gilts. *J Anim Sci* 100(5):skac051.

Feyera, T., et al. 2017. Dietary supplement rich in fiber fed to late gesating sows during transition reduces rate of stillborn piglets. *J Anim Sci* 95(12):5430–5438.

Feyera, T., et al. 2018. Impact of sow energy status during farrowing on farrowing kinetics, frequency of stillborn piglets, and farrowing assistance. *J Anim Sci* 96(6):2320–2331.

Feyera, T. and Theil, P.K. 2017. Energy and lysine requirements and balances of sows during transition and lactation: a factorial approach. *Livest Sci* 201:50–57.

Feyera, T., et al. 2021. Optimal feed level during the transition period to achieve faster farrowing and high colostrum yield in sows. *J Anim Sci* 99(2):skab040.

Filha, W.A., et al. 2009. Growth rate and age at boar exposure as factors influencing gilt puberty. *Livest Sci* 120(1–2):51–57.

Food and Nutrition Board, Institute of Medicine. 1998. *Dietary reference intakes: thiamin, riboflavin, niacin, vitamin B-6, vitamin B-12, pantothenic acid, biotin and choline.* National Academy of Sciences, Washington, DC. pp. 390–422.

Foxcroft, G.R., et al. 2009. Prenatal programming of the postnatal development in the pig. *Soc Reprod Fertil Suppl* 66:213–231.

Frank, G.R., et al. 1988. Riboflavin requirement of lactating swine. *J Anim Sci* 66(1):47–52.

Gaines, A.M., et al. 2006. The dietary valine requirement for prolific lactating sows does not exceed the National Research Council Estimate. *J Anim Sci* 84(6):1415–1421.

Gaillard, G. and Dourmad, J.Y. 2022. Application of a precision feeding strategy for gestating sows. *Anim Feed Sci Technol* 287:115280.

Genest, M. and D'Allaire, S. 1995. Feeding strategies during the lactation period for first parity sows. *Can J Anim Sci* 75:461–467.

Getty, C.M. and Dilger, R.N. 2015. Moderate perinatal choline deficieincy elicits altered physiology and metabolomic profiles in the piglet. *PLoS* One 10(7):e0133500.

Goncalves, M.A.D., et al. 2016a. Fact sheet – impact of increased feed intake during late gstation on reproductive performance of gilts and sows. *J Swine Health Prod* 24:264–266.

Goncalves, M.A.D., et al. 2016b. Effects of amino acids and energy intake during late gestation of high-performing gilts and sows on litter and reproductive performance under commercial conditions. *J Anim Sci* 94:1993–2003.

Gorr, S.C., et al. 2024. *Ad libitum* feeding systems for lactating sows: effects on productivity and welfare of sows and piglets. *Animal* 18(3):101093.

Gourley, K.M., et al. 2020. Effects of timing and size of meals prior to farrowing on sow and litter performance. *Transl Anim Sci* 4(2):724–736.

Graham, A.B., et al. 2014. Evaluation of full feeding sow pre-farrow in lactation. Carthage Veterianary Service 24th Annual Conference Proceedings.

Graham, A.B., et al. 2015. Impact of feeding level post-weaning on wean to estrus interval, conception and farrowing rates, and subsequent farrowing performance. *J Anim Sci* 93(Suppl_2):65.

Greiner, L., et al. 2016. The evaluation of increasing lysine or feed amounts in late gestation on piglet birth weights. *J Anim Sci* 94:123–124.

Greiner, L., et al. 2017. Evaluation of the optimal standardized ileal digestbile tryptophan:lysine ratio in lactating sow diets. *Transl Anim Sci* 1(4):526–532.

Greiner, L., et al. 2019. Evaluation of the optimal standardized ileal digestible valine:lysine ratio in lactating sow diets. *J Anim Sci* 97(1):2965–2971.

Greiner, L., et al. 2020. Lysine (protein) requirements of lactating sows. *Transl Anim Sci* 4(2):txaa072.

Hansen, L.L., et al. 2021. Precision feeding gestating sows: effects on offspring growth performance and carcass and loin quality at slaughter. *Transl Anim Sci* 5(4):txab227.

Hays, V.W. 1976. *Phosphorus in Swine Nutrition*. National Feed Ingredients Association, West Des Moines, IA.

Holen, J.P., et al. 2002a. A review of branched-chain amino acids in lactation diets on sow and litter growth performance. *Transl Anim Sci* 6(1):txac017.

Holen, J.P., et al. 2022b. Evaluation of essential fatty acids in lactating sow diets on sow reproductive performance, colostrum and milk composition, and pig survivability. *J Animal Sci* 100(6):skac167.

Hojgaard, C.K., et al. 2019. Optimal lysine in diets for high-yielding lactating sows. *J Anim Sci* 97(10):4286–4281.

Hoving, L.L., et al. 2011. An increased feed intake during early pregnancy improves sow body weight recovery and increases litter size in young sows. *J Anim Sci* 89(11):3542–3550.
Huang, S., et al. 2020. Effects of dietary fiber sources during gestation on stress status, abnormal behaviors and reproductive performance of sows. *Animals* 10(1):141.
Humphrey, D.C., et al. 2023. Evaluation of branched-chain amino acid interactions in 10 to 20 kg nursery pigs using a central composite design. *J Anim Sci* 101:skad253.
Jackson, S.C., et al. 2000. Cellular uptake of valine by lactating porcine mammary tissue. *J Anim Sci* 78(11):2927–2932.
Ji, F., et al. 2006. Characterization of mammary gland development in pregnant gilts. *J Anim Sci* 84(3):579–587.
Ji, F., et al. 2005. Changes in weight and composition in various tissues of pregnant gilts and their nutritional implications. *J Anim Sci* 83(2):366–375.
Jindal, R., et al. 1996. Effect of nutrition on embryonal mortality in gilts: association with progesterone. *J Anim Sci* 74(3):620–624.
Jindal, R., et al. 1997. Progesterone mediates nutritionally induced effects on embryonic survival in gilts. *J Anim Sci* 75(4):1063–1070.
Jo, H. and Kim, B.G. 2023. Effects of dietary fiber in gestating sow diets. *Anim Biosci* 36(11):1619–1631.
Jongbloed, A.W., et al. 1992. The effect of supplementary *Aspergillus niger* phytase in diets for pigs on concentration and apparent digestibility of dry matter, total phosphorus, and phytic acid in different sections of the alimentary tract. *J Anim Sci* 70:1159–1168.
Kanora, A. and Maes, D. 2009. The role of mycotoxins in pig reproduction: a review. *Vet Med – Czech* 54(12):565–576.
Kemme, P.A., et al. 1995. Apparent ileal digestibility of protein and amino acids from a maize-soybean meal diet with or without extrinsic phytate and phytase in pigs. p. 6 in *International Symposium on Nutrient Management of Food Animals to Enhance the Environment Abstracts*, June 4–8.
Ketaren et al. 1993. Phosphours studies in pigs. *Br J Nutr.* 70(1):289–311.
Kim, J.S., et al. 2015. Relationship between backfat thickness of sows during late gestation and reproductive efficiency at different parities. *Acta Agric Scand A Anim Sci* 65:1–8.
Klaaborg, J., et al. 2019. The effect of feeding strategy during rearing in a commercial setting on gilt body condition, lactation performance and culling rate in modern sows nursing large litters. *Livest Sci* 228:144–150.
Klevay, L.M. 2001. Iron overload can induce mild copper deficiency. *J Trace Elem Med Biol* 14(4):237–240.
Klindt, J., et al. 1999. Effect of prepubertal feeding regimen on reproductive development of gilts. *J Anim Sci* 77(8):1968–1976.
Knauer, M.T. and Baitinger, D.J. 2015. The sow body condition caliper. *Appl Eng Agric* 31(2):175–178.
Koketsu, Y., et al. 1996. Feed intake pattern during lactation and subsequent reproductive performance of sows. *J Anim Sci* 74(12):2875–2884.
Koning, D.B., et al. 2013. The influence of dietary restriction before and after 10 weeks of age on osteochondrosis in growing gilts. *J Anim Sci* 91:5167–5176.
Koning, D.B., et al. 2014a. The influence of floor type before and after 10 weeks of age on osteochondrosis in growing gilts. *J Anim Sci* 92:3338–3347.
Koning, D.B., et al.. 2014b. The influence of dietary restriction before and after 10 weeks of age on osteochondrosis in growing gilts. *J Anim Sci* 91:5167–5176.

Kornegay, E.R. 1985. Calcium and phosphorus in animal nutrition. pp. 1–106 in *Calcium and Phosphorus in Animal Nutrition*. National Feed Ingredients Association, West Des Moines, IA.

Kornegay, E.T. 1985. Biotin in swine nutrition. *Ann N Y Acad* Sci 447:112–121.

Krogh, U., et al. 2020. Impact of dietary protein to energy ratio and two different energy levels fed during late gestation on plasma metabolites and colostrum production in sows. *Livest Sci* 234:103999.

Langemeier, M. 2023. Long-term trends in pigs per litter. *Farmdoc daily* 13:19, Department of Agricultural and Consumer Economics, University of Illinois at Urbana-Champaign, February 3.

Langendijk, P. 2021. Latest advances in sow nutrition during early gestation. *Animals* 11(6):1720.

Lauridsen, C. and Danielsen, V. 2004. Lactational dietary fat levels and sources influence milk composition and performance of sows and their progeny. *Livest Prod Sci* 91:95–105.

Lawlor, P.G., et al. 2007. The influence of over feeding sows during gestation on reproductive performance and pig growth to slaughter. *Arch. Tierz, Dummerstorf* 70:82-91.

Leal, D.F., et al. 2019. Effects of post-insemination energy content of feed on embryonic survival in pigs: a systematic review. *Anim Reprod Sci* 205:70–77.

Lei, X.G., et al. 1993b. Supplementing corn–soybean meal diets with microbial phytase linearly improves phytate phosphorus utilization by weanling pigs. *J Anim Sci* 71:3359–3367.

Lents, C.A., et al. 2020. Dietary lysine-to-energy ratios for managing growth and pubertal development in replacement gilts. *Appl Anim Sci* 36(5):701–714.

Li, S., et al. 2023. Dietary fiber during gestation improves lactational feed intake of sows by modulating gut microbiota. *J Anim Sci Biotech* 14:65.

Libal, G.W., et al. 1997. Tryptophan needs of lactating sows fed diets supplemened with crystalline lysine. *J Anim Sci* 75(2):417–422.

Mallmann, A.L., et al. 2018. Two different feeding levels during late gestation in gilts and sows under commercial conditions: impact on piglet birth weight and female reproductive performance. *J Anim Sci* 96:4209–4219.

Mallmann, A.L., et al. 2019. Impact of feed intake during late gestation on piglet birth weight and reproductive. *J Anim Sci* 97:1262–1272.

Meegat, M.B., et al. 2019. *Kansas State University Swine Nutrition Guide: Water in Swine Nutrition*. Kansas State University Extension, KS.

Mendoza, S.M., et al. 2020. Sow performance in response to natural betaine fed during lactation and post-weaning during summer and non-summer months. *J Anim Sci Biotechnol* 11:69.

Miller, K. A., et al. 2023. Strategies to manage barn feed supply to prolong and hold late finishing pigs during a supply chain disruption. Transl Anim Sci, 7(1).

Miller, H.M., et al. 2000. Increasing food intake in late gestation improved sow condition throughout lactation but did not affect piglet viability or growth rate. *J Anim Sci* 71:141–148.

Miller, K.A., et al. 2023. Strategies to manage barn feed supply to prolong and hold late finishing pigs during a supply chain disruption. *Transl Anim Sci*. 7(1):1-8

Mroz, Z., et al. 1994. Apparent digestibility and retention of nutrients bound to phytate complexes as influenced by microbial phytase and feeding regimen in pigs. *J Anim Sci* 72:126–132.

Nasi, M. 1990. Microbial phytase supplementation for improving availability of plant phosphorus in the diet of the growing pig. *J Agric Sci Finland* 62:435–443.

National Research Council. 2012. *Nutrient Requirements of Swine*, 11th ed. National Academy Press, Washington, DC.

Noblet J, et al. 1985. Studies on the energy metabolism of the pregnant sow: 1. Uterus and mammary tissue development. *Br J Nutr* 53(2):251–265.

Oliviero, C., et al. 2009. Feeding sows with high fibre diet around farrowing and early lactation: impact on intestinal activity, energy balance related parameters and litter performance. *Res Vet Sci* 86:314–319.

Oliviero, C., et al. 2010. Environmental and sow-related factors affecting the duration of farrowing. *Anim Reprod Sci* 119(1–2):85–91.

Paiva, P.C., et al. 2023. Risk factors associated with sow mortality in breeding herds under one production system in the Mid-Western United States. *Preven Vet Med* 213:105883.

Pallauf, V.J., et al. 1992a. Effect of microbial phytase supplementation to a maize-soya-diet on the apparent absorption on Mg, Fe, Cu, Mn and Zn and parameters of Zn-status in piglets. *J Anim Physiol Anim Nutr* 68:1–9.

Pallauf, V.J., et al. 1992b. Effect of microbial phytase supplementation to a maize-soy-diet on the apparent absorption of phosphorus and calcium in piglets. *J Anim Physiol Anim Nutr* 67:30–40.

Peltoniemi, O.A.T., et al. 2010. Fertility of sows fed *ad libitum* with a high fibre diet during pregnancy. *Reprod Dom Anim* 45(6):1008–1014.

Peo, Jr E.R.. 1976. *Calcium in Swine Nutrition*. National Feed Ingredients Association, West Des Moines, IA.

Peo, Jr E.R. 1991. Calcium, phosphorus, and vitamin D in swine nutrition. pp. 165–182 in *Swine Nutrition*, E. R. Miller, D. E. Ullrey, and A. J. Lewis, eds. Butterworth-Heinemann, Stoneham, MA.

Peters, J.C. and Mahan, D.C. 2008. Effects of dietary organic and inorganic trace mineral levels on sow reproductive performances and daily mineral intakes over six parities. *J Anim Sci* 86(9):2247–2260.

Pettigrew, J.E., et al. 1996. Riboflavin nutrition of sows. *J Anim Sci* 74(9):2226–22230.

Quinn, A.J., et al. 2015. The effect of feeding a diet formulated for developing gilts between 70kg and ~140kg on lameness indicators and carcass traits. *Livest Sci* 174:87–95.

Ramirez-Camba, C.D., et al. 2020. Efficiency of standardized ileal digestible lysine utilization for whole body protein deposition in pregnant gilts and sows during early-, mid-, and late-gestation. *J Anim Sci* 98(11):skaa340.

Reese, D.E. and Hill, G.M. 2010. Trace minerals and vitmains for swine diets. pp. 37–52 in *National Swine Nutrition Guide*, Meisinger D.J., ed. U.S. Pork Center of Excellence.

Ren, P., et al. 2017. Effect of different feeding levels during three short periods of gestation on sow and litter performance over two reproductive cycles. *Anim Reprod Sci* 177:42–55.

Richert, B.T., et al. 1996. Valine requirement of the high-producing lactating sow. *J Anim Sci* 74(6):1307–1313.

Richert, B.T., et al. 1997. The effect of dietary lysine and valine fed during lactation on sow and litter performance. *J Anim Sci* 75(7):1853–1860.

Rodríguez, M., et al. 2023. Feed intake patterns of modern genetics lactating sows: characterization and effect of the reproductive parameters. *Porcine Health Manag* 31(1):6.

Rooney, H.B., et al. 2020. Effect of increasing dietary energy density during late gestation and lactation on sow performance, piglet vitality, and lifetime growth of offspring. *J Anim Sci* 98(1):skz379.

Rosero, D.S., et al. 2012. Response of the modern lactating sow and progen to source and level of supplemental dietary fat during high ambient temperatures. *J Anim Sci* 90(8):2609–2619.

Rosero, D.S., et al. 2015. Impact of dietary lipids on sow milk composition and balance of essential fatty acids during lactation in prolific sows. *J Anim Sci* 93(6):2935–2947.

Rosero, D.S., et al. 2016. Optimizing dietary lipid use to improve essential fatty acid status and reproductive performance of the modern lactating sow: a review. *J Animal Sci Biotechnol* 7:34.

Sampath, V., et al. 2023. Role and functions of micro and macro-minerals in swine nutrition: a short review. *J Anim Sci and Tech* 65(3):479.

Sawicki, C.M., et al. 2017. Dietary fiber and the human gut microbiota: application of evifence mappting methodology. *Nutrients* 9(2):125.

Seerley, R.W., et al. 1981. The influence of sow dietary lipids and choline on piglet survival, milk and carcass composition. *J Anim Sci* 52(3):542–550.

Shang, Q., et al. 2019. Effects of dietary fiber sources during late gestation and lactation on sow performance, milk quality, and intestinal health in piglets. *J Anim Sci* 97(12):4922–4933.

Shi, M., et al. 2016. Estimation of the standarized ileal digestible lysine requirement for primparous pregnant sows. *J Anim Physiol Anim Nutr.* 100(2):287–93.

Shelton, N.W., et al. 2009. *Effects of increasing feeding level during late gestation on sow and litter performance. Proc Kansas State University Swine Day.* Agricutlural Experiment Station and Cooperative Extension Service, Manhattan, Kansas. pp 38–50.

Simons, P.C.M., et al. 1990. Improvement of phosphorus availability by microbial phytase in broilers and pigs. *Br J Nutr* 64:525–540.

Soto, J., et al. 2011. Effects of increasing feeding levels in sows during late gestation on piglet birth weights (abstract). *J Anim Sci* 89:86.

Stalder, K.J., et al. 2003. Financial impact of average parity of culled females in a breed-to-wean swine operation using replacement gilt net present value analysis. *J Swine Health Prod* 11(2):69–74.

Strathe, A.V. and Bruun, T.S. 2023. 1. Main focus should be feeding levels and fiber supply when optimizing gestating and pre-farrowing feeding strategies for hyper prolific sows. *Anim Sci Proc* 14(6):707–710.

Strathe, A.V., et al. 2016. The effect of increasing the dietary valine-to-lysine ratio on sow metabolism, milk production, and litter growth. *J Anim Sci* 94(1):155–164.

Strathe, A.V., et al. 2017. Sows with high milk production had both a high feed intake and high body mobilization. *Animal* 11(11):1913–1921.

Strathe, A.V., et al. 2019. Effects of dietary protein level and energy intake from 50 to 120 kg on body weight, back fat thickness and body composition in gilts. *Livest Sci* 227:11–16.

Theil, P.K., et al. 2023. Feeding the modern sow to sustain high productivity. *Mol Reprod Dev* 90:517–532.

Thingnes, S.L., et al. 2015. The effect of different dietary energy levels during rearing and mid-gestation on gilt performance and culling rate. *Livest Sci* 172:33–42.
Thomas, L.L., et al. 2018a. Partitioning components of maternal growth to determine efficiency of feed use in gestating sows. *J Anim Sci* 96(10):4313–4326.
Thomas, L.L., et al. 2018b. Effect of parity and stage of gestation on growth and feed efficiency of gestating sows. *J Anim Sci* 96(10):4327–4338.
Tian, M., et al. 2020. Dietary fiber and microbiota interaction regulates sow metabolism and reproductive performance. *Anim Nutr* 6(4):397–403.
Tinkel, A., et al. Effects of feeding a mineral amino acid complex from nursery to second parity on sow productive and gait analysis. *J Anim Sci* 96(Suppl_2):133.
Tuong, N.D., et al. 2021. Effect of digestible lysine/metabolizable energy ratio in F1 (Landrace x Yorkshire) gilt diets on growth, age at puberty and reproductive performance in closed housing condition. *Adv Anim Vet Sci* 9(9):1347–1354.
United States Department of Agriculture, National Agricultural Statistics Service. Quick Stats. https://www.nass.usda.gov/Quick_Stats/, accessed February 2, 2021.
USDA-NASS 2024. Quarterly Hogs and Pigs Report (released 28th March, 2024). US Department of Agriculture – National Agricultural Statistics Service (USDA-NASS). ISSN: 1949–1921.
Vargas, A.J., et al. 2009. Factors associated with return to estrus in first service swine females. *Prev Vet Med* 89:75–80.
Wang, L., et al. 2022. A systematic review and meta-analysis of dietary fat effects on reproductive performance of sows and growth performance of piglets. *J Anim Sci Biotech* 13:Article 12.
Wedekind, K., et al. 2015. Efficacy of an equine joint supplement, and the synergistic effect of its active ingredients (chelated trace minerals and natural eggshell membrane), as demonstrated in equine, swine and an osteoarthritis rat model. *Open Access Anim Physiol* 7:13–27.
Weldon, W.C., et al. 1994. Postpartum hypophagia in primiparous sows: I. Effects of gestation feeding level on feed intake, feeding behavior, and plasma metabolite concentrations during lactation. *J Anim Sci* 72(2):387–394.
Wellington, M.O., et al. 2018. Impact of dietary fiber and immune stimulation on threonine requirement for protein deposition in growing pigs. *J Anim Sci* 96(12):5222–5232.
Wensley, M.R., et al. 2023. Feeding strategies to increase sow colostrum quality and yield. *J Swine Health Prod* 31(3):133–136.
West, J.W., et al. 1999. Effects of dietary fiber on intake, milk yield, and digestion by lactating dairy cows during cool or hot, humid weather. *JDairy Sci* 82(11):2455–2465.
Williams, N., et al. 2005. Nonnegotiables of gilt development. *Advan Pig Prod* 16:1–9.
Zhong, W., et al. 2021. Effects of dietary choline levels during pregnancy on reproductive performance, plasma metabolome and gut microbiota of sows. *Front Vet Sci* 8:771228.
Zinedine, A., et al. 2007. Review on the toxicity, occurrence, metabolism, detoxification, regulations and intake of zearalenone: an oestrongenic mycotoxin. *Food Chem Toxicol* 45(1):1–18.

Index

www.ingramcontent.com/pod-product-compliance
Ingram Content Group UK Ltd.
Pitfield, Milton Keynes, MK11 3LW, UK
UKHW021826100926
8556IPUK00003B/179